MODERN ATOMIC PHYSICS:

FUNDAMENTAL PRINCIPLES

MODERN ATOMIC PHYSICS: FUNDAMENTAL PRINCIPLES

B. CAGNAC
Faculty of Science, University of Paris VI

J.-C. PEBAY-PEYROULA
Faculty of Science, University of Grenoble

Translated by J. S. Deech
J. J. Thomson Physical Laboratory,
University of Reading

A HALSTED PRESS BOOK

JOHN WILEY & SONS
New York – Toronto

Authorised English language edition of
Physique atomique, tome 1, first published
by Dunod, Paris, in 1971

First published in the United Kingdom 1975 by
The Macmillan Press Ltd

Published in the U.S.A. and
Canada by Halsted Press, a
Division of John Wiley & Sons, Inc.,
New York

Printed in Great Britain

Library of Congress Cataloging in Publication Data

Cagnac, Bernard, 1931–
Modern atomic physics.

Translation of v. 1 of Physique atomique.
"A Halsted Press book."
Bibliography: p.
Includes index.
1. Nuclear physics. 2. Atomic theory. 3. Quantum theory. I. Pebay-Peyroula, Jean-Claude, 1930– joint author. II. Title.
QC174.12.C3213 1975 539.7 74–28161
ISBN 0 470–12920–4

Contents

PART 3 ANGULAR MOMENTUM AND MAGNETIC MOMENT

Preface

Atomic physics was responsible for the birth and development of quantum theory and indeed is still closely linked with it; one can go further and say that atomic physics cannot exist without quantum mechanics. A course on quantum mechanics that emphasised the experimental consequences of the mathematics would amount to a good course on atomic physics. However, experience has shown that students find great difficulty in coping with the mixture of terminologies that is inevitable when constantly switching from theoretical calculation to experimental description. Thus in order to facilitate the assimilation of new ideas, it would seem desirable to separate the whole subject into two distinct courses. These would comprise (a) the experimental justification of the principles of quantum mechanics and the consequences of these principles as they concern atomic structure, and (b) an account of quantum mechanics arranged, once the basic principles have been described, almost entirely on logical and deductive lines. It is to the first of these tasks that this volume is dedicated; the second task is undertaken in *Quantum Theory and its Applications*.

Our approach concentrates on describing experiments and comparing them with theoretical results; it has the objects of: (a) establishing the experimental basis of the fundamental principles of quantum mechanics, by describing the crucial experiments that demonstrate the limitations of classical theory and provide unchallengeable results which the theory must explain; (b) setting out the fundamental laws governing the internal structure of atomic systems, knowledge of which is indispensable now that they have such wide application; (c) surveying the experimental work currently being conducted in physics laboratories and thus appreciating its aims.

In discussing the experimental foundations of the principles of quantum mechanics, we think that the device of rediscovery is to be avoided. This device involves following the path of history from the beginning of the century with the object of proving for ourselves that the only way to progress would be to invent quantum mechanics. Quantum mechanics is now so solidly based and supported by such a wide range of experiments that further persuasion is unnecessary. Besides, the path of history has been winding and uneven and the course which it has taken is far from being the simplest and most direct. Certain ideas suddenly appeared during very prolific periods in the form of very complex deductions from work carried out simultaneously in different fields.

However, a certain amount of detailed and wide-ranging experimental evidence appears indispensable for an understanding of the significance of the essential ideas of quantum mechanics and for easing their assimilation; this is what we have tried to include. We have not forgotten some of the older experiments that are still of fundamental importance, but we have added many other more recent experiments; moreover it seemed sensible to group them in the same part of the book, even in the same chapter, without regard to their chronological order, as long as they supported the same essential idea.

It is a fact that when considering specific experiments different ideas often overlap, and the plan that we have adopted gives prominence to certain ideas at the expense of others. We have chosen to give general concepts priority over particular phenomena, and since the explanation of certain phenomena requires the use of several general concepts, we have often had to discuss the same phenomenon from different aspects. Optical resonance, for example, is a phenomenon involving an atom–photon collision, the explanation of which involves the principle of the conservation of energy (chapter 1), of the conservation of momentum (chapter 2) and the notion of transition probability or cross-section (chapter 3). To take another example, the interpretation of magnetic resonance brings into play the gyromagnetic ratio (chapter 9), spatial quantisation (chapter 10) the angular momentum associated with circular polarisation (chapter 11), and so on.

Our presentation has been guided by our concern both to graduate the difficulties and to allow for the possibility of a partial reading of the book. So we have gathered together in volume 1 those simple concepts that can be understood without any knowledge of the formal mathematics of quantum mechanics. Volume 1 is devoted to the description of experiments and to the development of those quantum mechanical ideas with the most straightforward experimental basis. It can be read before an explanation of quantum mechanics. Some more theoretically inclined readers might object that this first volume makes too much use of the language of classical physics; but it should not be forgotten that classical terminology remains indispensable for a rapid and convenient description of physical processes, and that the experimental physicist should develop the ability to change continually from quantum to classical language. Apart from which, it is precisely on the classical nature of measuring apparatus that the quantum theory of measurement is based. In any case, our concern to make the connection with quantum mechanics permeates this volume. It is this in particular which has led us to emphasise the chapters devoted to angular momentum; they promote a better understanding of the importance given in quantum mechanics to the theory of angular momentum.

We have, on the other hand, collected together ideas in the second volume, even some very old ones, that can be satisfactorily understood only by using quantum-mechanical calculations or their results. This volume can be read to best advantage by those who have already taken a course in quantum mechanics, and this assumption accounts for the alteration in pace and style. Nevertheless the first four chapters of the second volume do not make use of quantum mechanical formalism, a familiarity with the general ideas of quantum mechanics being the only prerequisite. It is only in the second volume that the description of atomic structure is actually completed. The final chapter surveys some experimental problems that are currently under consideration.

Writing this book has involved a number of compromises. The essential aim has been to give the reader the most precise and accurate review possible of the problems of atomic physics; it is this goal which has determined our choice of topics. The extent of the subject is obviously greater than an average student can absorb in a year in which he is taking courses in quantum mechanics and nuclear physics, perhaps in addition to being taught, for example, field theory or electronics. Our presentation makes a number of chapters relatively independent, so that the reader can design his own course and select accordingly. Also, by using small print, we have singled out certain passages in each chapter that are not essential to its understanding; they may be omitted on the first reading.

Another aim of this course is to show the reader how to work out the results of an experiment for himself and to give him a sense of orders of magnitude, necessary for justifying those approximations without which few calculations in physics would be possible. This prevented us from putting (in the way theoreticians do): $h = c = 1$, and so we had to confront the irritating problem of units. Since French students have for several years used rationalised MKS units, we did not want to interfere with this practice, and we have used rationalised formulae. However, most publications and major works in atomic physics, even the most recent, use the non-rationalised gaussian system and one should be able to convert from one to the other. To this end we have introduced into our formulae a coefficient κ defined by the relation

$$\kappa = \epsilon_0 \mu_0 c^2$$

(a) In the MKS system, $\kappa = 1$. The coefficient κ can be simply disregarded in all the formulae, which then become normal rationalised formulae, and the numerical values of the constants are

$$4\pi\epsilon_0 = \frac{1}{9 \times 10^9}; \qquad \frac{\mu_0}{4\pi} = 10^{-7}$$

(b) In the gaussian system, where electrical units from the electrostatic c.g.s. system and magnetic units from the electromagnetic c.g.s. system are used simultaneously, the coefficients are determined thus

$$4\pi\epsilon_0 = 1; \qquad \frac{\mu_0}{4\pi} = 1 \quad \text{and} \quad \kappa = c$$

Appendix 1 gives a set of electromagnetic formulae showing how the classical formulae should be modified to take account of the coefficient κ. We have taken these modified formulae as our starting point for all our calculations in atomic physics (see p. 301).

Before starting the course, we would like to thank all the staff and research workers in our laboratories, both in Paris and Grenoble, who have helped us to prepare this new presentation of atomic physics. Above all we must thank Professor Kastler and Professor Brossel for their inspiration, both in their teaching and in the everyday life of the laboratory.

B. Cagnac
J.-C. Pebay-Peyroula

Preface to the English Edition

The English edition of Cagnac and Pebay-Peyroula's *Physique Atomique* has been prepared with the aim of retaining as faithfully as possible both the spirit and the letter of their work. However, certain changes have been made.

(1) Wherever it was felt that English-speaking students would be unfamiliar with the French terminology, preference was given to the accepted English practice: for example, hyperfrequency transitions are always referred to as microwave transitions.

(2) The French edition is divided into two volumes, with the chapters numbered continuously throughout; the appendixes appear at the end of the second volume only. For the English edition it has been decided to publish two books that are as far as possible independent of one another. To this end, each book has been given a distinctive title—'Fundamental Principles' corresponding to volume 1 and 'Quantum Theory And Its Applications' corresponding to volume 2. The numbering of the chapters in the two volumes is independent and decimal notation has been used to subdivide each chapter. Reference to other chapters always imply the current volume unless otherwise stated. Each book now contains its own appendixes, some of which appear in both books.

(3) In consultation with the authors, some sections have been amplified to avoid any possible misunderstanding. Additionally, in an effort to ensure that the book is up to date at the time of writing, major developments in the past three years have been included where they have an important bearing on

the subject matter. Alterations, other than corrections and those due to changes of terminology, have been inserted as translator's footnotes.

(4) SI units (rationalised MKS) have been used throughout, though the coefficient κ (see preface to French edition) is retained in relevant formulae. Thus, despite some spectroscopists' aversion to changing, for example, from ångström units to a subunit of the metre or from gauss to tesla, numerical values are nearly always given in SI.

(5) Important names and dates are given for their historical interest only and accordingly precise references are usually omitted, as in the French edition. Where the French edition refers to textbooks that have not been translated into English, an English substitute has been given.

I should like to express my thanks to Dr M. H. Tinker for reading and commenting upon the entire translated manuscript, to my wife who sacrificed much of her time for several months in preparing the final typescript version, and finally to the publishers for their continued encouragement and helpfulness.

J. J. Thomson Physical Laboratory J.S.D.
Reading
July 1974

Glossary

1 Latin Alphabet

a	acceleration semi-major axis of an ellipse, or other length
A	magnetic vector potential amplitude of a sinusoidal function magnetic moment coupling constant probability of spontaneous emission atomic mass number
$\mathscr{A}$	atomic mass (approximately equal to A in CGS, or to $A/1000$ in MKS)
b	impact parameter in collision problems semi-minor axis of an ellipse, or other length
B	magnetic induction vector, usually called magnetic field (the excitation vector $\boldsymbol{H} = (1/\mu_0 \mu_r)\, \boldsymbol{B}$ is hardly ever used) absorption and induced emission probability coefficients (Einstein notation) amplitude of a sinusoidal function
$\mathscr{B}$	magnetic field vector
c	velocity of light

C	constant in a Coulomb law of force ($W(r) = C/r$) Curie constant of magnetic susceptibilities capacitance torsion constant of a wire
d	differential
d	symbol for the quantum number $l = 2$ (total orbital angular momentum of an electron)
D	electric induction vector probability density distance symbol for the quantum number $L = 2$ (total orbital angular momentum of an atom)
$\mathscr{D}$	intensity of a beam of particles per unit cross-sectional area ($N = \mathscr{D}tS$)
e	base of napierian logarithms e = 2·7183
e	elementary positive charge $e = 1{\cdot}6 \times 10^{-19}$ C
E	electric field vector algebraic value of the total energy of an atomic state
$\mathscr{E}$	complex function associated with the electric field of a wave
f	a function force vector oscillator strength symbol for the quantum number $l = 3$ (orbital angular momentum of an electron)
F	resultant vector for a system of forces symbol for the quantum number $L = 3$ (orbital angular momentum of an atom)
g	Landé factor symbol for the quantum number $l = 4$ (orbital angular momentum of an electron)
G	statistical weight or order of degeneracy symbol for the quantum number $L = 4$ (total orbital angular momentum of an atom)
h	Planck's constant ($\hbar = h/2\pi$)
H	hamiltonian operator
$\mathscr{H}$	hamiltonian function
i	square root of -1, the base of imaginary numbers
i	index number angle of incidence
I	magnitude of an electric current moment of inertia nuclear spin quantum number and the corresponding vector

j	the electric current density vector index number quantum number of angular momentum and the corresponding vector
J	total angular momentum quantum number of an atom, and the corresponding vector
k	Boltzmann's constant wave vector an integer
K	contact potential difference absorption coefficient symbol for the principal quantum number $n = 1$
l	a length orbital angular momentum quantum number and the corresponding vector
L	luminance the total orbital angular momentum quantum number of an atom and the corresponding vector symbol for the principal quantum number $n = 2$
$\mathscr{L}$	lagrangian function
m	mass, especially mass of the electron magnetic quantum number
M	molecular mass mass, especially the mass of an atom or a nucleus intensity of magnetisation vector symbol for the principal quantum number $n = 3$
$\mathscr{M}$	magnetic moment vector
n	number of particles per unit volume principal quantum number
N	an integer (dimensionless) unit normal vector symbol for the principal quantum number $n = 4$
$\mathscr{N}$	Avogadro's number
O	symbol for the principal quantum number $n = 5$
p	momentum or impulse vector electric dipole moment vector index number population of an energy level symbol for the quantum number $l = 1$ (orbital angular momentum of an electron)

P	power polarisation vector of a dielectric symbol for the quantum number $L = 1$ (total orbital angular momentum of an atom) symbol for the principal quantum number $n = 6$
$\mathscr{P}$	generalised impulse vector
q	algebraic electric charge (especially charge on the electron $q = -e$)
Q	electric charge quality factor of a resonant cavity or a resonant circuit reduced quadrupole moment
$\mathbf{Q}$	components of the quadrupole tensor
r	radius vector, distance between two points
R	distance between two points electric resistance Rydberg constant for atomic spectra gas constant for a perfect gas
$\mathscr{R}$	radial wave function
s	spin quantum number and the corresponding vector screening coefficient symbol for the quantum number $l = 0$ (orbital angular momentum of an electron
S	area total spin quantum number of an atom and the corresponding vector symbol for the quantum number $L = 0$ (total orbital angular momentum of an atom)
t	time
T	absolute temperature period spectral term
$\mathscr{T}$	work
u	energy density unit vector transverse component of the magnetisation $\boldsymbol{M}$ in a rotating frame
U	internal energy electrostatic potential
v	velocity vector transverse component of the magnetisation $\boldsymbol{M}$ in a rotating frame
V	electrostatic potential or electromotive force exceptionally, velocity vector

$\mathscr{V}$	volume
w	a small energy
W	energy in general, and especially potential energy $W(r)$
x, X	co-ordinate
y	co-ordinate
Y	co-ordinate angular wave functions of hydrogen (spherical harmonics)
z	co-ordinate
Z	co-ordinate atomic number of an element statistical partition function

2 Greek Alphabet

α	fine structure constant ($\alpha = e^2/4\pi\,\varepsilon_0\,\hbar c$) angle direction cosine name of a type of particle
β	Bohr magneton angle direction cosine name of a type of particle (β^- and β^+ rays)
γ	gyromagnetic ratio direction cosine name of a region of the electromagnetic spectrum
Γ	moment of a force or resultant moment of a system of forces
δ	increment, difference
Δ	laplacian operation
$\varDelta$	width of a line correction applied to an energy value following a perturbation calculation
ε	constant depending on units, ε_0, (in CGS ε_0 is replaced by $1/4\pi$) dielectric constant ε_r a conical eccentricity any very small quantity
η	efficiency, or other dimensionless fraction
θ	angle, especially latitude in spherical co-ordinates

Θ	angular wave function Debye temperature
κ	constant depending on units: $\varepsilon_0 \mu_0 c^2 = \kappa^2$ $\begin{cases}\kappa = 1 \text{ in MKSA system}\\ \kappa = c \text{ in gaussian system}\end{cases}$
λ	wavelength
Λ	Compton wavelength ($\Lambda = h/mc$)
μ	constant depending on units, μ_0, (in CGS, μ_0 is replaced by 4π) magnetic permeability reduced mass
ν	frequency
ξ	co-ordinate
ϖ	pressure probability
π	3·1416... symbol for a polarisation of a wave (parallel to a magnetic field)
Π	product
ρ	density electric charge density particular values of a radius vector
σ	angular momentum vector cross section symbol for a polarisation of a wave (circular or perpendicular to a field)
Σ	summation
τ	time constant (lifetime, relaxation time)
ϕ	angle, especially longitude in spherical co-ordinates
Φ	magnetic flux angular wave function
χ	angle magnetic susceptibility
ψ	angle total wave function
ω	angular velocity of rotation angular frequency of a sinusoidal function
Ω	solid angle exceptionally, angular velocity of rotation

Introduction

Atomic physics has its origins in the nineteenth century, but since then the microscopic character of the subject has only been revealed gradually. A course of atomic physics which claimed to be complete would have to examine the experiments and the different lines of thought which led to the molecular theory, from the discovery of the electron to the historic measurements of the fundamental constants: Avogadro's number, the charge e and mass m of the electron. Such material would contain little that is new for the student mastering the subject or for the reader already familiar with it, since the various methods of measuring e or e/m, for example, have been described in more elementary texts. From the start of this course, the microscopic nature of matter has been assumed; only in the first chapters do we describe and comment upon experiments which have had a fundamental role in the development of quantum physics. Nevertheless it seemed useful to draw up a brief preliminary table of the major historical developments in atomic physics in chronological order, (appendix 4). The main aim of this summary is to set out what we have taken as known, and to remind the reader of the main events in the history of the atom. This material can obviously be omitted by the reader eager to learn something new.

During the nineteenth century *the molecular theory* superseded the quantitative laws of chemistry (the law of definite proportions and the laws of multiple proportions). Dalton recognised that various chemical species were derived from the same basic unit, the molecule, the simple bodies present in these species appearing in the molecule in the form of indestructible particles called atoms. Avogadro's hypothesis in 1811 stated this very idea: equal volumes of a gas contain equal numbers of molecules. These proposals were of considerable importance in the development of chemistry and physics. In our study of atomic physics we shall use the concept of thermal agitation, a concept which originated with Brown's discovery of brownian motion in 1827, although analysis and appreciation of its implications continued throughout most of the nineteenth century. The kinetic theory of gases plays an important part in the explanation of brownian motion, and a comparison of the macroscopic properties of gases with the results of this theory led, around 1875, to the first calculation of Avogadro's number. Similarly the concept of thermal agitation applied to particles in a fine emulsion led Jean Perrin in 1908 to make the first accurate measurement of Avogadro's number. From then on, scientific progress was swift: various methods, relying on different physical principles, gave consistent values of Avogadro's number and thus verified the molecular theory (see appendix 4).

The discovery of *the electron* was also an important milestone. The analysis of electrolysis experiments in the light of molecular theory showed that an atom could carry an electric charge equal to, or a multiple of, a certain value—the deposit of one gram molecule corresponding to the passage of a quantity of electricity equal to one Faraday. Independently, and thanks to the availability of high voltages from a Ruhmkorff coil, the study of electric discharges in rarefied gases led in about 1860 to the initially very hazy idea of 'cathode rays' as the cause of the fluorescence observed from the walls of the bulb containing the rarefield gas. There was protracted controversy as to the nature of these 'cathode rays': they were deflected by a magnetic field, which seemed incompatible with a wave nature; they could pass through thin foils of metal, which contradicted the corpuscular theory. In the course of a series of celebrated experiments, Jean Perrin showed that cathode rays carried negative electricity which could be collected in a Faraday cage. The work which followed showed that, whatever the technique, the experimental results were consistent with the hypothesis of the existence of a particle of mass about two thousand times smaller than the hydrogen atom (Wiechert, 1897). Thenceforth the evolution of science was rapid, manifesting itself by measurements of the charge and mass of the electron, the interpretation of the photoelectric effect and the thermoelectric effect.

Knowledge of *the structure of the atom* was being gained contemporaneously with the discovery of the electron. In the experiments with discharges, interest centred on the positive rays which appeared behind the cathode on the opposite side to the anode and which were deflected by a magnetic field, manifested by a curvature in an inverse sense to that observed for cathode rays. Detailed study of these positive rays was carried out in the last decade of the nineteenth century and measurements of the curvature of their trajectories led to their interpretation as a beam of ions, that is to say atoms which have lost one or more electrons. In the case of hydrogen all the experiments showed that only one type of ion existed and the conclusion was that the hydrogen atom possessed only one electron. From this followed the identification of the atomic number Z of each atom (the order number in the Periodic classification) with the number of electrons it contained. Later, in 1913, an accurate measurement of the ratio q/m of the charge q to the mass m of the ions obtained in certain discharges led to the discovery of isotopes (J. J. Thomson) and the measurement of the mass number A of each isotopic species. The discovery of the electron and of positive ions had thus formed the first link with the understanding of the atom. Despite some indisputable successes of J. J. Thomson's atomic model, where the electron was situated in the midst of a cloud of positive charge exerting a force of attraction proportional to distance, we shall see how the Rutherford experiments on scattering of particles required a planetary model of the atom as suggested by Jean Perrin in 1901 (chapter 5).

Electromagnetism underwent great development at the beginning of the twentieth century. Hertz's experiments and the theoretical work of Hertz and Maxwell led rapidly to the formalism well known to all students who have taken a traditional course on electricity.

This development of electromagnetism made it possible to explain all the phenomena of wave optics (and of geometrical optics, which follows when all dimensions are very much greater than the wavelength). The connection between wave optics and atomic physics is usually rather weak. However, the importance of the phenomenon of X-ray diffraction by a crystal lattice should be stressed. X-Rays were discovered in 1895 by Röntgen; in the years that followed their main properties were analysed and their wave nature propounded. Clear confirmation of this was given by von Laue in 1912 when he showed that X-rays were diffracted because of periodicity in the structure of the crystals. The Bragg theory, by allowing a quantitative study, was the start of the microscopic description of the solid state.

On the other hand, electromagnetic theory is very closely concerned with atomic physics, in the study of the interactions between charged particles and electromagnetic fields. The electromagnetic radiation emitted by an accelerated charged particle is one of the most important consequences of this branch of physics. The simplest case is that of the oscillating dipole, which can be represented by a charged particle oscillating with very small amplitude; we have summarised the results of this study, which has fundamental implications, in appendix 2.

The connection with the microscopic theory of matter was made in this same period through the Thomson model mentioned above. The interaction between an electromagnetic field and an atom was interpreted in terms of the oscillatory forces acting on the elastically bound electrons. Results were obtained for absorption phenomena; absorption bands corresponding to resonance frequencies were related to the motion of the electron and the anomalous variation of refractive index in the neighbourhood of an absorption band was also interpreted by this model (see appendix 2). On the other hand the problems relating to the emission of light by the atom were only poorly described.

In addition, an electron forced into oscillations by the incident electromagnetic field radiates into space with a directionality which is the same as that of an oscillating dipole. The explanation of experiments on scattering of radiation by atoms has been another success of electromagnetic theory; the phenomena observed vary depending upon whether the incident wave has a frequency either close to or very different from the resonance frequencies of the elastically bound electrons (Thomson scattering, Rayleigh scattering). In the particular situation where the incident wave has a frequency much higher than the resonant frequencies of the electron, a simple calculation enables the result of measurements of scattering efficiency to be related to the number of electrons per atom. The measurements of Barkla (1909) have been of equal historical importance in confirming the significance attributed to the atomic number Z (see appendix 2).

Despite the simplicity of the model used, the theory of dipole radiation has provided a good approximation in accounting for scattering and absorption; we will come across it in other examples during the course of this book. These successes enable us to understand why modern theories of radiation use many results of classical electromagnetic theory as a starting point.

So we see that by the end of the first decade of this century physical phenomena could be considered in many respects to be understood; it was possible to comprehend the structure of matter intellectually. However, some difficulties persisted, particularly concerning the emission of light—thermal emission by a black body, emission of spectral lines—and in the course of attempts to resolve them, the quantum nature of the phenomena was recognised. It is this change in physicists' thinking which is essentially the subject matter of the first part of this book.

Part 1

Waves and Photons

1

Quantisation of Energy

This first chapter deals with applications of the general law of the conservation of energy. We shall use this law to explain a number of experiments and thus demonstrate the discontinuous nature of energy changes on a microscopic scale.

1.1 Review of Planck's Law

The thermal radiation emitted by a hot body is both a subject of current research and also a part of our everyday life (incandescent light). Historically it was through scientific study of this phenomenon that the idea of quantisation was introduced.

It is well known that the colour of radiation perceived by our eyes changes progressively from red to white when the temperature of a body is increased, and practical experience shows that the colour observed inside an oven is a measure of its temperature. So it is of interest to study the spectral distribution of the light intensity from the interior of an oven.

The luminous intensity received and measured by any detector depends greatly on the conditions of the experiment; that is why it is useful to relate experimental results to a single quantity having a simple theoretical interpretation: the energy density u of the electromagnetic radiation inside the oven. It is easy to relate this energy density inside the oven to the power P which is carried by a beam of light leaving the oven and which can be measured experimentally. (We use as an intermediate step the luminance L of the walls inside the oven: $u = (4\pi/c)L$ and $P = LS\Omega \cos i$ for a diverging beam of solid angle Ω emitted by the surface area S in a direction making an average angle i with the normal.) This radiation has a continuous spectrum, that is, its energy has a continuous distribution as a function of the frequency of the electromagnetic waves. In order to examine this distribution, we must define a differential energy density u_ν such that the product $u_\nu \mathrm{d}\nu$ represents the energy density for those waves whose frequencies are between ν and $\nu + \mathrm{d}\nu$. In other words, the total energy density u is obtained by integrating the function u_ν over all frequencies:

$$u = \int_0^\infty u_\nu \, \mathrm{d}\nu$$

From experimental measurements, a curve can be drawn representing the variation of u_ν as a function of ν for a fixed absolute temperature T, and the same curve is obtained for all ovens having the same temperature T. Figure 1.1 compares curves obtained for different temperatures: the energy is zero

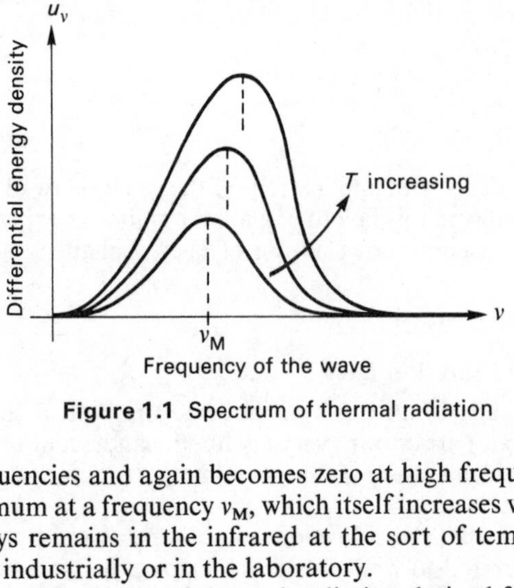

Figure 1.1 Spectrum of thermal radiation

at very low frequencies and again becomes zero at high frequencies; it passes through a maximum at a frequency ν_M, which itself increases with temperature but which always remains in the infrared at the sort of temperatures which can be achieved industrially or in the laboratory.

The thermodynamic theory of thermal radiation derived from the classical theory of electromagnetic radiation is incapable of reconstructing the form of these curves; it can explain the parabolic part of the curves observed at very low frequencies ($\nu \simeq 0$) but fails to explain the fall-off at high frequencies. To explain the shape of the curves the German physicist Planck proposed in 1900 an application of thermodynamic theory in which it is assumed that exchanges of energy between the hot walls and the electromagnetic waves are discontinuous. More precisely, he hypothesised that the energy exchanged between the walls and the wave of frequency ν is always a multiple of a certain minimum quantity $h\nu$ proportional to the frequency ν; this hypothesis allowed him to calculate the differential energy density

$$\boxed{u_\nu = \frac{8\pi h\nu^3}{c^3}\frac{1}{e^{h\nu/kT}-1}}$$

where c is the speed of light, k is Boltzmann's constant and h is Planck's constant.

A priori this formula accounts well for the form of the experimental curves. All the careful experiments done in 1901 by Lummer and Pringsheim were in complete agreement; they permitted a new measurement of the Boltzmann constant k which was in agreement with accepted values, as well as a first measurement of Planck's constant h. The presently accepted value is

$$h = 6{\cdot}6262 \times 10^{-27} \text{ erg s} = 6{\cdot}6262 \times 10^{-34} \text{ J s}$$

We shall not develop the statistical thermodynamic derivation of the Planck formula here, because it is usually treated in great detail in all courses on thermodynamics. We simply wish to recall how the phenomenon of thermal radiation, interpreted by statistical thermodynamics, gave us the first experimental proof of the quantisation of energy.

In the remainder of this chapter we shall describe in more detail some physical phenomena which demonstrate much more clearly the discontinuous character of energy exchange between matter and electromagnetic waves, and which firmly establish the existence of packets of energy $h\nu$ called photons. In the next chapter we shall identify the photon as a real relativistic particle.

This corpuscular description of radiation contradicts the classical theory of electromagnetic radiation as derived from Maxwell's equations since the latter always leads to continuous energy changes. Nevertheless the classical theory permits precise calculations to be made of the waves generated by radio and radar aerials and it completely explains the wave properties of radiation (interference, diffraction). Moreover that value $h\nu$ for the photon contains the frequency ν which is a wave quantity. The corpuscular description of the radiation is not a substitute for the classical wave picture which we shall continue to use. In chapter 4 we shall try to overcome the apparent contradiction between the two descriptions.

1.2 The Photoelectric Effect

1.2.1 Experimental Description

The photoelectric effect is particularly simple to demonstrate. One need only illuminate with a mercury lamp (an ultraviolet light source), a piece of zinc placed on a charged electroscope. If the electroscope is positively charged

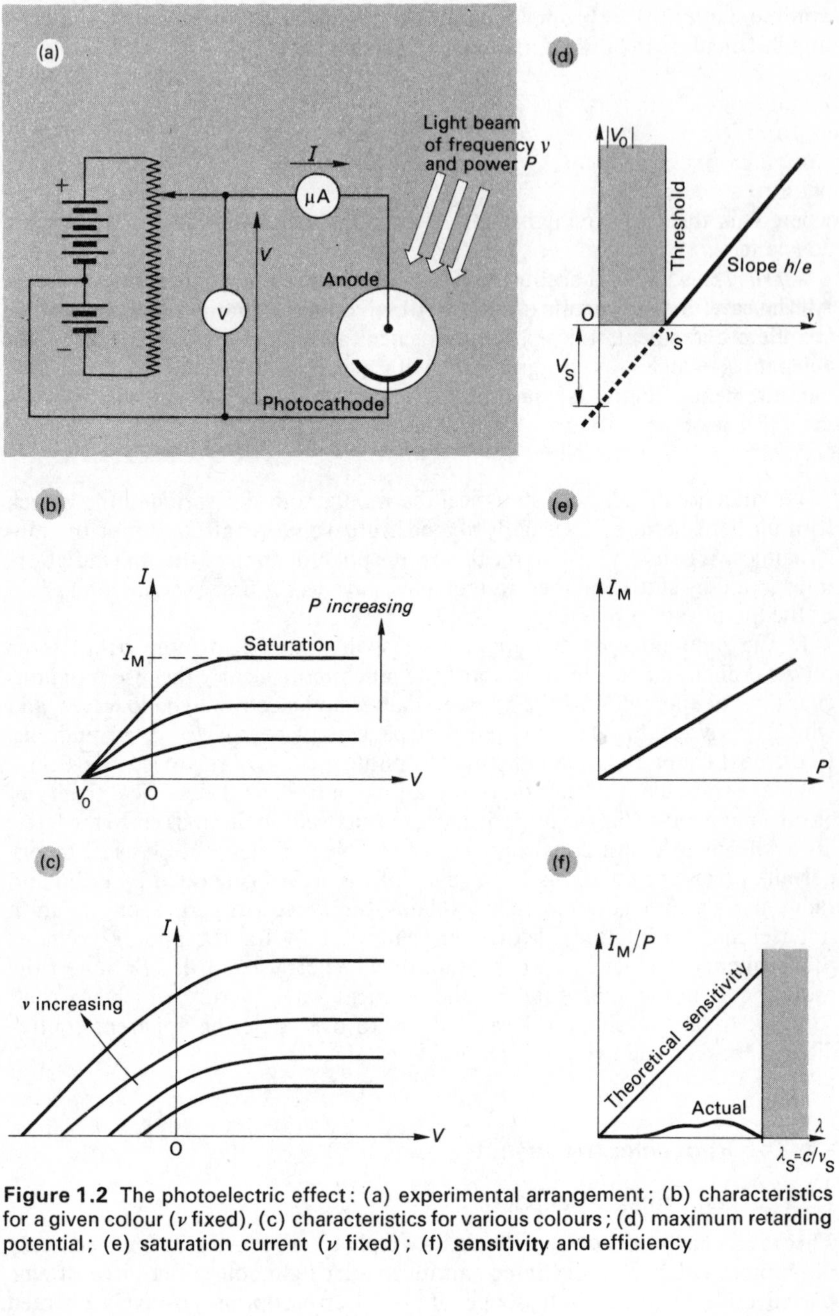

Figure 1.2 The photoelectric effect: (a) experimental arrangement; (b) characteristics for a given colour (ν fixed), (c) characteristics for various colours; (d) maximum retarding potential; (e) saturation current (ν fixed); (f) **sensitivity** and efficiency

nothing happens, but if it is negatively charged it slowly discharges; irradiation by the light allows the excess negative charge to leak away from the metal foil (Halbwachs, 1888). If a sheet of glass or perspex, transparent to visible light but which absorbes ultraviolet light, is placed between the light source and the piece of zinc the electroscope remains charged. This illustrates one of the fundamental characteristics of the photoelectric effect; it is produced only by electromagnetic waves which have a sufficiently short wavelength or high enough frequency. A more systematic study enables us to make an exact measurement of the *threshold frequency* ν_S, below which the effect does not occur. This threshold frequency is characteristic of the material used.

A more thorough study of this effect requires that the metal should be placed in a vacuum so that the negative charges that are ejected can be collected. By the use of a mass spectrograph one can measure the ratio q/m of the charge q to the mass m of these negative charges and thus identify them as electrons (Lenard, 1899).

A photoelectric cell is constructed by enclosing inside a vacuum tube a plate of metal sensitive to light, called the photocathode, and a threadlike electrode, called the anode, for collecting the electrons. Under normal operating conditions the collector electrode is at a positive potential V with respect to the plate. The relevant experimental facts concerning the operation of such a cell are given in figure 1.2.

Figure 1.2(a) illustrates the experimental apparatus used for studying the effect. The current I through the cell is measured as a function of the voltage V of the collector electrode with respect to the plate, a voltage which can be either negative or positive. Thus characteristic curves are obtained, as shown in figures 1.2(b) and 1.2(c). When the voltage V is positive and sufficiently high the current I has a maximum constant value I_M called the *saturation current*; but when the voltage V is reduced to near zero the current I is reduced as well; it does not, however, become zero at the same time as the voltage becomes zero, but only at a certain negative voltage V_0. Its modulus $|V_0|$ represents the *maximum retarding potential* beyond which no current flows because the electrode repels all the electrons.

From the point of view of using the technique in photoelectric cells, the important thing is the saturation current I_M which can be shown to be proportional to the power P of the incident light (figure 1.2(e)). This is what makes a photoelectric cell useful for measuring light intensity. However, for a proper understanding of the phenomenon the maximum retarding potential $|V_0|$ is more important. This obeys some very precise laws: (a) its value depends only on the frequency of the light used (if the power P of the beam of light is varied with ν fixed, V_0 does not change—see figure 1.2(b)); (b) its value increases with the frequency ν of the light (see figure 1.2(c)); and, more precisely, it is a linear function of the frequency. This is illustrated in figure 1.2(d), where one sees the straight line representing $|V_0|$ as a function of ν, $|V_0|$ being zero at the threshold frequency ν_S below which there is no photoelectric effect; (c) the slope of any such line is a constant independent of all experimental conditions and, in particular, of the photocathode material, whereas ν_S does depend on this material.

1.2.2 Interpretation of the Threshold and of the Maximum Retarding Potential

It is simple to understand the progressive increase of the photoelectric current I when V increases from a negative value V_0, as a space charge effect, common in all electronic tubes. Electrons that are being permanently produced and not collected quickly enough by the anode, accumulate in the neighbourhood of the cathode and form a negatively charged region within the vacuum, which tends to prevent the passage of further electrons. When the anode voltage V is large enough it pulls the electrons away sufficiently quickly and the space charge disappears. All the electrons detached from the photocathode by the light are then collected and so the maximum current I_M, the saturation current, is explained.

On the other hand the laws of classical physics cannot explain why the maximum retarding potential $|V_0|$ and the photoelectric threshold ν_S are both independent of the power P of the light wave. It might seem that the effect should depend upon the force exerted on the electrons by the electric field of the wave, and this would be correspondingly larger the greater the power P. However, these characteristic properties of the photoelectric effect can be explained quite easily if one supposes that energy exchange between radiation and matter is quantised, as suggested by Einstein in 1905.

The electrons are normally bound to the metal, so that in order to remove one of them from the metal, the opposition of attractive forces must be overcome. If an electron is successfully detached from the metal, these attractive forces will have done work, known as the extraction energy or work function $\mathscr{T}_S$, so that the electron will have been given an escape energy $W_S = |\mathscr{T}_S|$. In other words, in order to extract the electron from the metal, the potential energy of the metal–electron system must be increased by W_S. (For a fuller discussion of the significance of W_S consult a textbook on solid-state physics.) This escape energy can be measured by using the phenomenon of thermionic emission, since the theory of this effect enables one to calculate the saturation current of a diode as a function of the absolute temperature T of the heated metal filament from the formula

$$I_S = AT^2 \exp(-W_S/kT)$$

(see texts on statistical thermodynamics). The work function is a characteristic of each metal or alloy used.

This means that to detach an electron from a metal it must be given energy that can compensate for the work done $\mathscr{T}_S$ to overcome the forces which bind it to the metal; that is, an energy greater than W_S. If the energy in the light wave is transferred to the electrons in the form of separate photons each of energy $h\nu$, these can only detach electrons from the metal when

$$\boxed{h\nu \geqslant W_S = h\nu_S}$$

The threshold frequency of the photoelectric effect is thus explained; photons of threshold frequency ν_S have exactly the escape energy W_S.

When the frequency ν of the light wave is greater than ν_S the excess energy of the photon can be taken up by the electron in the form of kinetic energy, in such a way as to maintain energy balance

$$\boxed{h\nu = W_S + \tfrac{1}{2}mv^2}$$

This is Einstein's equation for the photoelectric effect. The kinetic energy $\frac{1}{2}mv^2$ thus calculated is the maximum kinetic energy of the photoelectrons when they are not prevented by space charge from leaving the metal. It is their initial kinetic energy which enables the electrons to reach the collector electrode even when its potential is negative and repels them. The maximum retarding potential $|V_0|$ that the electrons are able to overcome is determined by their maximum kinetic energy

$$e|V_0| = \tfrac{1}{2}mv^2 = h\nu - W_S = h(\nu - \nu_S)$$

where e is the elementary electric charge, equal to $1{\cdot}6 \times 10^{-19}$ C; the electron carries a charge $q = -e$.

This is a satisfactory explanation of why the maximum retarding potential $|V_0|$ is a linear function of the frequency ν, which is zero for the threshold frequency ν_S and whose slope is a positive constant h/e.

The validity of this interpretation is confirmed by numerical values measured experimentally:

(a) By measuring the slope of the straight line $|V_0| = f(\nu)$ the value of the constant h/e is obtained, and from it an experimental value of Planck's constant h may be deduced. Practically the same value is obtained as in measurements on thermal radiation.

(b) For different metals, escape potentials V_S may be determined, such that $eV_S = W_S = h\nu_S$ (see figure 1.2(d)). The value of V_S represents a measurement in electron volts of the escape energy W_S. This is in good agreement with thermionic emission measurements. Some values are set out below in order to give an idea of orders of magnitude

Metal	Cs	Rb	K	Na	Ca	Mg	Zn	Fe	Ni
Excape potential V_S in volts	2·1	2·2	2·4	2·5	2·3	2·4	3·4	4·8	5·0

V_S is easily related to the photoelectric threshold wavelength λ_S for each metal by the relation

$$\lambda_S = \frac{c}{\nu_S} = \frac{hc}{e}\frac{1}{V_S} = \frac{1240}{V_S\ \text{(in volts)}}\ \text{nm}$$

This wavelength is in the visible region for the alkali metals or the alkali earths but it is in the ultraviolet for Zn, Fe and Ni.

Comment These measurements are more difficult to obtain than might be supposed from the curves drawn in the diagram. The determination of the voltage V_0 is complicated in practice by a parasitic photoelectric effect at the anode collector, which can cause the current I to change sign. In the high precision measurements which he carried out between 1905 and 1916 Millikan made the anode of the photoelectric cell with a metal different from the photocathode, chosen so that its threshold frequency was much higher than that of the photocathode. By working with frequencies intermediate between the two thresholds, this parasitic photoelectric effect is suppressed.

But working with two different metals presents a new problem. It is known that contact potential differences exist between different metals. Their algebraic sum is zero in a closed and isothermal metallic chain, but this is not the case with a photoelectric cell where the metal chain is interrupted inside the cell between the two electrodes. It follows that the real potential V applied to the electrons in the vacuum is not equal to the potential U measured with the voltmeter; the difference between them is equal to the contact potential differences K between the metals which constitute the two electrodes. When the straight line $|V_0| = f(\nu)$ is drawn, it is displaced vertically by an unknown amount K. This does not change the slope h/e of the line, but prevents verification of the relation $h\nu_S = W_S = eV_S$ and prevents an accurate measurement of the escape potential V_S.

To overcome this difficulty Millikan used three metals, A, B and C with increasing photoelectric thresholds

$$\nu_S(A) < \nu_S(B) < \nu_S(C)$$

To study the photoelectric effect of metal A, an anode made from either metal B or metal C can be used, without being affected by the parasitic photoelectric effect of the anode. The voltmeter measures the following voltages when the photoelectric current is zero:

$$\begin{cases} \text{with anode B} \quad U_{OB}(A) = V_0(A) + K_{AB} \\ \text{with anode C} \quad U_{OC}(A) = V_0(A) + K_{AC} \end{cases}$$

$V_0(A)$ is the real potential existing in the vacuum such that $qV_0(A) = \frac{1}{2}mv^2$ and is unknown, as are the contact potentials K_{AB} between A and B and K_{AC} between A and C.

By subtracting the two equations term by term one obtains

$$U_{OC}(A) - U_{OB}(A) = K_{AC} - K_{AB} = K_{BC}$$

where K_{BC} is the contact potential between the two metals B and C (this last equality results from the sum of the contact potentials in a closed chain being zero). In this way the contact potential K_{BC} is measured.

The photoelectric effect for metal B can now be studied with an anode made from metal C. When the photoelectric current is zero the voltmeter measures the potential

$$U_{OC}(B) = V_0(B) + K_{BC}$$

Thus the real potential in the vacuum $V_0(B)$ can be calculated and enables the investigations we have discussed above to be carried out.

1.2.3 Sensitivity and Quantum Efficiency

We have seen that the saturation current I_M is for a given colour (a given frequency) proportional to the power P transported by the beam of light (see figure 1.2(e)). The sensitivity of the photocell is measured from the slope of this line, that is to say, the ratio I_M/P.

This ratio can be calculated by making the hypothesis that each incident photon frees an electron. The number of photons arriving in one second is $N = P/h\nu$; they then release N electrons, which give rise to a current $I_M = Ne$. One then deduces that

$$\frac{I_M}{P} = \frac{Ne}{Nh\nu} = \frac{e}{h\nu} = \frac{e}{hc}\lambda$$

The theoretical sensitivity thus calculated is proportional to the wavelength λ of the light; this is shown in figure 1.2(f). Also shown in this figure, on the same scale, is the behaviour of a curve measured with a real photocell; the actual sensitivity is seen to be a good deal less than the theoretical sensitivity calculated above.

This does not mean that the explanation of the photoelectric effect should be reconsidered, as there is no other way to explain the frequency threshold and the maximum retarding potential $|V_0|$. It is sufficient to make the following statement: the photons that successfully detach an electron from the metal are a small proportion of the incident photons; the majority of them are transformed into energy of thermal agitation (heat) in the metal plate, or else are scattered by the plate and reappear outside the cell. This proportion of effective photons in relation to the total incident number of photons is called the quantum efficiency; in other words the ratio $\eta = n/N$ between

$$\begin{cases} \text{the number } n \text{ of photoelectrons ejected in one second such that } I_M = ne; \\ \text{the number } N \text{ of incident photons, such that } P = Nh\nu. \end{cases}$$

The actual sensitivity of the photocell is now written as

$$\frac{I_M}{P} = \frac{ne}{Nh\nu} = \eta\frac{e}{hc}\lambda$$

It is equal to the product of the theoretical sensitivity and the quantum efficiency η, and it allows us to measure the latter.

To determine orders of magnitude, the current I_M and power P are expressed in SI units, and the wavelength in nanometres; the theoretical sensitivity is then calculated as

$$\frac{I_M}{P}\text{ (in amperes per watt)} = \frac{\lambda\text{ (in nanometres)}}{1240}$$

(Note that the numerical constant that appears in this formula is equal to that used in calculating the threshold wavelengths λ_S from the escape potentials.) Thus for the middle of the visible spectrum ($\lambda \approx 600$ nm) one calculates a theoretical sensitivity of 0·5 A/W. The sensitivities measured in modern photocells vary between 1 and 100 mA/W, so that typical values of quantum efficiency η vary between 1/5 and 1/500.

1.2.4 Photoionisation

In the experiments described above, the photoelectric effect is produced in an atomic assembly bound in the form of a solid body. However, the photoelectric effect can also be observed in isolated atoms such as exist in a monatomic vapour. By irradiating a vapour with ultraviolet light of sufficiently short wavelength, its bound electrons can be ejected and observed. The atoms from which these photoelectrons have been detached form positive ions; they can be simultaneously observed and identified by mass spectrographic techniques. This is why the photoelectric effect in atoms of a vapour is given the name photoionisation.

Photoionisation only occurs if the frequency of the light is sufficiently high, and a threshold frequency ν_i can be measured, below which ionisation does not occur. The threshold frequency is interpreted as above: a photon $h\nu$ in the incident wave can only detach an electron from an atom if it has an energy greater than the escape energy W_i of the electron from the atom

$$h\nu \geqslant W_i = h\nu_i = hc/\lambda_i$$

Threshold wavelengths λ_i for photoionisation are given in the table below for vapours of alkali metals and for the rare gases. It will be noted that thresholds always occur in the ultraviolet region and often in the far ultraviolet.

Atom	Cs	Rb	K	Na	Li	Xe	Kr	A	Ne	He
λ_i in nm	318·4	296·8	285·6	241·2	230·0	102·2	88·5	78·7	57·5	50·4

The escape energy W_i corresponding to the extraction of an electron from an isolated atom is called the ionisation energy of the atom; it will be discussed again in section 1.4.1 in relation to ionisation potentials. We limit ourselves here to a simple comparison: the wavelength λ_i corresponding to the ionisation of an alkali vapour is much shorter than the corresponding photoelectric threshold wavelength λ_S for a piece of metal composed of the same atoms. In other words, to detach an electron from an isolated atom requires an energy W_i much greater than the energy W_S necessary to extract an electron from a block of atoms bound to one another.

The comparison between experiments on the photoelectric effect carried out on isolated atoms, or on atoms bound to one another in the form of a block of metal, can be extended by considering the quantum efficiency η. We have seen that of the photons which strike a metal plate, the proportion η that produce a photoelectric effect is small. Most of the light energy undergoes absorption processes without the release of photoelectrons. These processes are related to the opaque properties of the solid metal body (in contrast to the transparency of glass for example), and one would not expect these to occur in a vapour. Measurements of the quantum efficiency of photoionisation have been made recently (1964) in the rare gases and they show that $\eta = 1$. This does not mean that all photons passed into the gas whose wavelength is above the threshold in the far ultraviolet are necessarily absorbed. If the pressure of the gas is very low, the number N of photons absorbed might be only a small fraction of the incident photons. However, it does mean that each photon absorbed has produced a photoelectron; the number N of the photons absorbed is equal in this case to the number of photoelectrons produced, and $\eta = n/N = 1$.

Comment In the case of photoionisation the measurement of quantum efficiency presents some difficulties. It is necessary to measure the number N of photons absorbed and the number n of photoelectrons produced simultaneously. By comparing the intensities of a beam of light before and after traversing the vapour, the fraction x of the power which is absorbed by the vapour can readily be determined and thus the number of photons absorbed in one second is found to be $N = xP/h\nu$. But the photoelectrons are produced within the vapour and it is difficult to collect all of them in order to measure the corresponding electric current.

To extract the electrons from the vapour it is confined between the two plates of a condenser, to which a potential difference V is applied. The electrons are attracted to the positive plate, but the number of electrons collected depends on the potential V. If V is too small they are not all successfully collected and the current I measured is too small (some of them recombine with positive ions within the vapour); if on the other hand V is too large, an electric discharge starts in the vapour and the discharge current is added to the current of the photoelectrons. However, experiments show that there exists an intermediate plateau in the values of V where the current I collected on the plates is constant: that is to say, all the photoelectrons produced in the vapour are successfully collected without initiating a discharge and the number of photoelectrons produced in one second $n = I/e$ can then be calculated.

We have discussed in this section only the photoelectric effect under the action of ordinary visible or near visible light, so that the energies concerned are of the order of several electron-volts; but the photoelectric effect occurs with X-ray photons whose energies are measured in thousands of electron-volts. The magnitude of the energy involved enables the production of photoelectrons corresponding to different escape energies W_S, some of which can be very large; thus evidence is provided of the existence of more or less firmly attached electrons inside the atom. The photoelectric effect with X-rays is described at the beginning of chapter 7—section 7.1, X-ray absorption spectra and section 7.2, the velocity spectrum of X-ray photoelectrons, may be found appropriate here and could be read at this stage.

The photoelectric effect in metals proves that exchanges of energy between electromagnetic waves and matter take place in minimum quantities equal to $h\nu$. The photoelectric effect with X-rays confirms this and also provides information about the structure of atoms. That is why we have devoted a chapter entirely to X-rays.

1.3 Optical Spectra

Light sources commonly used fall into two main categories as follows.

(*a*) *Thermal sources*, in which a refractory material is at a high temperature (this can be a metal conductor as in the tungsten filament of an electric light bulb or an insulator such as the oxide of cerium found in the Auer mantles of old gas lamps). These thermal sources emit radiation distributed continuously over all frequencies; this *continuous spectrum* depends largely on the temperature and little on the material used. It is the thermodynamic explanation of this radiation which led Planck to envisage discontinuous energy changes between radiation and matter (see section 1.1).

(*b*) *Light sources from a discharge*, in which an electric current passes through a vapour of a given chemical element (this may be a vapour in equilibrium, occupying a closed space and heated as in a discharge lamp, or it may be a short-lived but constantly replenished vapour formed by atoms detached from the electrodes of an electric arc).

In contrast to thermal sources, discharge sources only emit radiation of certain well-defined frequencies, evidenced by the *line spectra* observed in a spectrograph. The frequencies corresponding to each of these lines are defined with extreme precision (relative precision of the order of 10^{-6}) and are characteristic of the different atoms or ions in the vapour present in the discharge.

Comment Spectroscopists sometimes distinguish between the arc spectra emitted by neutral atoms and the spark spectra emitted by ions produced in the more severe conditions of an electric discharge. To be more exact, they are called for example

a spectrum of Fe^{I}, the spectrum emitted by neutral iron atoms
a spectrum of Fe^{II}, the spectrum emitted by Fe^{+} ions
a spectrum of Fe^{III}, the spectrum emitted by Fe^{2+} ions
a spectrum of Fe^{IV}, the spectrum emitted by Fe^{3+} ions, and so on.

The lines of the different ionic spectra Fe^{II}, Fe^{III} and so on can be identified by the fact that they appear only under certain discharge conditions, simultaneously with the presence of ions Fe^{+}, Fe^{2+} and so on observed in a mass spectrograph.

1.3.1 The Combination Principle and the Bohr Hypothesis

The existence of these emission frequencies, characteristic of each type of atom, is an extremely important experimental fact. Its interpretation was possible only after the discovery of the fundamental law which relates the emitted frequencies from a particular atom.

In the case of the hydrogen atom, this law is called the Balmer–Rydberg law: the various wavelengths λ emitted by the hydrogen atom are given by the formula

$$\frac{1}{\lambda} = R\left(\frac{1}{n^2} - \frac{1}{p^2}\right)$$

where R is an experimentally determined constant called the Rydberg constant and n and p are two integers.

If the wavelengths are expressed in centimetres so that their reciprocals, called *wavenumbers* are in cm^{-1}, the Rydberg constant has the value $R = 109\,677$ cm^{-1} (to establish orders of magnitude, note that the visible wavelength $\lambda = 500$ nm corresponds to the wavenumber $1/\lambda = 20\,000\ \mathrm{cm}^{-1}$).

The wavenumbers of the various lines emitted by hydrogen can thus be written $1/\lambda_{np}$ with two indices n and p, and they are equal to the differences which can be calculated between the terms R/n^2 of a series dependent on only one index.

Historical Comment Balmer in 1885 was aware only of the lines of hydrogen appearing in the visible spectrum corresponding to a fixed value $n = 2$, and he wrote the law in the form $\lambda = \lambda_0 p^2/(p^2 - 4)$. This expression was recast in 1889 by Rydberg, who wrote

$$\frac{1}{\lambda} = \frac{4}{\lambda_0}\left(\frac{1}{4} - \frac{1}{p^2}\right) = R\left(\frac{1}{4} - \frac{1}{p^2}\right)$$

This form of the law could readily be generalised when hydrogen lines corresponding to other values of n were discovered in the ultraviolet and infrared at the beginning of this century.

The law obeyed by the hydrogen lines is particularly straightforward because hydrogen with its single electron is the simplest atom. But this law can be generalised to some extent for all atoms by means of the combination principle stated by Ritz in 1908: the wavenumbers $1/\lambda$ of radiation emitted by a particular atom can be labelled with two indices and expressed as the differences between terms of a series dependent on a single index

$$\boxed{\frac{1}{\lambda_{np}} = T_n - T_p}$$

Thus, *one can determine from experimental values measured on a particular atom, a series of numbers T_n called spectral terms, such that every wavenumber corresponding to a spectral line of this atom is equal to the difference of two spectral terms.*

The Danish physicist Niels Bohr interpreted this law in 1913 in terms of photons: the wavenumber $1/\lambda$ is proportional to the frequency ν, and therefore to the energy $h\nu$ of the corresponding photons

$$h\nu_{np} = hc/\lambda_{np} = hcT_n - hcT_p$$

so that the combination principle stated for the wavenumbers applies equally well to the energy $h\nu_{np}$ of the various photons that can be emitted by the same atom.

If we suppose that the emission process takes place independently for each isolated atom, so that each photon is emitted by a single atom, the energy $h\nu_{np}$ represents the loss of energy suffered by an atom during the emission process and the law of conservation of energy requires that

$$h\nu_{np} = E_{\mathrm{i}} - E_{\mathrm{f}}$$

where E_{i} is the initial energy of the atom before emission and E_{f} the final energy of the atom after the emission of a photon.

Comparison of the two preceding equations leads to an identification of the numbers hcT_n and hcT_p with the values of energy E stored in the atom before and after emission. Since the energy of an emitted photon cannot have a value other than the difference between the two terms hcT_n, it may be concluded that the atom cannot possess values of energy other than the values hcT_n. Thus, by taking into account the existence of photons, Bohr eventually explained the Ritz combination principle in his hypothesis '*the energy stored in an atom can take only particular values forming a discontinuous series*'.

The atomic states corresponding to these particular values of energy are called the energy levels of the atom. Electromagnetic waves corresponding to the spectral lines are emitted when the atom makes a transition between two energy levels. For reasons to be given subsequently (see section 1.3.2 below and chapter 6) these energy values are defined to be negative $E_n = -hcT_n$. (It should not be forgotten that these energies are defined only by their differences and are not defined to within an additive constant.)

The atom emits a wave of frequency λ_{np} when it goes from the initial energy state $E_p = -hcT_p$ to the lower energy state $E_n = -hcT_n$ and the Bohr hypothesis is written as

$$\boxed{h\nu_{np} = E_p - E_n}$$

All that we have said is summarised with reference to hydrogen in figure 1.3. The values of the spectral terms T_n (expressed in cm^{-1}) are plotted on the vertical axis increasing towards the bottom, and for each value a straight horizontal line has been drawn. The number n of the spectral term is shown in the column on the right, increasing towards the top with an appropriately changing scale so that the horizontal lines represent the energies corresponding to each atomic state (this is the relation $E_n = -hcT_n$). The transitions between the energy levels are represented by the vertical arrows. The energy of the photon emitted in each transition is represented by the length of the arrow. Alongside some of these arrows the wavelength measured in nanometres has been indicated.

Hydrogen is an exceptional case. It is not usual to observe in the spectrum of an atom all the frequencies ν_{np} that can be calculated by associating any number n with any number p; transitions between any particular level E_n and any other level E_p are not all possible.

Comment I (historical) The Bohr hypothesis is a simple generalisation applied to an isolated atom of the law relating to quantised energy changes between radiation and matter, already described in the case of a red-hot oven (Planck's Law) or the free electrons within an illuminated piece of metal (photoelectric effect). However, this law completely contradicts the results of classical electromagnetic theory which in many respects is a very satisfactory theory of electromagnetic waves. In fact, Bohr was inspired to make this generalisation only after Rutherford's experiment had demolished Thomson's classical model of the atom (elastically bound electrons) and showed the impossibility of explaining by means of classical theory the mechanism for the emission of a wave by an atom (see section 5.3).

Comment II Atoms also emit spectral lines in the X-ray region (see section 7.3.2). These X-ray lines obey the same combination principle as lines in the visible spectrum and in the same way they enable an interpretation of spectral terms as energy

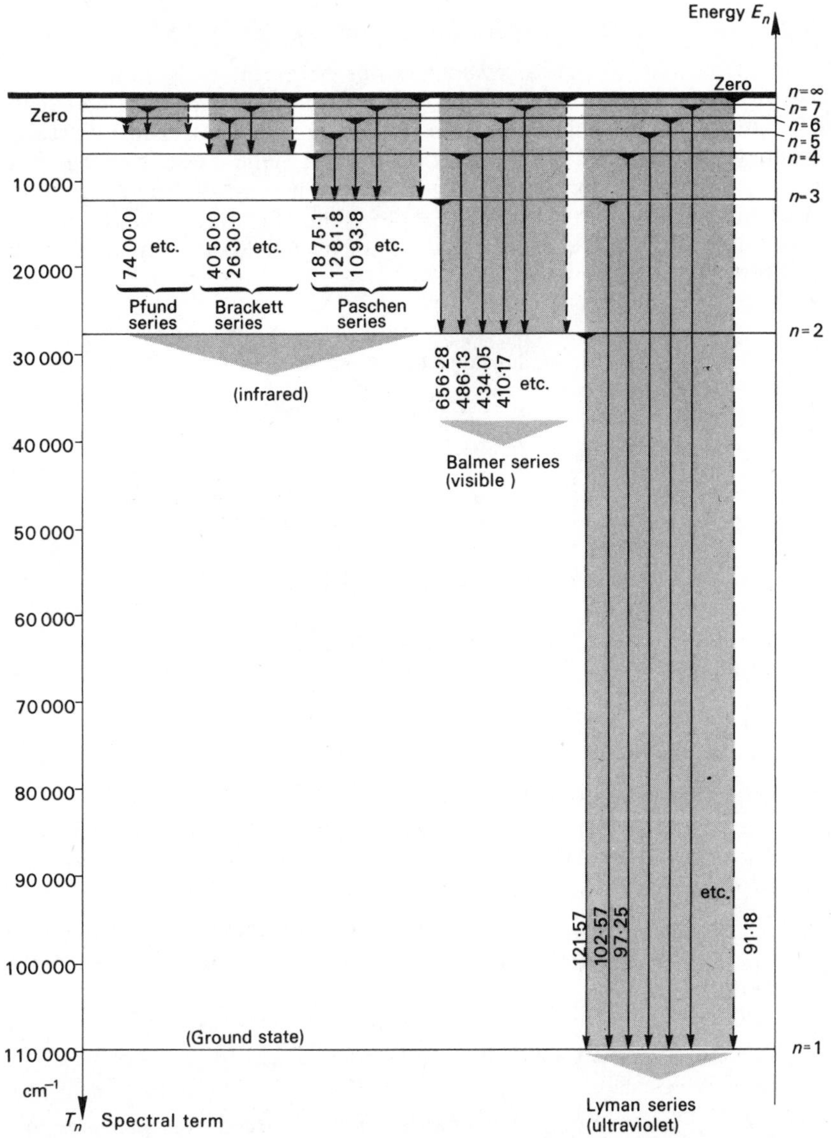

Figure 1.3 Energy diagram and spectral transitions of atomic hydrogen. Alongside some arrows the value in nanometres of the corresponding wavelength has been indicated

levels. The comparison of the X-ray emission lines with the X-ray photoelectric effect promotes further understanding of the internal structure of atoms, and is the reason for gathering together everything concerned with X-rays in a separate chapter.

1.3.2 Optical Resonance Experiments — the Ground State of an Atom

The phenomenon of optical resonance was observed on sodium atoms in 1905 by Wood, an American. Observation of this phenomenon is quite easy with sodium atoms. Discharge lamps containing sodium are manufactured nowadays and are often used to light streets. They emit a yellow-orange light which is practically monochromatic; its spectrum consists almost entirely of the sodium D line, of wavelength $\lambda = 589{\cdot}3$ nm (the other spectral lines of sodium occurring in the visible range are of negligible intensity under these conditions).

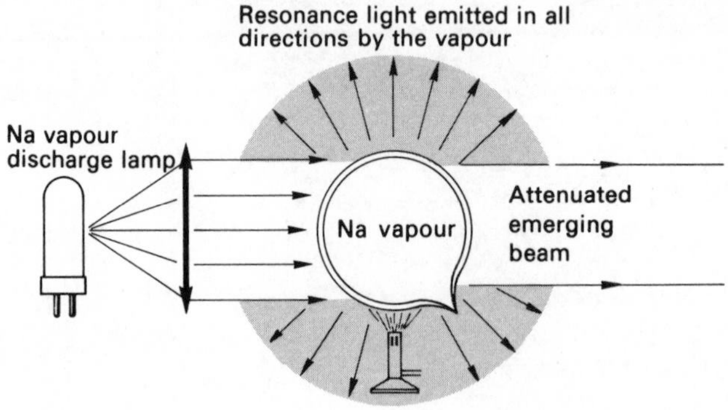

Figure 1.4 An optical resonance experiment

This light can be used to illuminate a glass bulb containing sodium vapour. (For this a bulb containing a small piece of sodium is evacuated and then heated with a Bunsen burner to establish a sufficiently high vapour pressure.) The following observations can then be made (see figure 1.4).

(a) The beam of light emerging from the bulb after passing through the vapour is attenuated, part of its intensity having been absorbed in the vapour.

(b) The vapour through which the incident beam passes, itself becomes a source of light. It emits in all directions light of the same wavelength, called fluorescent light or, to be more precise, resonance fluorescence.

The phenomenon of optical resonance can be regarded as a *particular case of the phenomenon of fluorescence*. The term fluorescence refers in a general way to the emission of light by a body after it is excited by irradiation with a beam of primary light. The light emitted in fluorescence is generally of lower frequency than the primary light which caused the effect. For example, illumination in darkness uses fluorescence which produces visible light under the action of a primary beam of ultraviolet light. Optical resonance is a particular case of fluorescence where the light emitted by the body has the same wavelength as the primary beam used to excite the emission. This *fluorescence without change of frequency* only arises in the case of certain frequencies characteristic of the atom in the vapour.

The same kind of experiment can be carried out on other monatomic vapours (in which the particles constituting the vapour are single atoms not grouped in molecules) and confirms the following general laws:

(a) a monatomic vapour can strongly absorb a beam of light only if the frequency of this light coincides with that of a spectral line of the atoms in the vapour;

(b) only certain spectral lines have frequencies permitting the observation of this intense absorption which is accompanied by the emission of fluorescence of the same frequency. These spectral lines are called resonance lines.

Comment The phenomenon of optical resonance is clearly distinguishable from the phenomenon of scattering, which also occurs without change of frequency, by two essential characteristics: first, a much greater intensity, and secondly a high selectivity of frequency and the fact that the appropriate frequencies are characteristic of the atom used.

These experimental facts may be readily interpreted by the Bohr hypothesis concerning the existence of atomic energy levels.

(i) If the exchange of energy between a single atom and an electromagnetic wave occurs only in the form of photons, and if the energies corresponding to the possible states of the atom can only take discrete values $E_1, E_2, \ldots E_n, \ldots$ then the atom can only absorb photons having the exact energy difference between two possible states E_n and E_p, thereby allowing a transition to be made from a lower level E_n to a higher level E_p; in other words photons obey the Bohr condition $h\nu_{np} = E_p - E_n$.

This is the reason the atom can only absorb waves having the same frequency as one of its characteristic spectral lines.

(ii) It remains to explain why absorption only occurs for some frequencies and not for others. As is generally the case in physics, the most stable states of a system are those for which its energy is a minimum. The atom can be regarded as being stable only when in the state corresponding to the minimum value E_1 of its possible energy values. This state is called the *ground state of the atom.* If the vapour normally contains only atoms in the ground state, it can only absorb those photons capable of inducing a transition from this ground state, that is to say, corresponding to certain special spectral lines

$$h\nu_{1p} = E_p - E_1$$

This is why there are so few resonance lines in relation to all the spectral lines of the atom.

This interpretation is confirmed if the two spectral terms corresponding to the wavenumbers of each of the resonance lines are considered. It can be seen that the first of these spectral terms remains the same for all the resonance lines of the atom

$$\frac{1}{\lambda_{1p}} = T_1 - T_p$$

The fact that the spectral term T_1 common to the various resonance lines is the largest of the spectral terms of the atom helps to justify the choice of negative energies: the maximum spectral term T_1 corresponds to the energy minimum $E_1 = -hcT_1$ of the ground state.

For hydrogen (as in the diagram) the resonance lines are the lines of the Lyman series, all falling in the ultraviolet.

To conclude, the radiation emitted or absorbed by an atom is accounted for by the hypothesis of energy levels: the atom is normally in the ground state of minimum energy E_1 but an external influence (radiation at the resonance frequencies or electric current in a discharge) can raise the atom to a higher energy state E_n; the atom is said to be excited or in an excited state. The excited atom tends to return spontaneously to states of lower energy by giving up excess energy in the form of a photon; it can make either a transition to another excited state of lower energy or a transition direct to the ground state (by emitting a resonance frequency). The experiments described in the next section will confirm this hypothesis by analysing the excitation mechanism of atoms in an electric discharge.

Comment I By using a spectrometer with sufficient dispersion the sodium D line is resolved into the two components D_1 at $\lambda = 589{\cdot}6$ nm and D_2 at $\lambda = 589{\cdot}0$ nm. Each of these lines behaves separately as a resonance line and using the two simultaneously gives rise to the same effects.

Comment II In the classical theory of radiation the fixed frequencies of the spectral lines were explained by supposing the existence of elastically bound electrons in the atom having their own well-defined frequencies of free oscillation (Thomson model: see section 5.2 and appendix 2). This did not succeed in explaining the Ritz combination principle but, on the other hand, it did explain the optical resonance experiments described above.

The electric field of the incident wave exerts an alternating force on each electron which has the same frequency ν as the wave; this sets it into forced oscillation at the frequency ν. The results of the general theory of elastically bound oscillators are well known: if the frequency of the alternating force differs greatly from the natural frequency ν_0 of the oscillator, the amplitude of the forced oscillation is very small and the stored energy of the oscillator does not change; but if the frequency of the force is equal to the natural frequency ν_0 of the oscillator, the amplitude of the forced oscillation is extremely large and the oscillator therefore stores a large amount of energy. This is called a resonance phenomenon from which the term optical resonance experiment is derived.

In contrast to the classical description, note that all that has been said in this section relies solely on energy conservation and does nothing to prejudge the internal make-up of the atom.

Comment III Specular reflection. When an optical resonance experiment is carried out as we have described above, the density of the vapour (that is the number of atoms per unit volume n) can be varied by changing the temperature of the bulb containing it. The density n increases with vapour pressure, which in turn is determined by the temperature of the coldest point of the container (principle of cold walls). When the density of the vapour increases the attenuation of the incident beam increases as well as the intensity of the fluorescent light re-emitted by the vapour. If the density is greatly increased, the incident beam is completely absorbed inside the bulb; it can even be totally absorbed before completely traversing the bulb. In this case, only the vapour near the entrance face of the bulb is illuminated and the vapour on the other side of the bulb is unaffected; the fluorescent light then comes only from that part of the bulb which is nearest to the point of entrance of the incident beam. If the distance over which the incident beam is totally absorbed becomes less than a millimetre, the fluorescent light appears to come from the walls themselves; then it seems as if the bulb is painted the colour of the incident beam, but the phenomenon has not changed fundamentally.

On the other hand a significant change is produced in the optical resonance phenomenon if the vapour density becomes sufficiently high so that the incident beam is totally absorbed in a distance from the entrance window smaller than the wavelength of the beam. Instead of being re-emitted in all directions, the light is transmitted through the vapour in a special direction; it appears as if the incident beam is reflected at the boundary of the vapour, as at the surface of a metal conductor. This is called specular reflection. Note that this reflection occurs only when the wave has a wavelength exactly equal to that of the resonance line; all other waves of different wavelengths continue to pass normally through the bulb of vapour.

This phenomenon of specular reflection can be explained by using wave language and assuming that the wavelets emitted by each excited atom have a precise phase relation to the incident wave that excited the atom (in agreement with the classical model of an elastically bound electron). The emitting atoms are enclosed within a volume element of linear dimension which is small compared with the wavelength λ, and the dephasing of a wave propagating over this distance within the volume element may be disregarded. That is to say, the emitting atoms may be thought of as distributed over a surface. If the wavelets emitted from this surface by different atoms all have the same phase relation to the incident wave, this becomes a problem of wave optics similar to the problem of metallic reflection. The wavelets are all in phase for a particular direction, symmetric with the direction of incidence in relation to the normal. In this direction the total amplitude resulting from interference is very large. In other directions, the wavelets have various phases and their interference gives rise to a zero resultant amplitude.

This experiment with specular reflection shows that photon language does not make wave language redundant. The problem of compatibility between the two languages therefore arises and is taken up again in chapter 4.

1.3.3 Width of Spectral Lines

Throughout this course it will be seen that many experiments involve the study of a transition between energy levels of an atomic system. This transition, which may be either in the optical or radiofrequency regions, is not perfectly monochromatic, but has a certain frequency width $\Delta\nu$. Several effects can contribute to this width.

(*a*) *Doppler effect.* Atoms emitting electromagnetic radiation are not usually stationary. In a majority of cases they are in the form of a gas (which is excited by an electric discharge). The emitting atoms then have a certain velocity of thermal agitation. For an observer in the laboratory frame of reference, the emission must be considered as coming from a source in motion, and it is necessary to take account of the Doppler effect. The speed v of an atom is small compared to the speed of light c, so it is possible to use the classical expression for the Doppler effect. If we let θ be the angle between the direction of observation and the velocity vector, the change of frequency $\delta\nu$ between the exact frequency ν_0 of the transition and the frequency ν seen by the observer is

$$\frac{\delta\nu}{\nu_0} = \frac{\nu - \nu_0}{\nu_0} = \frac{v\cos\theta}{c} = \frac{v_x}{c}$$

where v_x is the component of the velocity in the direction of the observer. If we assume that the temperature of the gaseous source of light is uniform, the distribution of speeds of the atoms is a maxwellian distribution and therefore we know the number of atoms in the source whose component of velocity v_x is between v_x and $v_x + \mathrm{d}v_x$

$$\mathrm{d}N = Nf(v_x)\,\mathrm{d}v_x$$

$f(v_x)$ being the probability density for the component v_x

$$f(v_x) = \sqrt{\left(\frac{M}{2\pi RT}\right)}\exp\left[-\frac{M}{2RT}v_x^{\ 2}\right]$$

where N is the total number of atoms in the source, M the molecular weight of the gas and R the perfect gas constant.

Let $P_\nu \mathrm{d}\nu$ be the power emitted in the band of frequencies between ν and $\nu + \mathrm{d}\nu$ (where ν is close to ν_0). It is proportional to the number of atoms emitting in this same band of frequencies, that is to say, having a component of velocity v_x between v_x and $v_x + \mathrm{d}v_x$ such that

$$v_x = c\frac{\nu - \nu_0}{\nu_0} \quad \text{and} \quad \mathrm{d}v_x = c\frac{\mathrm{d}\nu}{\nu_0}$$

If K, K' and K'' are constants of proportionality

$$P_\nu \mathrm{d}\nu = KNf\left(c\frac{\nu - \nu_0}{\nu_0}\right)c\frac{\mathrm{d}\nu}{\nu_0}$$

so that

$$P_\nu = K'f\left(c\frac{\nu - \nu_0}{\nu_0}\right) = K''\exp\left[-\frac{M}{2RT}\frac{c^2(\nu - \nu_0)^2}{\nu_0^{\ 2}}\right]$$

Figure 1.5 shows how the intensity varies as a function of ν. It is easy to determine $\Delta\nu_D$, the width at half-height of the curve, by finding the difference $\nu - \nu_0$ such that the exponential term equals $\frac{1}{2}$. It is then found that

$$\delta\nu = \frac{\nu_0}{c}\sqrt{\left(\frac{2RT}{M}\ln 2\right)}$$

and the width at half-height will be twice this difference

$$\Delta\nu_D = \frac{2\nu_0}{c}\sqrt{\left(\frac{2RT}{M}\ln 2\right)}$$

Thus Doppler broadening is evidently proportional to the square root of the temperature, proportional to the frequency ν_0 and inversely proportional to the square root of the molecular weight. The reader may verify numerically that for the yellow lines of sodium, in a gas at 500 K the Doppler broadening in wavenumbers would be $\Delta(1/\lambda) = 0{\cdot}057\ \text{cm}^{-1}$.

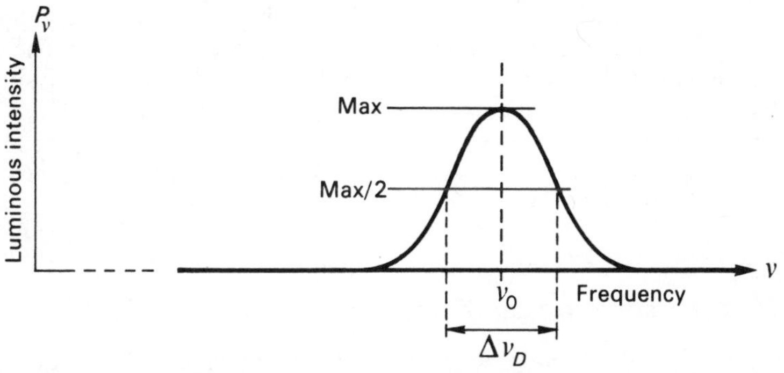

Figure 1.5 Doppler width of a spectral line

(*b*) *Broadening due to the Stark effect.* Under the influence of an electric field atoms are perturbed and their usual energy levels are replaced by several closely spaced energy levels. This is known as the Stark effect. The observed spectrum becomes complex, a line splitting into many closely spaced components. In order to resolve these components, when studying the influence of the electric field, it is generally necessary to use high voltages of the order of many kilovolts, giving rise to electric fields as high as 10^7 V/m. Light sources contain constant or alternating electric fields created between the electrodes; in addition electric fields are created by the various ions produced in the discharge and influence the levels of each atom. These different fields are, on the one hand, very small compared with the value of 10^7 V/m mentioned previously, but on the other hand their influence on a level of an excited atom fluctuates, both in magnitude and direction. The average effect observed over the whole source does not permit the various components to be resolved; only a broadening of the lines is observed.

(*c*) *The natural width and broadening by collisions.* The natural width is related to the time during which an isolated atom can remain in an excited state. The study of the natural width involves fundamental concepts and we shall return to them in chapter 4. In practice, this cause of broadening is negligible in comparison with the other causes which have been described. However, an atom is not isolated within the source and different atoms are continually colliding with each other. When such a collision involves an excited atom, it generally loses its excitation energy in the course of the collision and consequently the lifetime of the excited state is reduced. It will be seen in chapter 4 that the natural width is inversely proportional to the lifetime of the excited state so that the effect of collisions is to increase the natural width of transitions. For clarification, it should be remembered that at a pressure of the order of one atmosphere, collision broadening is comparable to Doppler broadening. In order to reduce the effect of collisions the emitting atoms must be in a gas at low pressure; this is usually the case for luminous sources used in spectroscopy (pressures of the order of a torr, or millimetre of mercury). When the gaseous source is at high pressure (many atmospheres) or is strongly ionised, the shapes of the lines become complex and displacement can occur. The study of the broadening and shifts of lines thus becomes a method of analysing the source region, and it is frequently used in plasma physics.

Another cause of broadening, appropriate to resonance lines, is the phenomenon of self-absorption which will be touched on in chapter 3 (see comment at the end of section 3.1.4).

1.4 Electronic Excitation of an Atomic Vapour

When an electric discharge is struck in a monatomic vapour, two main effects can be observed: (a) the formation of positive ions from atoms of the vapour, which can be identified by mass spectrographic techniques; (b) the emission of light of characteristic spectral frequencies from neutral atoms of the vapour or from the positive ions that have been produced. In both cases, the function of the electric discharge is to supply extra energy to the atoms of the vapour. It is the mechanism for supplying this energy that we shall study in the experiments described below.

1.4.1 Ionisation Potential

The first experiments of this kind were carried out in 1902 by Lenard, who studied the phenomenon of ionisation. To produce a controlled discharge, Lenard used an arrangement of three electrodes which resembled a vacuum triode tube but which contained a vapour at low pressure (see figure 1.6(a)). The first electrode is a heated filament from which electrons are emitted: their speeds are very low and can be considered as almost zero (this is thermionic emission, the theory of which can be found in a course on statistical thermodynamics). The second electrode is a coarsely meshed grid. If a positive potential difference V_g is established between the grid and the filament, the electrons are attracted by the grid. When the electrons of charge $q = -e$ reach the surface of the grid and pass through its mesh, the electric field that has been applied to them has provided an amount of work closely equal to eV_g (see figure 1.6(b) showing the equipotential surfaces in the tube between

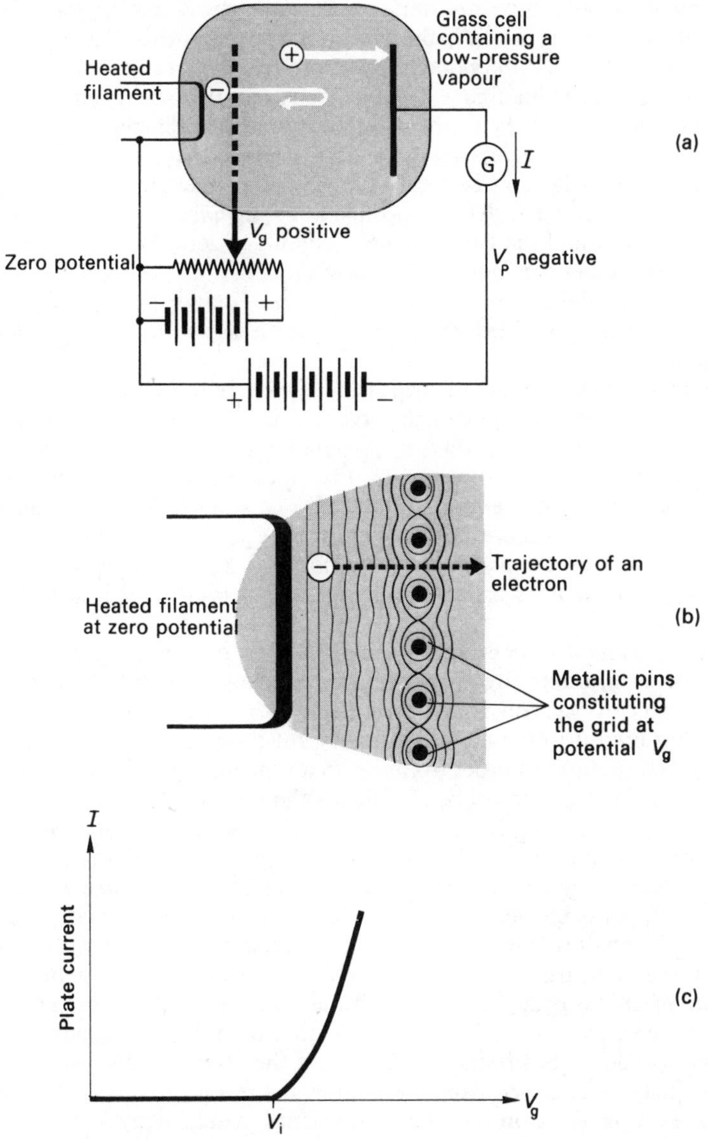

Figure 1.6 Ionisation potential: (a) experimental arrangement; (b) equipotential lines within the triode; (c) characteristic curve observed

the filament and the grid). Since the electrons start with zero velocity, they have acquired a kinetic energy $\frac{1}{2}mv^2 \approx eV_g$. The potential difference V_g can be varied and thus one can control the velocity v of the electrons that cross the vapour and that are liable to enter into collision with the atoms.

On the other hand, by establishing a sufficiently negative potential difference V_P between the plate, which is the third electrode, and the filament, the plate repels all the electrons and none of them can reach it (see the trajectory of an electron symbolised in figure 1.6(a)). However, if some positive ions are formed in the vapour they will be strongly attracted by the plate and will strike it. This flow of positive charges to the plate leads to the passage of a current I and thus to the deflection of a galvanometer G connecting the plate and the voltage source. In the absence of positive ions, the plate receives nothing, the current I is zero and the galvanometer remains at zero. We therefore have a means of detecting the formation of positive ions.

(*Note.* In the definitions of the grid voltage V_g and the plate voltage V_P, both have signs opposite to the signs chosen in the normal use of triodes in electronics.)

The experiment consists of measuring the current I as a function of the voltage V_g applied to the grid. The curve in figure 1.6(c) shows the form of the results obtained: when the voltage V_g is small, the current I is zero. A current I flows only if the voltage V_g is greater than a certain minimum value V_i; the current I then increases very rapidly with V_g. This shows that the formation of positive ions in the vapour only occurs when $V_g \geqslant V_i$, or when the electrons that collide with the atoms of the vapour possess a kinetic energy $\frac{1}{2}mv^2 = eV_g \geqslant eV_i$. This voltage V_i is called the *ionisation potential* of the corresponding atoms.

This experiment is interpreted using the idea of energy of extraction or energy of separation, already introduced in dealing with the photoelectric effect: when an electron is detached from an isolated atom, the attraction forces that bind it to the atom do an amount of work, $\mathscr{T}_i$; this is called the energy of extraction. In order to ionise an atom, energy must be given to it to compensate for the work done $\mathscr{T}_i$; this is the ionisation energy $W_i = |\mathscr{T}_i|$.

The electrons accelerated by the grid may have the necessary energy when they take part in inelastic collisions with the atoms. An inelastic collision is a collision between particles in the course of which a part of the kinetic energy of the particles is converted into another form of energy (in contrast to an elastic collision during which the sum of the kinetic energies of the particles is conserved, see the next section). An electron transfers the maximum energy to an atom when it is brought to rest after the collision and therefore gives up all its kinetic energy $\frac{1}{2}mv^2 = eV_g$. If this is smaller than the ionisation energy W_i, the collision does not produce ionisation of the atom. On the other hand an electron may be detached from an atom as the result of an inelastic collision with an external electron possessing sufficient kinetic energy

$$\tfrac{1}{2}mv^2 = eV_g \geqslant W_i = eV_i$$

From the last equality ($W_i = eV_i$) the experiment described above is easily explained ($V_g \geqslant V_i$), and thus the significance of the ionisation potential V_i measured experimentally is emphasised. The table below gives values of V_i observed in alkali metal vapours and in the rare gases. These are also the measured values in electron-volts of the ionisation energies W_i of the different atoms

Atom	Cs	Rb	K	Na	Li	Xe	Kr	A	Ne	He
Ionisation potential V_i (in volts)	3·89	4·18	4·34	5·14	5·39	12·1	14·0	15·8	21·6	24·6

These experiments with electron bombardment provided the first measurements of the ionisation energy W_i of atoms, and these measurements were confirmed later by photoionisation experiments. We have already discussed photoionisation (section 1.2.4) and have used the concept of ionisation energy W_i to explain the characteristic threshold wavelengths λ_i of this phenomenon. There must be the same relationship between the ionisation potentials V_i and the threshold wavelengths λ_i of photoionisation as between the escape potentials V_S of metals and the threshold wavelengths λ_S of the photoelectric effect (see section 1.2.2)

$$\lambda_i = \frac{c}{\nu_i} = \frac{hc}{W_i} = \frac{hc}{e} \times \frac{1}{V_i} = \frac{1240}{V_i(\text{in volts})} \text{ nm}$$

The tables of numerical values which we have set out enable the accuracy of this relationship to be verified. This agreement confirms the validity of the ideas introduced in the interpretation of the experiments.

Comment I As already noted in relation to photoionisation, it can be ascertained that the ionisation energies of atoms $W_i = eV_i$ (see table above) are about twice the escape energies $W_S = eV_S$ of the corresponding metals (see table in section 1.2, photoelectric effect). That is to say the electrons are more strongly attached to an individual atom than to a metallic block made of the same atoms. This property confirms the hypothesis of free electrons belonging to no particular atom, by means of which electric conduction in metals is explained.

Comment II In his experiments Lenard used a mercury pump to create a vacuum, and the low pressure gases that he used were contaminated by mercury vapour originating from the pump with the result that he always measured the ionisation potential of the mercury atom $V_i = 10{\cdot}5$ V. However, Lenard's experiments were repeated by Franck and Hertz under improved conditions and they measured real ionisation potentials which did depend upon the atoms studied.

1.4.2 Elastic and Inelastic Collisions

The interpretation of electron bombardment experiments is based on the concept of an inelastic collision, that is, a momentary interaction between two particles in the course of which part of their kinetic energy is transformed irreversibly into other forms of energy. These inelastic collisions contrast with elastic collisions during which the total work done by the interaction forces is zero, so that the total kinetic energy is conserved.

In appendix 3 we review the treatment of collision problems as generally presented in mechanics courses using the laws of conservation of energy and of momentum. In order to be able to use these results one must first establish some orders of magnitude. The speeds of thermal agitation of atoms in a gas can be calculated from the laws of statistical thermodynamics: the root-mean-square velocity v for the atoms, mass M, of the gas is given as a function of the absolute temperature T by the equation $Mv^2 = 3kT$, where k is Boltzmann's constant; thus one calculates atomic speeds of the order of 100 to 1000 m/s. At first sight these speeds seem to be large, but from the kinetic energy $\frac{1}{2}mv^2 = eV_g$ of an electron, one calculates an electron speed $v = 400\,000$ m/s for a grid voltage $V_g = 1$ V. The electrons accelerated by voltages V_g of several volts therefore have speeds far in excess of the atoms and to a first approximation we can consider the atoms as quasi-stationary in relation to the electrons. Because of this the collisions that we are studying are a special case: *where a very light projectile* (an electron accelerated by the grid) *hits a very heavy quasi-stationary target* (an atom).

In this special case it may be shown quite generally that the heavy target remains virtually stationary after the collision, and that its kinetic energy also remains practically zero. That is to say that the heavy target can absorb the momentum of the projectile and yet only take up a negligible kinetic energy. The following consequences can thus be deduced.

(a) *If the collision is elastic* the projectile (the electron) conserves its kinetic energy. Only the direction of the velocity is altered (compare the rebounding of a billiard ball).

(b) *If the collision is inelastic* the kinetic energy lost by the projectile (the electron) compensates completely for the resistive work $\mathscr{T}$ of the interaction forces; that is to say, it is converted into potential energy $W = |\mathscr{T}|$.

In the contrasting situation where the projectile and the target are of comparable mass, the target would be given momentum by the collision and so would take up a significant amount of kinetic energy. In an elastic collision the kinetic energy of the projectile is no longer conserved. In an inelastic collision the kinetic energy lost by the projectile is greater than the work done $|\mathscr{T}|$, in other words, for a gain in potential energy W, a part of the projectile's kinetic energy is transformed into the kinetic energy of the target.

The large difference between the masses of the atom and of the electron supports the interpretation of the ionisation experiments as due to collisions if we write the ionisation limit as $eV_i = \frac{1}{2}mv^2 = W_i$. These results for elastic or inelastic collisions promote a better understanding of the experiments described in the next section.

1.4.3 Resonance Potentials—the Experiments of Franck and Hertz

In ionisation experiments such as Lenard's, one observes what happens to atoms when they are ionised by collisions with electrons, but one does not discover what happens to the electrons. After repeating Lenard's experiments, the German physicists Franck and Hertz in 1913 designed an experiment to observe the behaviour of the electrons during collisions.

(*a*) *Evidence of inelastic collisions.* The experimental arrangement (see figure 1.7(a)) is similar to the gas triode used by Lenard, but the collector plate of this triode is now maintained at a positive potential V_P so that it collects the electrons and not the ions. This flow of negative charges to the plate gives rise to the passage of a current I (which is in the opposite direction to the current produced in Lenard's experiment by positive ions). This positive potential

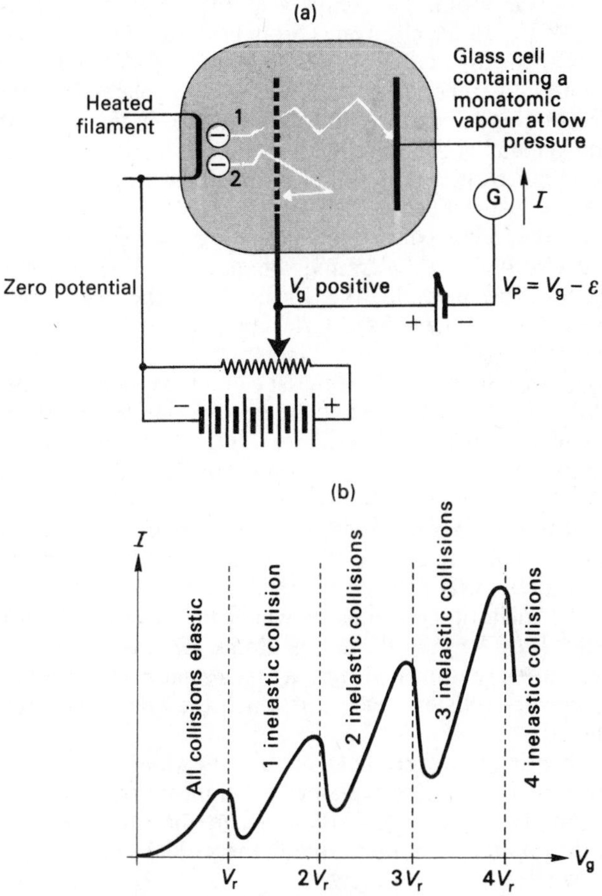

Figure 1.7 The resonance potential (Franck and Hertz experiment): (a) experimental arrangement; (b) characteristic curve observed

V_P does not have an arbitrary value; it is very slightly lower than the grid potential, so that $V_P = V_g - \varepsilon$, where ε is a positive constant with a value of a fraction of a volt.

In these conditions, it may be deduced that:

(i) The electrons that cross the grid with kinetic energy $\frac{1}{2}mv^2 = eV_g$ are subjected to a repulsive force between the grid and the plate which slows down their motion. Nevertheless, they reach the plate with a reduced kinetic energy

$\frac{1}{2}mv^2 = eV_g - e\varepsilon$. This applies for electrons that undergo only elastic collisions with gas atoms so that their kinetic energy is conserved, allowing them to overcome the opposing potential difference ε.

(ii) On the other hand, if an electron undergoes an inelastic collision its kinetic energy is suddenly reduced by an amount equal to the potential energy W gained by the atom. The velocity of the electron is completely annulled if $\frac{1}{2}mv^2 = W$ and so the electron is brought to rest. Once the electron is brought to rest by an inelastic collision, it is no longer capable of overcoming the opposing potential difference ε, and cannot reach the plate. Two electron trajectories, corresponding to cases (i) and (ii), are traced in figure 1.7(a). The experimental arrangement conceived by Franck and Hertz amounts to a means of separating out from a group of electrons those that have lost their kinetic energy in an inelastic collision.

Once again, the experiment consists of measuring the current I received by the plate as a function of the accelerating potential V_g. Figure 1.7(b) shows a typical shape of an experimental curve. In the curve, the following effects are observed in succession as the voltage V_g increases.

(1) While V_g remains small, the current I increases steadily. This portion of the curve is analogous to the diode characteristic and is explained in the same way in terms of the space charge formed by the electrons emitted by the heated filament and accumulated near it. The increase of the current I represents the gradual diminution of this space charge in proportion as the electrons are attracted more rapidly by the grid.

(2) When the potential V_g exceeds a certain threshold value V_r, a sharp drop of current is observed, indicating the occurrence of a new phenomenon: most of the electrons that previously reached the plate are now stopped in transit. This is explained by inelastic collisions in which the electrons give up all their kinetic energy $\frac{1}{2}mv^2 = eV_r = W_r$ to the atoms. To explain the sudden change in shape of the curve it must also be assumed that below the threshold potential V_r only elastic collisions occur, in which the kinetic energy of the electrons is conserved.

(3) When the potential V_g appreciably exceeds the value V_r the current I starts to increase again. This may be explained by supposing that the electrons continue to give up the same energy $W_r = eV_r$ to the atoms by inelastic collisions. They then retain part of their kinetic energy and remain capable of reaching the plate with reduced a velocity.

(4) Once again a sharp drop of current I is observed when the potential V_g reaches the value $2V_r$; that is to say most of the electrons are stopped again. The electrons can lose all their kinetic energy $\frac{1}{2}mv^2 = 2eV_r = 2W_r$ if they make two inelastic collisions with two different atoms in succession.

The repetition of the effect for $V_g = 3V_r$ or $V_g = 4V_r$ is explained similarly by the total loss of kinetic energies $\frac{1}{2}mv^2 = 3W_r$ or $\frac{1}{2}mv^2 = 4W_r$ as a result of three or four successive inelastic collisions. This interpretation is confirmed by the observed variation of the curve $I = f(V_g)$ with the pressure of the gas. When the pressure—that is the number of atoms per unit volume—is greatly reduced the probability that an electron will collide with the gas atoms is reduced correspondingly. The probability of an electron undergoing several successive inelastic collisions during its passage to the plate then becomes very small and it is clearly established that the current minima at $4V_r$, $3V_r$ or $2V_r$ become much less pronounced than the first minimum at V_r, in contrast to figure 1.7(b).

To sum up, the explanation of this experiment is entirely consistent with the following hypothesis: the atom can take from the electron only an exactly determined quantity of energy $W_r = eV_r$.

(*b*) *Emission of light and the agreement with the Bohr hypothesis.* In the experiments carried out on mercury vapour the measured threshold potential is $V_r = 4{\cdot}9$ V whereas the ionisation potential of the mercury atom is $V_i = 10{\cdot}5$ V. The energy W_r given to the atom in this experiment is much less than the ionisation energy W_i necessary to divide it into two parts (electron and positive ion). We observe another effect in which the atom preserves its identity but passes from its normal state to another state where it stores additional potential energy W_r. In other words, the atom makes a transition from its ground state of energy E_1 to an excited state of higher energy $E_2 = E_1 + W_r$.

The fact that the electron cannot transfer a quantity of energy less than W_r *to the atom is strong support for the idea of discontinuous energy states introduced by the Bohr hypothesis.*

This agreement with the Bohr hypothesis can be taken further. When this electron bombardment experiment is carried out, it is established that the potential V_r is equally a threshold for the emission of light by the vapour: as long as the grid voltage V_g is less than V_r no radiation comes from the vapour, but once V_g exceeds the threshold V_r the vapour becomes a source of light. The light emitted is monochromatic; for mercury it is ultraviolet light of wavelength $\lambda_r = 253{\cdot}7$ nm corresponding to a resonance line of the mercury spectrum.

The simultaneous occurrence of light emission and of inelastic collisions transferring the atoms to the excited energy state E_2, agrees with the Bohr hypothesis: excited atoms return spontaneously to lower energy states by giving up the excess energy in the form of photons. If an intermediate energy level does not exist, the atom in an excited state E_2 returns directly to the ground state E_1 and the photons emitted correspond to what is called a resonance line. In these conditions the conservation of energy may be written

$$\frac{hc}{\lambda_r} = h\nu_r = E_2 - E_1 = W_r = eV_r$$

that is to say, there exists the same relation between the wavelength λ_r of this resonance line and the threshold potential V_r measured experimentally as between the threshold wavelength λ_S of the photoelectric effect and the escape potential V_S of the metal used

$$\lambda_r = \frac{hc}{e} \times \frac{1}{V_r} = \frac{1240}{V_r(\text{in volts})}\ \text{nm}$$

The potential V_r is called the resonance potential: its value also represents the measurement in electron-volts of the energy difference $W_r = E_2 - E_1$ between the ground state E_1 and the excited state E_2 corresponding to this resonance line.

So we establish agreement between the values of the energy levels determined in completely independent ways:

(i) *from the wavenumbers* $1/\lambda$ *measured in spectroscopic experiments*;
(ii) *from resonance potentials measured in electron bombardment experiments.*

1.4.4 Critical Potentials

From spectroscopic experiments we know that the same atom can be found in a number of different excited states; and yet in the experiment described above we have shown only the existence of a single excited state $E_2 = E_1 + W_r$. This can be explained as follows: the electrons emitted from the heated filament take part in a number of collisions with gas atoms before reaching the grid and, consequently, before acquiring their maximum kinetic energy $\frac{1}{2}mv^2 = eV_g$. As soon as the electron

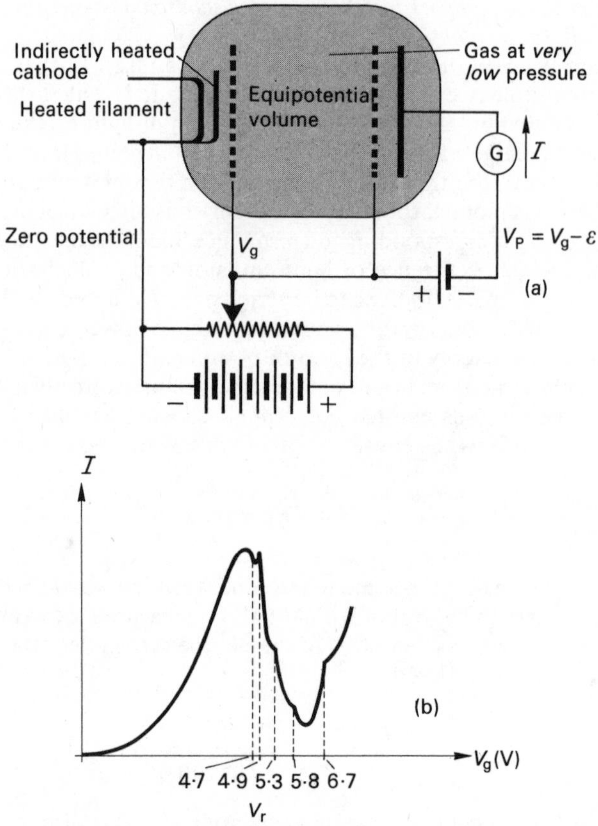

Figure 1.8 Franck and Hertz critical potentials

between the filament and grid, has acquired a sufficient part of this kinetic energy such that $\frac{1}{2}mv^2 = eV_r$, it is able to undergo an inelastic collision in which the atom can make only a transition to the lowest excited level E_2. In other words, the electron never has the time between collisions to acquire a kinetic energy much greater than W_r and therefore it can never excite the atom to an energy level greater than E_2. (The light that is observed corresponds to the emission of the resonance line having the lowest frequency or the longest wavelength.)

In order to observe atomic transitions to other excited levels, Franck and Einsporn repeated the same experiment in 1920 with an improved arrangement as follows (see figure 1.8(a)).

(a) They used an indirectly heated cathode which permitted a more accurate definition of the potential difference V_g between the grid and the cathode.

(b) They reduced simultaneously the vapour pressure and the grid–cathode distance so that there was only a very small probability of the electrons meeting an atom between the cathode and the grid (the mean free path of the electrons is greater than the grid–cathode distance).

(c) They placed a second grid connected to the first just in front of the collector electrode; the two grids bounded an equipotential volume where electrons that did not undergo a collision had a constant kinetic energy. This volume was large enough for the electrons to have a reasonable probability of making a collision in spite of the low pressure of the vapour (strictly speaking it is necessary to maintain a slight potential difference between the two grids to prevent the formation of a space charge which could cause malfunctioning).

To sum up, this new arrangement enables one to distinguish within the cell (i) a region of acceleration between cathode and grid where there are practically no collisions, and (ii) a collision region between the two grids where the electrons are not accelerated.

Figure 1.8(b) shows a typical example of a curve obtained in these conditions, when the plate current I is measured as a function of the grid voltage V_g (for the case of mercury vapour). A sudden drop of current is always observed at the resonance potential V_r but the curve also shows other less pronounced discontinuities indicating a sudden increase in the number of inelastic collisions when V_g exceeds certain critical values.

It can be shown that these critical potentials correspond well with the energy levels E_n determined from optical spectra, that is, one can associate with each of them a particular energy E_n such that $E_n - E_1 = eV_c$. Furthermore, the emission of additional wavelengths can be observed for certain critical potentials; in the case of mercury, for example, the emission of another resonance line of wavelength $\lambda = 185$ nm occurs when the critical potential of 6·7 V is exceeded.

Comment In the case of mercury there is no radiative transition (by absorption or emission of a photon) between the ground state and the excited state corresponding to the critical potential 4·7 V (this is a 'metastable' level, see volume 2, section 4.3.3). A transition between these same two levels induced by electron collisions is possible but the probability is very small. This is why the experiment shows up principally the transition to the resonance level whose energy is a little higher ($V_r = 4{\cdot}9$ V) and from which the resonance wavelength $\lambda_r = 253{\cdot}7$ nm is emitted.

To conclude, the experiments on electron bombardment allow the quantities of energy exchanged between electrons and atoms in inelastic collisions to be measured and show that they can take only certain values, forming a discontinuous series of discrete terms. These experiments also confirm the idea of discrete or quantised energy levels introduced by Bohr to explain spectroscopic experiments.

The numerical values measured for the energy differences between these levels are the same in spectroscopic experiments (hc/λ) and in electron bombardment experiments (eV).

The improvement in experimental techniques has led to an improvement in the precision of energy loss measurements of electrons in inelastic collisions (precision of the order of one hundredth of a volt has been achieved). This method of studying atoms and molecules is still in use at the present time (see volume 2, section 7.5).

1.5 Units of Energy

The whole of this first chapter amounts to a simple application of the law of conservation of energy. Its systematic use in the interpretation of each experiment leads to the concepts of an energy quantum of radiation (the photon) and of energy levels of atoms. In considering the experiments, we have had to calculate energies from measurements of a variety of physical quantities.

(a) In spectroscopic experiments, frequencies or wavenumbers $1/\lambda = \nu/c$ are measured and from them values of energy are deduced using the formulae

$$W = h\nu = hc \times \frac{1}{\lambda}$$

(b) In experiments where the motion of electrons is controlled (the photoelectric effect, ionisation, Franck and Hertz) the energy is calculated from electrostatic potentials V using the formula $W = eV$. Since the electric charge e is the same in all the calculations, it is not necessary to multiply V by e in order to compare the different values of energy. These are all proportional to the potential V so that the value of V represents the measurement of energy W in units of electron-volts. In the same way, to compare various energies it is not necessary to multiply a frequency ν by the constant h or a wavenumber $1/\lambda$ by the constant hc. Measurement of the frequency or of the wavenumber $1/\lambda$ is sufficient to specify the value of the corresponding energy. This is why physicists usually quote energies measured in frequency units (cycles/second, or hertz) or in wavenumbers (cm^{-1}, or keysers, obtained by expressing the wavelength in cm). Knowledge of the fundamental constants h, c and e allow comparisons of different energy measurements either with one another or with the traditional units, the erg and joule.

Other physical quantities can be used to measure energies when studying other phenomena; we can extend the list already started as follows.

(c) When the Boltzmann formula of statistical thermodynamics is used, one compares energies with a reference energy $W = kT$ proportional to the absolute temperature T (k being Boltzmann's constant). This reference energy can be specified simply by the value of the temperature T, which amounts to measuring energy in units of temperature (degrees Kelvin).

(d) The study of nuclear reactions and the operation of large accelerators always confirms Einstein's celebrated formula attributing to a mass m

Table 1.1 Practical units for expressing energies[†]

Formulae for definition of these practical energy units: $W = mc^2 = eV = hc\frac{1}{\lambda} = kT = h\nu$

When the unit is used →	joule	erg	a.m.u.	eV	cm^{-1}	degree	Hz
1 g is		$8{\cdot}98755 \times 10^{20}$ (c^2 in CGS)	**$6{\cdot}0222 \times 10^{23}$ ($\mathcal{N}$ for $^{12}C = 12$)**				$1{\cdot}3564 \times 10^{47}$ (c^2/h in CGS)
1 joule is	1	10^7		$0{\cdot}6241 \times 10^{19}$ ($1/e$ in $coulomb^{-1}$)		$0{\cdot}72431 \times 10^{23}$ ($1/k$ in MKS)	$1{\cdot}5092 \times 10^{33}$ ($1/h$ in MKS)
1 erg is	10^{-7}	1	$0{\cdot}67006 \times 10^3$ ($\mathcal{N}/c^2$ in CGS)		$0{\cdot}5034 \times 10^{16}$ ($1/hc$ in CGS)	$0{\cdot}72431 \times 10^{16}$ ($1/k$ in CGS)	$1{\cdot}5092 \times 10^{26}$ ($1/h$ in CGS)
1 a.m.u. is		$1{\cdot}49224 \times 10^{-3}$ ($c^2/\mathcal{N}$ in CGS)	1	$0{\cdot}93148 \times 10^9$			
1 eV is	**$1{\cdot}6022 \times 10^{-19}$ (e in coulomb)**		$1{\cdot}07356 \times 10^{-9}$	1	$0{\cdot}80655 \times 10^4$	$1{\cdot}1605 \times 10^4$ (e/k in MKSA)	$2{\cdot}4180 \times 10^{14}$ (e/h in MKSA)
1 cm^{-1} is		$1{\cdot}9865 \times 10^{-16}$ (hc in cgs)		$1{\cdot}2399 \times 10^{-4}$	1	$1{\cdot}4388$ (hc/k in CGS)	**$2{\cdot}99793 \times 10^{10}$ (c in CGS)**
1 degree is	**$1{\cdot}3806 \times 10^{-23}$ (k in MKS)**	**$1{\cdot}3806 \times 10^{-16}$ (k in CGS)**		$0{\cdot}8617 \times 10^{-4}$ (k/e in MKSA)	$0{\cdot}6950$ (k/hc in CGS)	1	$2{\cdot}0836 \times 10^{10}$ (k/h)
1 Hz is	**$6{\cdot}6262 \times 10^{-34}$ (h in MKS)**	**$6{\cdot}6262 \times 10^{-27}$ (h in CGS)**		$0{\cdot}4136 \times 10^{-14}$ (h/e in MKSA)	$0{\cdot}33356 \times 10^{-10}$ ($1/c$ in CGS)	$0{\cdot}47994 \times 10^{-10}$ (h/k)	1

[†] The values of the five fundamental constants are taken from a compilation: Taylor, B. N., Parker, W. H., and Langenberg, D. N. (1969). *Rev. Mod. Phys.*, **41**, 375.

an energy that is proportional to it, $W = mc^2$. Specifying the mass of a physical system involves specifying the energy that it contains; that is to say, energies can be measured in units of mass. When dealing with a macroscopic system the common unit, the gram, can be used; when dealing with a system on an atomic scale a unit called 'the atomic mass unit', abbreviated to a.m.u., is used. This unit is the gram divided by Avogadro's number. (The mass of a hydrogen atom is very nearly equal to 1 a.m.u.)

Table 1.1 allows conversion to be made between these various units, by showing the values that each of them has in terms of the others. These units have been classified in order of size. It will be noticed that the cm^{-1} and the degree are of exactly the same order of magnitude. The five fundamental constants, the elementary charge e, Planck's constant h, the speed of light c, Boltzmann's constant k and Avogadro's number $\mathcal{N}$, around which the table is established, are shown in bold type.

There is some value in comparing two other units of energy frequently used with the various energy units used by physicists:

(i) in industry, the kilowatt-hour = 3600 joules = $3{\cdot}6 \times 10^{13}$ ergs;

(ii) in chemistry, the kilocalorie per mole which must be divided by Avogadro's number, $\mathcal{N}$, to relate it to an actual molecule. 1 kcal/mole would be 7×10^{-21} J = 1/23 eV per molecule.

Comment The value given for Avogadro's number corresponds to a system of atomic masses which has now been universally adopted by physicists and chemists. In this system, the atomic mass of the carbon 12 isotope is $M = 12{\cdot}000\,000$; in other words the mass of this carbon atom is exactly 12 a.m.u. In these circumstances, the mass of the proton is $M_0 = 1{\cdot}007\,277$ a.m.u. and that of the electron

$$m = \frac{M_p}{1836{\cdot}1} = \frac{1}{1823}\ \text{a.m.u.} = 0{\cdot}910\,9 \times 10^{-27}\ \text{g}$$

2

The Momentum of Radiation

2.1 Classical Treatment and Radiation Pressure

Chapter 1 was devoted to the study of energy exchange between atoms and electromagnetic radiation; but this radiation has properties other than energy —in particular it transports momentum. Detailed texts on electromagnetism give very generalised but inconveniently abstract reasons in trying to establish the origin of this property of radiation. Here we have chosen to introduce the idea of momentum of radiation on the basis of an experimentally observable fact: the force exerted by a beam of light on a surface, called radiation pressure.

Section 2.1.1 gives an exact calculation of the radiation pressure using the fundamental laws of electromagnetism. However, the details of the calculation are not essential for an understanding of the discussion that follows; only the results shown at the end of section 2.1.1 are necessary. What is important is the interpretation of radiation pressure in terms of the momentum, which is described in section 2.1.2.

2.1.1 Calculation of Radiation Pressure by Classical Electromagnetism

The main hypothesis on which this calculation is based is the opacity encountered by an electromagnetic wave at a wall. A wave can be reflected, scattered or absorbed at any boundary, but which of these is not important. We need only suppose that the wave is not transmitted at the boundary. In addition we suppose that the boundary has a sufficiently large radius of curvature so that locally it is in a plane. Let xOy be the boundary between the vacuum (in the negative z direction) and the material that forms the wall (in the positive z direction) (see figure 2.1).

To carry out the calculation, it is necessary to take into account the fact that the wave penetrates a very small surface layer of the wall of thickness ε and exerts forces on the elementary electric charges, stationary or in motion, contained in the wall. We suppose that the motion reaches a steady-state condition and in this case the forces exerted on the elementary charges are transmitted to the whole solid body making up the wall. Let ρ and $\boldsymbol{j}$ respectively be the volume density of charge and the vector current density for the elementary charges. Let $\boldsymbol{E}$ and $\boldsymbol{B}$ respectively be the

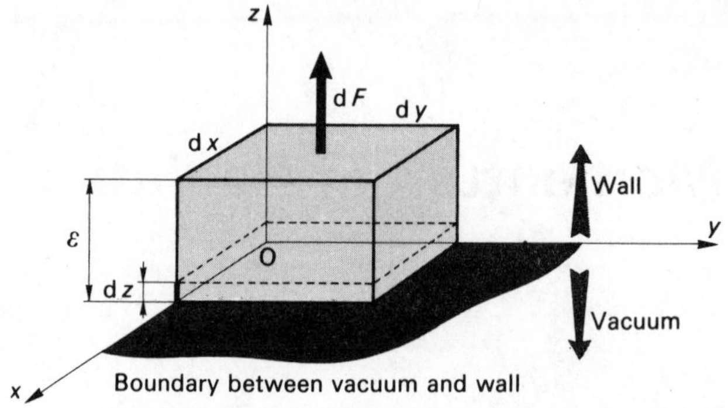

Figure 2.1 Radiation pressure

electric and magnetic field vectors of the wave, which must depend upon the three co-ordinates x, y and z and on time. The force exerted by the wave on a small element of volume $\mathrm{d}x\mathrm{d}y\mathrm{d}z$ in this surface layer can then be written

$$\left(\rho \boldsymbol{E} + \frac{1}{\kappa}\boldsymbol{j} \times \boldsymbol{B}\right)\mathrm{d}x\,\mathrm{d}y\,\mathrm{d}z$$

(for the meaning of the coefficient κ see the preface and appendix 1).

The force $\mathrm{d}\boldsymbol{F}$ exerted on the small surface area $\mathrm{d}x\mathrm{d}y$ of the wall is obtained by adding the contributions of all the elements of thickness $\mathrm{d}z$ within the surface layer of thickness ε, that is, by integrating the expression above with respect to z.

In this way the pressure exerted on the wall is deduced

$$\frac{\boldsymbol{F}}{S} = \frac{\mathrm{d}\boldsymbol{F}}{\mathrm{d}x\,\mathrm{d}y} = \int_0^{\varepsilon}\left(\rho \boldsymbol{E} + \frac{1}{\kappa}\boldsymbol{j} \times \boldsymbol{B}\right)\mathrm{d}z\,\mathrm{d}t$$

This pressure, like the fields $\boldsymbol{E}$ and $\boldsymbol{B}$, will alternate as a function of time with the same period T as the electromagnetic wave. We are interested only in the time average value (calculated over a whole number of periods) of the vector

$$\varpi = \frac{1}{T}\int_0^T \frac{\boldsymbol{F}}{S}\,\mathrm{d}t = \frac{1}{T}\int_0^T\int_0^{\varepsilon}\left(\rho \boldsymbol{E} + \frac{1}{\kappa}\boldsymbol{j} \times \boldsymbol{B}\right)\mathrm{d}z\,\mathrm{d}t$$

The volume densities $\boldsymbol{\rho}$ and $\boldsymbol{j}$ are related to the fields $\boldsymbol{E}$ and $\boldsymbol{B}$ through Maxwell's equations

$$\text{(I)} \begin{cases} \operatorname{div} \boldsymbol{B} = 0 \\ \operatorname{curl} \boldsymbol{E} + \dfrac{1}{\kappa}\dfrac{\partial \boldsymbol{B}}{\partial t} = 0 \end{cases} \qquad \text{(II)} \begin{cases} \operatorname{div} \boldsymbol{E} = \dfrac{\rho}{\varepsilon_0} \\ \operatorname{curl} \boldsymbol{B} - \dfrac{\varepsilon_0 \mu_0}{\kappa}\dfrac{\partial \boldsymbol{E}}{\partial t} = \dfrac{\mu_0}{\kappa}\boldsymbol{j} \end{cases}$$

Using equations (II) we deduce

$$\rho \boldsymbol{E} + \frac{1}{\kappa}\, \boldsymbol{j} \times \boldsymbol{B} = (\varepsilon_0 \operatorname{div} \boldsymbol{E})\boldsymbol{E} + \left(\frac{1}{\mu_0}\operatorname{curl} \boldsymbol{B} - \frac{\varepsilon_0}{\kappa}\frac{\partial \boldsymbol{E}}{\partial t}\right) \times \boldsymbol{B}$$

Using equations (I) we carry out the following transformation

$$\frac{\partial \boldsymbol{E}}{\partial t} \times \boldsymbol{B} = \frac{\partial}{\partial t}(\boldsymbol{E} \times \boldsymbol{B}) - \boldsymbol{E} \times \frac{\partial \boldsymbol{B}}{\partial t} = \frac{\partial}{\partial t}(\boldsymbol{E} \times \boldsymbol{B}) + \boldsymbol{E} \times \kappa \operatorname{curl} \boldsymbol{E}$$

and we obtain

$$\rho \boldsymbol{E} + \frac{1}{\kappa}\, \boldsymbol{j} \times \boldsymbol{B} = \underbrace{(\varepsilon_0 \operatorname{div} \boldsymbol{E})\boldsymbol{E} - \varepsilon_0 \boldsymbol{E} \times \operatorname{curl} \boldsymbol{E}}_{\mathscr{E}} - \frac{\varepsilon_0}{\kappa}\frac{\partial}{\partial t}(\boldsymbol{E} \times \boldsymbol{B}) - \underbrace{\frac{1}{\mu_0}(\boldsymbol{B} \times \operatorname{curl} \boldsymbol{B})}_{\mathscr{B}}$$

We can change the order of integration with respect to the two independent variables z and t. The integration over time of the term $(\partial/\partial t)(\boldsymbol{E} \times \boldsymbol{B})$ is straightforward, and the definite integral over one period T is zero. So this term can be dropped and we have thus reduced the integrand to a vector $\mathscr{E}$ which only depends on the electric field $\boldsymbol{E}$ and a vector $\mathscr{B}$ which only depends on the magnetic field $\boldsymbol{B}$.

For simplicity, we confine ourselves to calculating the normal component of the mean pressure ϖ_z, in other words, to the normal components $\mathscr{E}_z$ and $\mathscr{B}_z$ of the two vectors in the integrand

$$\varpi_z = \frac{1}{T}\int_0^{\varepsilon}\int_0^{T} (\mathscr{E}_z + \mathscr{B}_z)\, \mathrm{d}z\, \mathrm{d}t$$

where

$$\mathscr{E}_z = \varepsilon_0 \left(\frac{\partial E_x}{\partial x} + \frac{\partial E_y}{\partial y} + \frac{\partial E_z}{\partial z}\right) E_z - \varepsilon_0 E_x \left(\frac{\partial E_x}{\partial z} - \frac{\partial E_z}{\partial x}\right) + \varepsilon_0 E_y \left(\frac{\partial E_z}{\partial y} - \frac{\partial E_y}{\partial z}\right)$$

or, by rearranging the terms

$$\mathscr{E}_z = \frac{\varepsilon_0}{2}\frac{\partial}{\partial z}(E_z^2 - E_x^2 - E_y^2) + \varepsilon_0 \frac{\partial}{\partial x}(E_x E_z) + \varepsilon_0 \frac{\partial}{\partial y}(E_y E_z)$$

Equally we can change the order of partial differentiation with respect to a space variable and of integration with respect to time. This allows us to obtain immediately for the orthogonal terms

$$\int_0^T \frac{\partial}{\partial x}(E_x E_z)\,\mathrm{d}t = \frac{\partial}{\partial x}\int_0^T (E_x E_z)\,\mathrm{d}t = 0$$

If we assume that the wall is homogeneous and that the electromagnetic wave has an almost constant amplitude in the xOy plane, the mean value in time of $E_x E_z$ and $E_y E_z$ must depend only on the z co-ordinate and hardly at all on the x and y co-ordinates. Therefore the contribution of the electric vector to the normal pressure ϖ_z reduces to the first partial derivative with respect to z whose integration over z is straightforward. The result is expressed as a function of the components of the electric field, on the one hand at a depth $z = \varepsilon$ (where they are zero by definition) and on the other hand in the xOy plane, at the external surface of the wall

$$\frac{1}{T}\int_0^T \mathrm{d}t \int_0^\varepsilon \mathscr{E}_z\,\mathrm{d}z = \frac{1}{T}\int_0^T \mathrm{d}t\frac{\varepsilon_0}{2}(E_x^2 + E_y^2 - E_z^2) = \frac{\varepsilon_0}{2}(\overline{E_x^2} + \overline{E_y^2} - \overline{E_z^2})$$

where $\overline{E_x^2}$, $\overline{E_y^2}$ and $\overline{E_z^2}$ are time average values, calculated in the vacuum close to the wall.

To complete the calculation, it is necessary to account for the contribution of the vector $\mathscr{B}$. By using the first of the equations (I) we can write $\mathscr{B}$ in a form analogous to that of $\mathscr{E}$

$$\mathscr{B} = \frac{1}{\mu_0}(\mathrm{div}\,\boldsymbol{B})\,\boldsymbol{B} - \frac{1}{\mu_0}\dot{\boldsymbol{B}} \times \mathrm{curl}\,\boldsymbol{B}$$

The calculation for the magnetic term is therefore identical to that for the electric term and we can write the result straight away

$$\boxed{\varpi_z = \frac{\varepsilon_0}{2}(\overline{E_x^2} + \overline{E_y^2} - \overline{E_z^2}) + \frac{1}{2\mu_0}(\overline{B_x^2} + \overline{B_y^2} - \overline{B_z^2})}$$

where the symbols $\overline{B_x^2}$, $\overline{B_y^2}$ and $\overline{B_z^2}$ have the same meaning as $\overline{E_x^2}$, $\overline{E_y^2}$ and $\overline{E_z^2}$: *they represent the time averaged values of the squares of the components of the electric and magnetic fields calculated in the vacuum in the immediate neighbourhood of the wall.*

We finish this section by seeking an expression for the radiation pressure in terms of the energy density. It is shown in texts on electromagnetism that the energy density exists in the space through which the electromagnetic wave passes and has a mean value

$$u = \frac{\varepsilon_0}{2}\overline{E^2} + \frac{1}{2\mu_0}\overline{B^2} = \frac{\varepsilon_0}{2}(\overline{E_x^2} + \overline{E_y^2} + \overline{E_z^2}) + \frac{1}{2\mu_0}(\overline{B_x^2} + \overline{B_y^2} + \overline{B_z^2})$$

From the energy density u in a beam of light of cross-section S, the power that it transports can be calculated: $P = Suc$.

We can immediately transform the expression for ϖ_z as follows

$$\varpi_z = \frac{\varepsilon_0}{2}\overline{E^2} - \varepsilon_0\overline{E_z^2} + \frac{1}{2\mu_0}\overline{B^2} - \frac{1}{\mu_0}\overline{B_z^2} = u - \left(\varepsilon_0\overline{E_z^2} + \frac{1}{\mu_0}\overline{B_z^2}\right)$$

The calculation of E_z and B_z is carried out by restricting ourselves to *the particular case of a plane wave*, and this plane wave can be in any direction. Let i be the angle of incidence, in other words, the angle between the normal to the walls Oz and the normal to the plane of the wave ON (see figure 2.2). We choose the axis Oy to be in the plane of these two normals NOz; the plane of the wave passing through O contains the line Ox and cuts the plane NyOz in a line OY which makes an angle i with Oy. Let us suppose that the electric field $\boldsymbol{E}$ of the wave makes an angle α with this line OY. We can calculate immediately from figure 2.2 the normal component of $\boldsymbol{E}$ as $E_z = E\cos\alpha\sin i$. By letting β be the angle between the magnetic field $\boldsymbol{B}$ and the line OY, the same argument gives us the normal component of $\boldsymbol{B}$ as $B_z = B\cos\beta\sin i$.

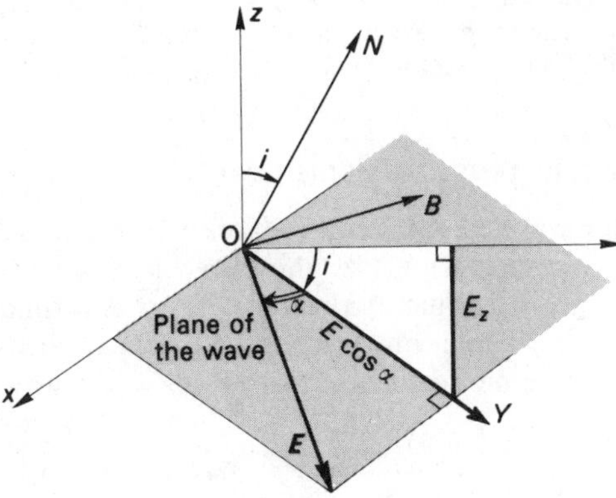

Figure 2.2 Diagram for the calculation of E_z and B_z

Again the relationships between the two fields $\boldsymbol{E}$ and $\boldsymbol{B}$ for a plane wave must be used

$$\begin{cases} \text{in magnitude:} & B = E\sqrt{(\varepsilon_0\mu_0)} \\ \text{in direction:} & \text{the vectors } \boldsymbol{E} \text{ and } \boldsymbol{B} \text{ are perpendicular, so} \\ & |\cos\beta| = |\sin\alpha|, \end{cases}$$

$$\varepsilon_0 E_z^2 + \frac{1}{\mu_0}B_z^2 = \left(\varepsilon_0 E^2\cos^2\alpha + \frac{1}{\mu_0}B^2\cos^2\beta\right)\sin^2 i = \varepsilon_0^2 E^2(\cos^2\alpha + \sin^2\alpha)\sin^2 i$$

and so

$$\varepsilon_0 \overline{E_z^2} + \frac{1}{\mu_0}\overline{B_z^2} = u\sin^2 i$$

We finally obtain an expression for the normal component ϖ_z of the radiation pressure as a function of the energy density u and of the angle of incidence i

$$\boxed{\varpi_z = u\cos^2 i}$$

It will be noticed that ϖ_z is positive, so that the wave tends to push the wall. We consider two important special cases:

(a) the case of a normal wave—the radiation pressure is equal to the energy density: $\varpi_z = u$;

(b) the case of isotropic radiation, that is, a superposition of waves travelling in all directions with an energy distributed in proportion to the solid angle. It is necessary to find the mean value of $\cos^2 i$ within a solid angle of 2π, corresponding to half of space

$$\varpi_z = u \int_0^{\pi/2} \cos^2 i \sin i \, di = u/3$$

This is the formula used in the theory of thermal radiation in equilibrium within a closed space.

Comment We have not attempted to calculate the tangential components of the pressure ϖ_x and ϖ_y. They are zero in the two special cases above for symmetry reasons, but it would be incorrect to generalise this result.

2.1.2 Interpretation in Terms of Momentum

The electromagnetic wave that strikes a wall of surface area S exerts a force on it $\boldsymbol{F} = S\varpi$, and sets it in motion (see the experiment described in the next section); that is, it gives it some momentum. There is a fundamental law of dynamics that expresses a relation between the forces applied to the wall and the rate of change of the momentum vector $\boldsymbol{p}=m\boldsymbol{v}$ of the wall

$$\frac{d\boldsymbol{p}_{\text{wall}}}{dt} = \boldsymbol{F} = S\varpi$$

Ignoring the source emitting the wave (which in many problems is too far away and unknown) we can take the wave and the wall as forming an isolated system, and apply to this system the general law of conservation of momentum. To ensure conservation of momentum, it must be assumed that the wave also possesses a momentum vector $\boldsymbol{p}_{\text{wave}}$ and that its change is opposite to that of the wall, so that

$$\boldsymbol{p}_{\text{wall}} + \boldsymbol{p}_{\text{wave}} = \text{constant}$$

We shall consider only the normal components just as we have calculated only the normal component of the pressure. For an experiment of short duration t, we calculate the change of the momentum of the wall

$$\delta p_{z\,\text{wall}} = F_z t = S\varpi_z t = Su \cos^2 i \times t$$

We shall examine the consequences for a wave train of length ct striking the wall for a time t, distinguishing two particular cases.

(*a*) *The case of an absorbing wall.* The wall absorbs the incident wave train and at the same time takes up its momentum $\boldsymbol{p}_{\text{wave}}$, so that $p_{z\,\text{wave}} = \delta p_{z\,\text{wall}}$.

The light beam, incident at an angle i, strikes the surface S of the wall and has a cross-sectional area of $S \cos i$; since the energy density there is u, it transports an energy $W = Suct \cos i$. The relationship between the momentum and the energy of the wave train is therefore: $p_{z\,\text{wave}} = W \cos i/c$.

(*b*) *The case of a reflecting wall.* The wave train initially possesses a momentum $\boldsymbol{p}_{\text{incident}}$, and leaves the wall with a different momentum $\boldsymbol{p}_{\text{reflected}}$, thus undergoing a change

$$\delta \boldsymbol{p}_{\text{wave}} = \boldsymbol{p}_{\text{reflected}} - \boldsymbol{p}_{\text{incident}} = -\delta \boldsymbol{p}_{\text{wall}}$$

But the wave train is reflected by the mirror, and the energy density u near the wall is the sum of the energy densities of the incident wave and the reflected wave respectively. These two contributions to the total energy density u are equal, so that each is half the total. So actually the energy density in the incident beam is only half the total energy density, and the energy carried by the incident wave train in a time t (and carried away by the same wave train after reflection) must be calculated using a density $u/2$ instead of u; that is to say, $W = S(u/2)ct \cos i$.

From this we find the relation

$$p_{z\,\text{incident}} - p_{z\,\text{reflected}} = 2\frac{W}{c}\cos i$$

The results obtained for the normal components p_z in cases (a) and (b) for any angle of incidence i, may be understood completely, together with their signs, if a wave train carrying an energy W is also allowed to possess a momentum vector

$$\boldsymbol{p}_{\text{wave}} \begin{cases} \text{—parallel to the direction of propagation and in the same sense} \\ \text{—of magnitude } |\boldsymbol{p}_{\text{wave}}| = W/c \end{cases}$$

Comment Concerning the comparison of cases (a) and (b) above, if we use a beam of light carrying a certain power $P = W/t$ in order to perform an experiment, this beam produces an energy density u near a reflecting wall twice that produced near an absorbing wall. Therefore it exerts twice the pressure on a reflecting wall (a mirror) as it exerts on an absorbing wall (a black wall):

on an absorbing wall, it exerts a normal force $F_z = \dfrac{P}{c}\cos i$;

on a reflecting wall, it exerts a normal force $F_z = 2\dfrac{P}{c}\cos i$.

(Care must be taken not to confuse the power, denoted by P, with the momenta always denoted by p.) In other words, the reflected wave exerts a pressure equal to that of the incident wave.

2.1.3 Experimental Verification

(*a*) *The first proof of the existence of radiation pressure* was obtained indirectly from measurements on thermal radiation in equilibrium inside a closed space, which is equivalent to radiation emitted by a black body (see section 1.1). In 1879, Stefan measured the power P radiated over all frequencies by a black body, by directing this radiation onto a perfectly absorbing blackened plate and measuring the heat gained by this plate. By varying the absolute temperature T of the black body, he discovered the law later named after him which can be formulated as $P = \sigma T^4$.

Some years later (1884) Boltzmann explained this law theoretically by assuming a radiation pressure $\varpi = u/3$ exerted normally on the walls by isotropic radiation of energy density u (see the end of section 2.1.1). Stefan's law and its theoretical interpretation given by Boltzmann were already indirect proof of the existence of this pressure.

Comment A brief review of Boltzmann's reasoning is given below. The thermodynamic system consists of radiation in equilibrium within a closed space of volume $\mathscr{V}$, where there is an energy density u; this depends only on T. If the volume of the enclosure is changed, the radiation does an amount of work $\varpi \mathrm{d}V$ and therefore some work $\mathscr{T} = -\varpi \mathrm{d}\mathscr{V} = -(u/3)\mathrm{d}\mathscr{V}$ is done on the system. The internal energy of the system $U = u\mathscr{V}$. The heat exchanged with the walls is given by

$$\mathrm{d}Q = \mathrm{d}U - \mathscr{T} = u\mathrm{d}\mathscr{V} + \mathscr{V}\mathrm{d}u + \frac{u}{3}\mathrm{d}\mathscr{V}$$

and the entropy change is

$$\mathrm{d}S = \frac{\mathrm{d}Q}{T} = \frac{4}{3}\frac{u}{T}\mathrm{d}\mathscr{V} + \frac{\mathscr{V}}{T}\mathrm{d}u$$

Since $\mathrm{d}S$ is an exact total differential, then

$$\frac{\partial}{\partial u}\left(\frac{4}{3}\frac{u}{T}\right) = \frac{\partial}{\partial \mathscr{V}}\left(\frac{\mathscr{V}}{T}\right)$$

or

$$\frac{4}{3T} - \frac{4u}{3T^2}\frac{\mathrm{d}T}{\mathrm{d}u} = \frac{1}{T}$$

and so $\mathrm{d}u/u = 4\mathrm{d}T/T$, so that $u = CT^4$. From this, Stefan's law is deduced since the radiation power P is proportional to u.

(*b*) *Direct experimental evidence of radiation pressure* was more difficult to obtain since it is very small. The intense beams of light that can now be obtained in the laboratory (a beam produced from an electric arc or from a powerful laser) have a power P of the order of a watt. If such a beam is focused onto a plate of surface area $S = 1\ \mathrm{cm}^2$, an energy density $u = P/Sc$ is produced nearby and the radiation pressure can be calculated as $\varpi = u = 0{\cdot}25 \times 10^{-6}$ torr (millimetre of mercury).

For the effect of this to be observable, it is necessary to suspend the receiving plate within an evacuated enclosure where the residual pressure of air is very small; the heat produced by the passage of the light beam is a source of parasitic forces called radiometric forces, whose order of magnitude is a sizeable fraction of the residual pressure. Certain types of pump, now available commercially, can readily achieve residual pressures smaller than 10^{-7} torr, and it is possible to do much better than this. However, these techniques were not sufficiently advanced in 1900 when Lebedew was able for the first time to produce evidence for, and measure, the recoil momentum of a plate due to radiation pressure.

Radiation pressure in the laboratory gives rise to only very small forces which are extremely difficult to observe; but this is not the case in stars where very high energy densities often occur. Radiation pressure plays an essential role in the theories put forward to explain the structure of certain stars.

Comment It is possible to buy in novelty shops small radiometers which reproduce the old experiments of the last century, and which give a good illustration of the radiometric forces we have discussed. Two small vanes are mounted symmetrically on a light pivot, each of them having a black surface and a polished surface. The device thus constructed is enclosed by a glass cover within which there is a partial vacuum (see figure 2.3). When the apparatus is placed beside a lamp, the vanes rotate in a direction such that the polished surfaces advance and the black surfaces recede.

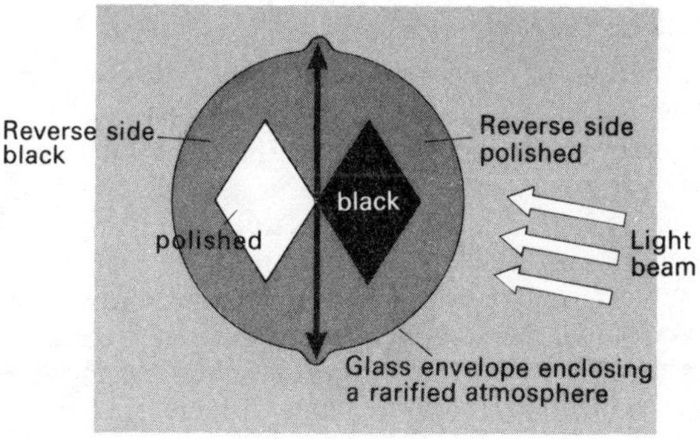

Figure 2.3 Radiometric forces

The explanation is simple once it is realised that the elastic collisions of elementary gas kinetic theory do not occur: experiments have shown that a beam of molecules all moving in the same direction are more or less scattered in every direction when they meet a wall; we must assume that the molecules adhere to the wall for a short time before leaving it.

The black surface of the vanes absorbs the light and is strongly heated. The molecules that strike and adhere to it for a short time are also heated, so that they leave the blackened surface with a mean kinetic energy greater than their mean energy in the gas.

On the other side of the same vane the polished surface reflects most of the light and is heated very little; it returns the molecules with a normal mean kinetic energy for a gas at ordinary temperatures. Under these conditions, the molecules that leave the black surface carry away a total momentum greater than those leaving the polished surface and conservation of total momentum can be assured only by recoil of the black surface.

These radiometric forces can be much stronger than the forces due to radiation pressure, but they cannot be confused because their effects are in opposite directions. If there is a good vacuum to suppress the radiometric forces, and if a fibre suspension is used (more sensitive than the pivot from the point of view of friction) the uniformly illuminated vanes turn in the opposite direction and the polished surface recoils. The latter is subjected to a radiation pressure twice as great as the blackened surface, because the energy density near the polished surface is doubled due to the reflected light or, if preferred, because it receives the momentum of the reflected beam in addition to the momentum of the incident beam (see comment at the end of section 2.1.2).

2.2 The Momentum of a Photon

2.2.1 From the Pressure of Radiation

Considerations of radiation pressure, in the preceding section, have resulted in attributing a momentum vector of magnitude $|\boldsymbol{p}| = W/c$ to a wave train which transports energy W. Therefore a photon which transports energy $W = h\nu$ must also possess a momentum vector

$$\begin{cases} \text{of magnitude } |\boldsymbol{p}| = h\nu/c = h/\lambda \\ \text{parallel to the direction of propagation and in the same sense.} \end{cases}$$

In this book we have chosen to introduce the notion of radiation momentum of a classical wave train (see section 2.1.2) so that it appears unrelated to quantum hypotheses. However, this approach is inappropriate to our present study, the properties of photons. We may postulate straight away that photons have momentum as indicated above and rederive the expression for the radiation pressure exerted on a wall by applying the conservation law for the momentum carried by the photons. More exactly, we may calculate the pressure ϖ_z normal to the wall from the normal components of the momenta of the photons $p_z = \pm(h\nu/c)\cos i$ (i is the angle of incidence; the sign is + or − according as the photon is incident or either reflected or scattered). In the case of isotropic radiation, we may average over all possible angles of incidence. This calculation would be entirely analogous to the calculation in the kinetic theory of gases for finding the pressure of a gas from the momenta of the molecules, and we refer to texts on thermodynamics where this calculation is carried out. However, we wish to make two remarks about this.

(a) In the kinetic theory of gases the calculation is usually carried out by assuming that each molecule makes an elastic collision with the wall, from which it is 'reflected' like a light ray by a mirror; we have already mentioned that in practice this does not occur (see comment at the end of section 2.1.3). However, the calculation is valid because the mean distribution of the velocities of the molecules arriving at the wall is the same as the mean distribution of the molecules leaving it; the situation is exactly as if each molecule were 'reflected' elastically.

(b) To find the pressure of a gas, it is necessary to average over the velocities v of the molecules. The calculation of radiation pressure is simpler, because all the photons have the same momentum $h\nu/c$, and we can assume that they move with the speed of the wave, that is to say, they all have the same speed $v = c$.

The result of the calculation for the two cases can be expressed as a function of the energy density u existing near the wall. For a gas, the energy density is the sum of the kinetic energies $W = \frac{1}{2}mv^2$ of the n molecules contained in a unit volume. It is here that a small difference between the two calculations arises:

(i) *in the case of radiation*
for each photon $|\boldsymbol{p}| = W/c = W/v$ and the pressure is calculated to be $\varpi = u/3$;

(ii) *in the case of a gas*
for each molecule $|\boldsymbol{p}| = mv = 2W/v$ and the pressure is calculated to be $\varpi = 2u/3$.

For a gas, one ends up by using the mean translational kinetic energy of a molecule $\frac{3}{2}kT$, which gives

$$u = n\frac{3}{2}kT \quad \text{and} \quad \varpi = \frac{2u}{3} = nkT$$

2.2.2 From Relativity

(a) First we shall recall some classical formulae of relativity theory. The general laws of conservation of energy and of momentum can be extended to relativistic particles providing that one defines

$$\begin{cases} \text{their energy } W = mc^2, \\ \text{their momentum } \boldsymbol{p} = m\boldsymbol{v}, \end{cases}$$

with the relativistic mass $m = m_0/\sqrt{[1-(v^2/c^2)]}$. The proportional relationship between this and the energy allows one to write the momentum vector of each particle in the form

$$\boxed{\boldsymbol{p} = \frac{W}{c^2}\boldsymbol{v}}$$

In addition, W and the three components of $\boldsymbol{p}$ form the four components of a four-vector in Minkowski's four-dimensional space–time. The modulus of this impulse–energy four-vector is invariant under changes of reference frame, and from this follows the invariance of the quantity $W^2 - p^2c^2$ for any system of particles (p is the modulus of $\boldsymbol{p}$). If the system is confined to a single particle, a more exact calculation from the framed equation above gives

$$\boxed{W^2 - p^2c^2 = m_0^{\,2}c^4}$$

(b) In chapter 1, we treated the photon as a simple quantity of energy, but in the preceding section we have seen that, like any other particle, it must also possess momentum. This is why we have introduced the photon as a relativistic particle. The photon propagates with the speed of the wave, in other words its speed $v = c$. Such a speed normally leads to an infinite mass m and therefore to an infinite energy W. The only way of reconciling a speed $v = c$ with a finite energy is to assume a rest mass m_0 of zero. Taking these factors into account, the two formulae above may be applied

$$\begin{cases} \text{with } |\boldsymbol{v}| = c \text{ the first gives } |\boldsymbol{p}| = \dfrac{W}{c^2}c = \dfrac{W}{c} \\ \text{with } m_0 = 0 \text{ the second gives } W^2 - p^2c^2 = 0, \text{ so that } p = W/c. \end{cases}$$

Therefore the same value for the momentum is obtained as in the calculation from radiation pressure. The rest energy of a particle m_0c^2 must be distinguished from its kinetic energy $W - m_0c^2$; for a photon, all its energy is kinetic energy. Consideration of the photon as a relativistic particle is justified in the following section; the conclusions drawn there are the only way of explaining certain experiments.

2.3 Elastic Collisions of Photons—the Compton Effect

2.3.1 Compton's Experiments on X-Ray Scattering

The scattering of X-rays passing through a block of material was studied in 1909 by Barkla. He interpreted these experiments with the help of Thomson's classical theory: the scattered wave is emitted by the electrons bound to the atoms when the alternating electric field of the electromagnetic X-rays forces them into small oscillations at a frequency much greater than their natural frequency. (On the other hand when the frequency of forced oscillation is much smaller than the natural frequency, this is called Rayleigh scattering; this is the theory that accounts to a first approximation for the scattering of visible light by molecules in the air.) Barkla's measurements of the scattered intensity were in reasonable agreement with Thomson's theory. At the same time he was able to make a fairly accurate estimate of the number of electron oscillators and, as a result, of the number of electrons bound to each atom. These were the first direct measurements of the numbers of electrons in atoms and succeeded in confirming the atomic numbers Z corresponding to their positions in the Periodic Table (see appendix 2).

However, Barkla noticed certain divergences from classical theory, particularly when he used hard X-rays. X-Rays were classified by their hardness, that is, by their ability to pass through matter, because there was no way of measuring their wavelength. These measurements became possible after the work of Laue (1912) and Bragg (1914) on the diffraction of X-rays by crystals. In 1923 Compton repeated Barkla's scattering experiments and measured the wavelength of the scattered radiation with a crystal spectrometer.

More precisely, Compton passed a very nearly monochromatic pencil of X-rays, of wavelength λ_0, and having a well-defined direction (let this be the Oz direction) through a block of material. He collected the fraction of the radiation scattered in a particular direction making an angle θ with the direction of incidence Oz (figure 2.4(a)), on the entrance slit of the crystal spectrometer.

Figure 2.4(b) shows the type of spectra obtained in this way for various values of the angle θ. Each curve represents the variation of the intensity P

as a function of the wavelength λ. The three lower curves corresponding to the scattered radiation should be compared with the upper curve representing the spectrum of the incident radiation. The spectrum of the scattered radiation is composed of two lines: (a) a component at the incident wavelength λ_0, called the Thomson component since from Thomson's classical theory, the scattered radiation has the same frequency as the incident radiation; (b) a

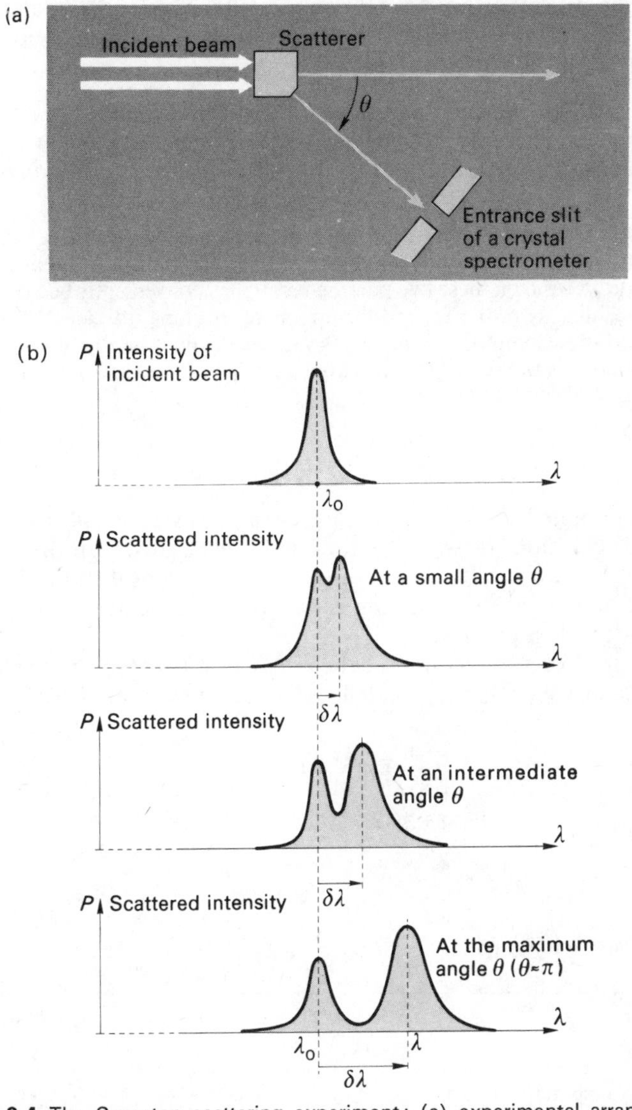

Figure 2.4 The Compton scattering experiment: (a) experimental arrangement; (b) spectra obtained

component of different wavelength $\lambda = \lambda_0 + \delta\lambda$, not accounted for by classical theory and called the Compton component. The measurements made by Compton showed in addition that:

(i) the difference $\delta\lambda = \lambda - \lambda_0$ is positive—the Compton component always has a longer wavelength than the Thomson component;
(ii) the difference $\delta\lambda$ is an increasing function of the angle θ between the direction of incidence and the direction of scattering;
(iii) the difference $\delta\lambda$ depends only on the angle θ. It is completely independent of the incident wavelength λ_0 and the composition of the material used for scattering.

It may be seen that the component scattered with a change of wavelength has some very characteristic properties. Compton explained them by considering the photon as a particle and by treating the problem as a collision between this particle and an electron; this is what we shall do now.

Comment While the wavelength λ of the Compton component is independent of the incident wavelength λ_0 and the composition of the scatterer, its intensity on the other hand does depend on these two parameters. It may conveniently be measured in relation to the intensity of the Thomson component: for long wavelengths ($\lambda_0 > 100$ pm, soft X-rays) the Compton component is very weak and is hardly observable, but its intensity increases rapidly when the wavelength λ_0 becomes smaller and it dominates when hard X-rays are used ($\lambda_0 \approx 1$ pm; $h\nu \sim 1$ MeV).

2.3.2 Theory of an Elastic Collision Between a Photon and a Free Electron

Compton interpreted the scattering with change of wavelength as an effect due to elastic collisions between photons and free electrons. In this problem all electrons that are weakly bound to atoms can be considered as free electrons; it is enough that their binding energy is negligible compared with the energy of the incident photons $h\nu_0$.

Whatever the exact mechanism of the interaction between the photon and the free electron, they form an isolated system during the very brief instant of

Table 2.1

		Before the collision	After the collision
Photon	Energy	$h\nu_0 = hc/\lambda_0$	$h\nu = hc/\lambda$
	Momentum	modulus $h\nu_0/c = h/\lambda_0$ direction along Oz	modulus $h\nu/c = h/\lambda$ direction making angle θ with Oz
Electron	Energy	$m_0 c^2$	$W = mc^2 = \sqrt{(p^2 c^2 + m_0{}^2 c^4)}$
	Momentum	zero (electron at rest)	modulus p direction making angle ϕ with O_z

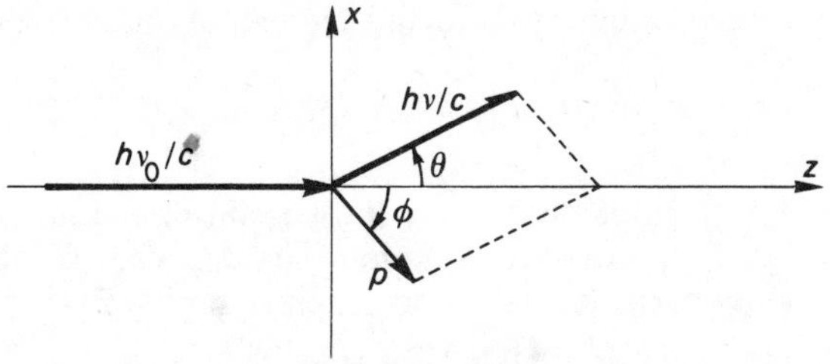

Figure 2.5 Vector diagram of the collision between a photon and a free electron

collision to which the general laws of conservation of energy and momentum can be applied. In examining the consequences of these two laws in detail, we shall not have described the phenomenon completely, but we shall obtain some important information about it. Table 2.1 and figure 2.5 show the notation we use for this calculation. (We have taken into account the fact that p and W are related by relativistic invariance $W^2 - p^2c^2 = m_0{}^2c^4$.)

To ensure conservation of momentum, the geometrical sum of the two momentum vectors after the collision must be equal to the single momentum vector before the collision. This shows that figure 2.5 is planar, and it is sufficient to project the vectors on to the two axes Oz, and Ox perpendicular to Oz, shown in the diagram. With the above notation we can write the conservation equations

Energy $$h\nu_0 + m_0c^2 = h\nu + \sqrt{(p^2c^2 + m_0{}^2c^4)}$$

Momentum
$$\begin{cases} \text{projection on O}z\text{:} \; \dfrac{h\nu_0}{c} = \dfrac{h\nu}{c}\cos\theta + p\cos\phi \\ \text{projection on O}x\text{:} \; 0 = \dfrac{h\nu}{c}\sin\theta + p\sin\phi \end{cases}$$

Three equations are thus obtained relating the incident frequency ν_0 and the four unknowns ν, θ, p and ϕ. Therefore the problem is not completely determined, but we can calculate three of the unknowns as a function of the fourth, chosen as a parameter. In an experimental measurement, a particular value of θ is used, so we choose θ as a parameter and calculate the other unknowns as a function of θ. Initially we are interested in calculating the frequency ν of the scattered photon in order to make a comparison with experimental results, so we must eliminate the two unknowns relating to the electron, p and ϕ.

First we eliminate ϕ by rewriting the two momentum equations in the form

$$p\cos\phi = \frac{h}{c}(\nu_0 - \nu\cos\theta)$$

$$p\sin\phi = -\frac{h}{c}\nu\sin\theta$$

Squaring each of these equations and adding them term by term we obtain

$$p^2 = \frac{h^2}{c^2}(\nu_0{}^2 + \nu^2 - 2\nu_0\,\nu\cos\theta)$$

We can also find p^2 from the first equation by isolating the square root in the second term and then squaring the equation thus obtained. We then eliminate p by equating the two expressions found for the quantity p^2c^2

$$p^2c^2 = [h(\nu_0 - \nu) + m_0c^2]^2 - m_0{}^2c^4 = h^2(\nu_0{}^2 + \nu^2 - 2\nu_0\,\nu\cos\theta)$$

This equation may be simplified considerably and after dividing by m_0c^2

$$\nu_0 - \nu = \frac{h}{m_0c^2}(1 - \cos\theta)\,\nu_0\,\nu$$

(When ν is nearly equal to ν_0, the product $\nu_0\,\nu$ can be replaced to a first approximation by $\nu_0{}^2$, and the asymptotic value of the difference $\nu_0 - \nu$ is obtained.)

To make a comparison with experiment, it is convenient to finish the calculation by replacing frequencies with wavelengths. This is done simply by dividing the equation above by the product $\nu_0\,\nu$

$$\frac{1}{\nu} - \frac{1}{\nu_0} = \frac{h}{m_0c^2}(1 - \cos\theta)$$

whence

$$\boxed{\delta\lambda = \lambda - \lambda_0 = \frac{h}{m_0c}(1 - \cos\theta)}$$

The difference $\delta\lambda$ thus calculated gives a good account of the three properties observed experimentally.

(a) $\delta\lambda$ is positive. This is a consequence of conservation of energy. The energy of a photon becomes smaller because a part of its energy is transformed into kinetic energy of the electron. We can regard all the photon energy as kinetic energy, because its rest mass is zero, the general expression for the relativistic kinetic energy being $W - m_0c^2$. Therefore, all the energy exchanged in the collision is kinetic energy; the collision conserves kinetic energy and is therefore called an elastic collision (see section 1.4.2).

(b) $\delta\lambda$ is an increasing function of the angle θ, because the latter can vary only between 0 and π where the cosine function is monotonic. Alternatively, one can write

$$1 - \cos\theta = 2\sin^2\theta/2$$

(c) $\delta\lambda$ is completely determined once the angle θ is known. It is totally independent of the composition of the scatterer, as well as of the incident wavelength λ_0. This very characteristic property of the Compton effect is thus explained: the effect gives rise to an absolute difference of wavelength. The calculation of this difference $\delta\lambda$ reduces to a constant appearing in front of the factor $(1-\cos\theta)$, having the dimensions of wavelength and called

$$\boxed{\text{the Compton wavelength } \Lambda = \frac{h}{m_0 c} = 2{\cdot}426 \text{ pm}}$$

This is the wavelength for which the energy of the photons is equal to the rest energy of the electron: $h\nu = hc/\Lambda = m_0 c^2 = 0{\cdot}511$ MeV.

Numerical agreement between the theoretical and experimental values of the Compton wavelength confirms the interpretation of the effect. (Compton's rough measurements were refined in 1930 by the more accurate measurements of Gingrich.)

The Compton wavelength Λ is the wavelength difference for a wave scattered at right angles. The maximum difference, observed for back-scattered waves $(\theta = \pi)$ is 2Λ.

In any case, the absolute difference $\delta\lambda$ is of the order of magnitude of Λ. From this absolute difference it is easy to find relative differences and consequently the relative importance of the energy exchanges between the photon and the electron may be evaluated as follows

(1) $h\nu_0 \ll m_0 c^2$ or $\lambda_0 \gg \Lambda \rightarrow \delta\lambda \ll \lambda_0$ the photon gives up very little energy

(2) $h\nu_0 > m_0 c^2$ or $\lambda_0 < \Lambda \rightarrow \delta\lambda > \lambda_0$ the photon gives up most of its energy

This illustrates a very general result of collision theory: if the projectile has a total energy (mass + kinetic) much smaller than that of the stationary target, the latter can take up the momentum of the projectile while retaining practically zero kinetic energy. (We have already seen this in section 1.4.2 with regard to collisions of electrons on nearly stationary and much heavier targets in the form of atoms, and we shall return to this in a general way in section 2.5.). In the particular example of *elastic* collisions, it means that very little energy is exchanged between the two particles. On the other hand, if the projectile has a total energy greater than that of the target, the energy exchanged between the two particles is far greater.

2.3.3 Observations of Compton Electrons

The calculation carried out in the last section was concerned only with the characteristics of the radiation; but now that we have determined the relationship between λ and θ, it is easy to find—also as a function of θ—the behaviour of the electron set into motion by the collision

$$\text{Kinetic energy:}\quad W - m_0 c^2 = h(\nu_0 - \nu) = \frac{h\nu_0}{1 + [m_0 c^2/h\nu_0(1-\cos\theta)]}$$

$$\text{Direction:}\quad \tan\phi = \frac{-\sin\theta}{\dfrac{\nu_0}{\nu} - \cos\theta} = \frac{-\sin\theta}{[1-\cos\theta][1+(h\nu_0/m_0 c^2)]} = \frac{-\cot\theta/2}{1+(h\nu_0/m_0 c^2)}$$

From the expression for the kinetic energy of the electron we can confirm the results of the discussion at the end of the last section: the electron takes only a very small part of the energy $h\nu_0$ of the incident photon when $h\nu_0 \ll m_0c^2$.

With regard to the direction of the electron, the minus sign in the expression for $\tan\phi$ indicates simply that the photon and the electron fly off on opposite sides of the Oz axis. Disregarding the signs, when θ increases from 0 to π, that is to say $\theta/2$ increases from 0 to $\pi/2$, ϕ decreases from $\pi/2$ to 0; but ϕ is never greater than $\pi/2$, so *the electron always goes off in the forward direction.*

Comment In the case where $h\nu_0 \ll m_0c^2$, that is where the electron takes up very little energy, one can write approximately $\tan\phi = -\cot\theta/2$. From this one sees that the trajectory of the electron is perpendicular to the interior bisector of the angle θ (see figure 2.6; when the photon loses very little energy, the two vectors $h\nu_0/c$ and $h\nu/c$ are nearly equal). This is exactly what happens in a collision between a light ball (the photon) and a heavy stationary ball (the electron).

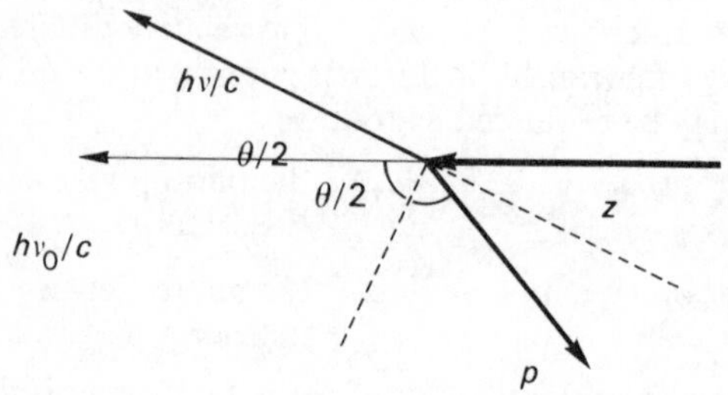

Figure 2.6 Scattering of a free electron by a photon when $h\nu_0 \ll m_0c^2$

To verify these formulae experimentally Compton and Simon (1925) passed a beam of X-rays into a cloud chamber and observed the trajectories of the electrons as a result of collisions. The Compton electrons can be distinguished from photoelectrons, which are also produced by the X-rays, by the fact that the photoelectrons go off perpendicular to the direction of propagation of the photons (parallel to the electric field $\boldsymbol{E}$ of the incident wave). The trajectory of a scattered photon is not observable in the cloud chamber, but in certain circumstances it can be reconstructed from its point of scattering (which is the point from which the Compton electron comes) and from its point of arrival at a place where a photoelectron is produced. In this way the angle θ can be measured. Compton photoelectrons have also been observed simultaneously with scattered photons by coincidence techniques commonly used in nuclear physics.

2.3.4 Elastic Collision Between a Photon and Bound Electron—Thomson Scattering

We have assumed in the theory of the last two sections that the target electron is stationary and free, but the theory is also valid when the electron is weakly bound to an atom with a binding energy of a few electron-volts (see section 1.4, ionisation energy). This additional energy which has not been taken account of in the theory is in fact negligible compared with the energy of the X-ray photons, which commonly varies between 10 keV and 1 MeV (that is to say between $\lambda \approx 100$ pm and $\lambda \approx 1$ pm). (To be precise, the collision is no longer completely elastic since a small part of the energy of the photon has been used to compensate for the binding energy and is not available in the form of kinetic energy.)

However, the atom also contains other electrons that are very strongly bound to it and whose binding energy cannot be disregarded (see chapter 7, X-ray spectra). In an elastic collision with a photon the binding energy of the electron remains unchanged, that is, the electron remains bound to the atom. Ultimately, it is the whole atom which takes part in the collision with the photon and plays the role of a target. The conservation equations which we have written down are generally useful whatever the target; they apply equally well to this new problem together with all the conclusions drawn from them, providing that one adjusts the meaning of the quantities W, p and m_0 which must now be related to the whole atom rather than to one of its electrons. All the formulae that have been derived in the preceding sections remain valid simply by changing the numerical value of the rest mass m_0 of the target: m_0 now represents the mass of the whole atom and is 10^4 to 10^5 times greater than the mass of the electron. The calculated difference of wavelength is now divided by this factor and one obtains $\delta\lambda \lesssim 10^{-4}$ pm, a value which is practically negligible over the whole range of X-ray wavelengths.

If the scatterer is a solid body, the atom itself is embedded in a rigid crystalline lattice, and the whole sample of material is subjected to collisions. The mass m_0 of the specimen must now be substituted in all the formulae; that is m_0 is now multiplied by a number of the order of Avogadro's number $\mathcal{N} = 6 \times 10^{23}$. The calculated differences $\delta\lambda$ are now so small as to defy all possibility of measurement and are no longer significant. There is no longer any difference between the result of the collision theory and Thomson's classical theory. The Thomson component in the spectrum of the scattered light is therefore explained.

In other words, the energy of the X-ray photon projectile is now very small compared with the rest energy of the target: $h\nu_0 \ll m_0 c^2$, and under these conditions the two particles can exchange a large quantity of momentum (the projectile can be scattered backwards) without exchanging much energy. This is the general property of *elastic* collisions which was reviewed at the end of section 2.3.2.

2.4 Inelastic Collisions of Photons

The phenomena of Compton or Thomson scattering which have been described are based on the conservation of total *kinetic energy* of the system under study, and this is why they are classified under the title of elastic collisions. On the other hand we have described two phenomena in chapter 1, namely the photoelectric effect and the phenomenon of optical resonance, where exchanges occur between kinetic energy of the photons and another form of energy. We have assumed the conservation of *total energy* of the system under study, quite naturally, but we have not taken the momentum of the photons into consideration at all. We shall now treat the phenomenon of optical resonance as an inelastic collision process.

2.4.1 Absorption of Photons

We adopt the notation used when studying optical resonance in section 1.3.2: $h\nu_{12}$ is the energy difference between two states of an atom with internal energies E_1 (the ground state) and E_2 (the excited state). The more rigorous investigation given below significantly modifies the description of the phenomenon.

After the collision, the atom and the photon effectively form one particle and the initial momentum of the incident photon must be preserved by this single particle. In other words, the atom must be in motion after the collision and must therefore possess a certain kinetic energy that it did not have previously. This kinetic energy of the atom can only have been taken from the energy of the incident photon, and this is only possible if the energy $h\nu$ of the incident photon is greater than the energy difference $E_2 - E_1 = h\nu_{12}$. The problem then is to determine the exact relationship between $h\nu$ and $h\nu_{12}$.

To perform this calculation we use, as in section 2.3.2, the properties of relativistic particles, and this requires a preliminary remark: the relativistic relationship between mass and energy is completely general and so an excited atom, which has more internal energy than a normal atom in the ground state, must also be heavier. We must therefore distinguish between the rest mass m_1 of the ground state and the rest mass m_2 of the excited state. We summarise the notation in Table 2.2.

Table 2.2

		Before the collision	After the collision
Photon	Energy	$h\nu$	non-existent
	Momentum	magnitude $h\nu/c$ direction along Oz axis	
Atom	Total energy	E_1	$W = \sqrt{(p^2 c^2 + m_2^2 c^4)}$
	Momentum	zero (atom stationary)	magnitude p direction along Oz
	Rest mass	$m_1 = E_1/c^2$	$m_2 = E_2/c^2 = (E_1 + h\nu_{12})/c^2$

The table indicates that there is only a single momentum vector both before and after the collision. To assure conservation of momentum these two vectors must be in the same direction. The problem does not then include an unknown angle and this considerably simplifies the writing of the conservation equations

Energy: $$h\nu + E_1 = \sqrt{(p^2 c^2 + m_2{}^2 c^4)}$$

Modulus of the momentum: $h\nu/c = p$

Eliminating p immediately by substituting the expression for p from the second equation into the first gives

$$(E_1 + h\nu)^2 = (h\nu)^2 + m_2{}^2 c^4 = (h\nu)^2 + (E_1 + h\nu_{12})^2$$

This expression simplifies to

$$\boxed{\nu = \nu_{12}\left(1 + \frac{h\nu_{12}}{2m_1c^2}\right)}$$

We find indeed, as foreseen, that the frequency ν of the incident photon must be greater than the theoretical frequency of the spectral transition $\nu_{12} = (E_2 - E_1)/h$.

However, this does not throw doubt on what was said in chapter 1. The energy involved in the optical transitions that have been discussed is very small compared with the rest energy of the atom

$$\left.\begin{aligned} h\nu_{12} &\sim 1 \text{ to } 10 \text{ eV} \\ m_1 c^2 &\sim 10 \text{ to } 100 \text{ GeV} \end{aligned}\right\} \rightarrow \frac{h\nu_{12}}{m_1 c^2} \sim 10^{-10}$$

The width of optical resonance lines observed experimentally is usually determined by the Doppler effect arising from the velocity v of thermal agitation of the atoms in the vapour (see section 1.3.3); thus one obtains for the relative width of spectral lines $\delta\nu/\nu \sim v/c \sim 10^{-6}$. It may be seen that the displacement of lines due to conservation of momentum is much smaller than their width and is therefore not observable.

Finally, we consider once again the general property of collision processes which has already been stressed in relation to elastic collisions (and to which we return in section 2.5): when the total energy of the projectile is far smaller than that of the target at rest, the target can absorb a significant quantity of momentum while only taking up negligible kinetic energy. In the case of *inelastic* collisions, this means that for practical purposes all the energy absorbed by the target goes to increase its internal energy. This justifies in retrospect the simplified arguments of chapter 1 where no account is taken of momentum.

2.4.2 Emission of Photons

We conclude this new description of optical resonance, which takes momentum into account, by dealing with the other aspect of the phenomenon, that of the emission of a photon from an atom in an excited state. The notation is

Table 2.3

			Before emission	After emission
Photon	Energy			$h\nu$
	Momentum		non-existent	magnitude $h\nu/c$; direction along Oz axis
Atom	Energy		E_2	$W = \sqrt{(p^2c^2 + m_1{}^2c^4)}$
	Momentum		zero stationary	magnitude p; direction along Oz axis
	Rest mass		$m_2 = E_2/c^2$	$m_1 = E_1/c^2 = (E_2 - h\nu_{12})/c^2$

the same as in the last section, so in table 2.3 we need only state the assumptions of this new problem without explanation.

The total momentum before emission is zero; the two momentum vectors after emission must be colinear and it is enough to equate their algebraic values in this direction. The conservations equations are

Energy: $E_2 = h\nu + \sqrt{(p^2c^2 + m_1{}^2c^4)}$

Momentum: $0 = h\nu/c + p$

The elimination of p is as simple as in the preceding derivation and leads to

$$\boxed{\nu = \nu_{12}\left(1 - \frac{h\nu_{12}}{2m_2c^2}\right)}$$

The result obtained resembles the earlier one, but m_1 has been replaced by m_2 to which it is nearly equal, and above all the sign of the correction term has changed. This time the frequency displacement is in the opposite direction. The atom must take up momentum opposing that of the emitted photon, that is to say, it must preserve in the form of kinetic energy part of the internal energy lost $h\nu_{12}$. The photon therefore cannot carry off all this energy $h\nu_{12}$.

As before, it is clear that in the case of optical transitions this frequency displacement is negligible. The derivation again serves only to justify simplifications made in chapter 1, but the consequences of these formulae are different for transitions of large energy, as we see in the next section.

2.4.3 Application to γ-rays—the Mössbauer Effect

The γ-rays emitted by naturally or artificially radioactive atoms are known to be only electromagnetic waves of particularly high frequency. The region of γ-rays covers part of the X-ray region but extends to higher frequencies.

The process of emission of a γ-ray photon is interpreted as a radiative transition between two energy states of the nucleus, totally analogous to our discussion of optical spectroscopy. However, the energy $h\nu_{12}$ of nuclear transitions is several orders of magnitude greater, commonly varying between

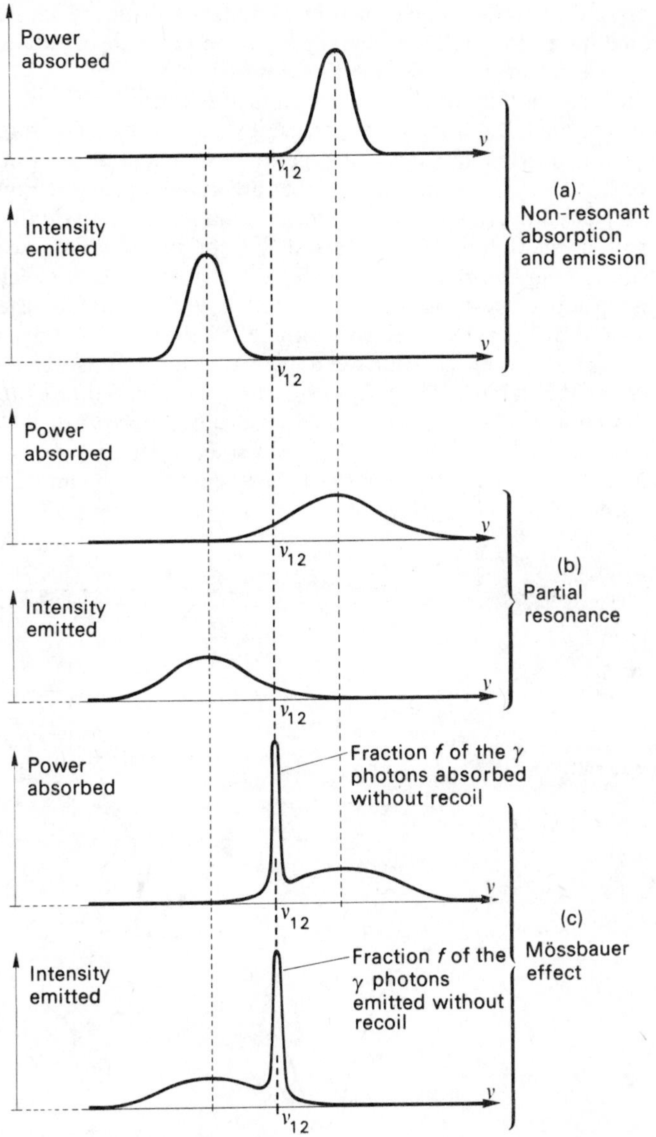

Figure 2.7 Nuclear radiative transitions

10 keV and 10 MeV. The ratio $h\nu_{12}/m_0c^2$ remains small compared with unity (we write m_0 to represent both m_1 amd m_2, as the distinction between them is unimportant), but the frequency displacement from the formulae derived above become greater than the linewidth. Thus a situation arises (shown diagrammatically in figure 2.7(a)) where the absorption line and the emission

line corresponding to the same transition no longer coincide. That is, the photons emitted have too small a frequency for them to be absorbed, and in this case an optical resonance experiment is impossible.

Nevertheless, the linewidth is determined essentially by the Doppler effect (see section 1.3.3) and is an increasing function of temperature. Thus with γ-rays it is possible to carry out experiments analogous to optical resonance as long as the source and the absorber are at a temperature high enough to ensure a partial overlap of the Doppler profiles of the emission line and of the absorption line (figure 2.7(b)). Partial absorption can then take place.

In 1958, a young German physicist, R. Mössbauer, used this technique to study absorption corresponding to the 129 keV radiation of the nucleus ^{191}Ir. He discovered that by cooling the source and the absorber the absorption increased, contrary to expectation. The interpretation of this phenomenon is as follows: at low temperatures, below a certain temperature characteristic of the solid containing the emitting and absorbing nuclei (called the Debye temperature; see texts on thermodynamics) some of the γ-rays are emitted or absorbed without recoil or Doppler broadening. The phenomenon can be

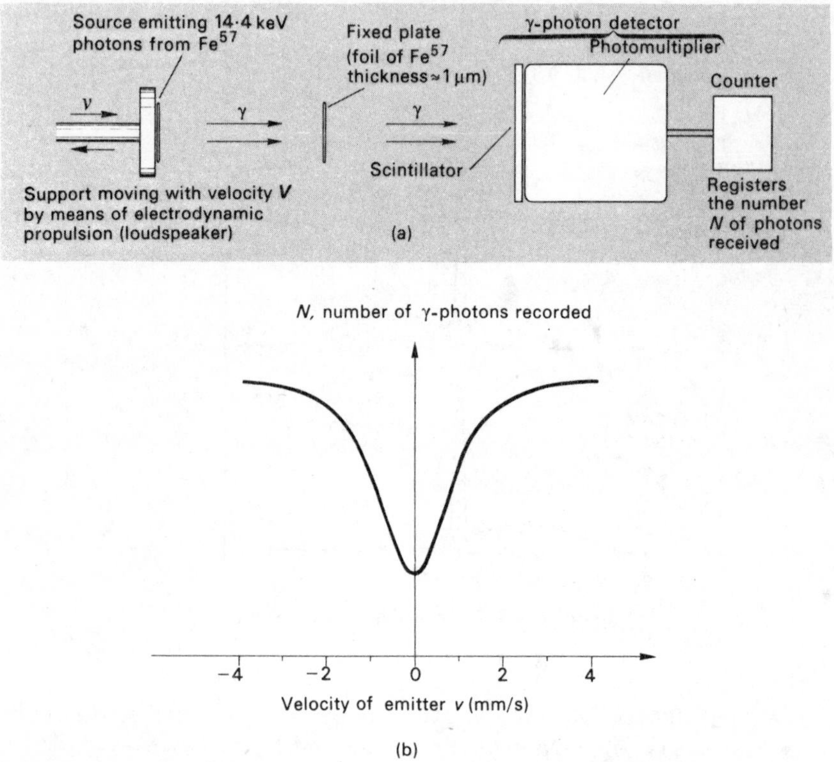

Figure 2.8 Mössbauer effect: (a) diagram of a Mössbauer experiment; (b) Mössbauer spectrum

summarised by saying that the emitted or absorbing nucleus is embedded in a crystalline lattice and it is the crystal as a whole which recoils. As the mass of the crystal is considerably greater than the mass of the nucleus (by a factor of the order of Avogadro's number $\mathcal{N} = 6 \times 10^{23}$) the velocity of recoil is negligible. An exact but complicated calculation using the techniques of solid-state theory, shows that only a proportion f of the γ-photons is emitted or absorbed without recoil, the factor f depending essentially on the ratio T/Θ (T is the actual temperature, Θ is the Debye temperature of the solid). In such a situation, figure 2.7(c) shows the profile of the lines.

The width of the recoilless emission or absorption line is very narrow and in simple cases is the natural width of the line, related to the radiative lifetime of the nuclear level (see section 4.2.3). In frequency units, this natural width is $\Delta\nu \sim 10^6$ Hz. Since γ-ray frequencies are of the order of 10^{19} Hz, the relative width $\Delta\nu/\nu \sim 10^{-13}$ is smaller than in any other physical phenomenon.

To demonstrate these extremely narrow recoilless absorption and emission lines, it is necessary to use a rather unusual experimental technique. With the absorber fixed, the emitter is moved with velocity v parallel to the direction of propagation of the γ-rays (figure 2.8). A detector counts the number N of γ-photons that have passed through the absorber and in this way the absorption of γ-rays as a function of v is studied. From the Doppler effect, the frequency seen by the absorbing atoms is not ν_{12} but

$$\nu_{12} \pm \nu_{12}\, v/c$$

the $\pm$ sign depending on the direction of the velocity vector. With $\Delta\nu$ being the width of the recoilless line, it can be seen immediately that if

$$v/c > \Delta\nu/\nu_{12}$$

the conditions for resonant absorption are no longer satisfied. With the numerical value of $\Delta\nu/\nu$, very small velocities of the order of 1 cm s^{-1} are sufficient to suppress resonant absorption. The curve $N = f(v)$, N being the total number of counts, then represents the profile of the absorption and emission lines. This curve is usually known as a 'Mössbauer spectrum' (see figure 2.8).

The nuclear energy levels can have complex structures, particularly in magnetic substances; the profile of the emission or absorption line will reflect the properties of the solid. Mössbauer spectroscopy has advanced considerably in recent years and has made important contributions in fields as varied as magnetism, chemical bonding and crystallography.

Comment It is impossible to excite a nucleus into an excited state in the same way as an atom. Only certain nuclear reactions starting from a radioactive nucleus B lead to excited states of another nucleus A

$$\text{B} \rightarrow \text{excited A} \rightarrow \text{ground state of A} + \gamma$$

For example

$$^{57}\text{Co} \rightarrow \text{excited } ^{57}\text{Fe} \rightarrow \text{ground state of } ^{57}\text{Fe} + 14{\cdot}4 \text{ keV}$$

$$^{191}\text{Os} \rightarrow \text{excited } ^{191}\text{Ir} \rightarrow \text{ground state of } ^{191}\text{Ir} + 129 \text{ keV}$$

This kind of reaction is common for heavy nuclei, but hardly ever occurs for light nuclei. Absorption experiments can be carried out on about 60 nuclei, corresponding to about 40 distinct elements for example

$$^{40}Ca,\ ^{57}Fe,\ ^{61}Ni,\ ^{67}Zn,\ ^{73}Ge,\ ^{83}Kr,\ ^{99}Ru,\ ^{107}Ag,\ ^{109}Ag,\ ^{117}Sn,\ ^{119}Sn, \text{etc.}$$

The analogy existing between the Mössbauer effect and Thomson scattering (see section 2.3.4) should be noted, although they correspond to different types of collision, one inelastic the other elastic. In both cases the emitting atom is rigidly bound to a crystal lattice and the latter can absorb all the momentum without taking up kinetic energy. *Thomson scattering has the same relationship to the Compton effect as the Mössbauer effect has to the phenomenon of absorption and emission of non-resonant γ-photons.*

2.5 Review of the Various Types of Collision Processes

This chapter has examined only those collision processes involving two particles because they lead to simple systems of equations; but there are other photon collision phenomena involving a larger number of particles.

(a) The photoelectric effect or, to be more exact, the phenomenon of photo-ionisation, is an inelastic collision involving three particles, as an independent electron appears after the collision between the incident photon and the atom.

(b) In relation to X-ray spectra, we shall discuss *Bremsstrahlung* radiation emitted by fast electrons when they strike a piece of metal (see section 7.3.1). This is an elastic collision involving three particles: when the incident electron collides with the nucleus of an atom part of its kinetic energy is transformed into a photon.

(c) In nuclear physics the creation of γ-photons from an electron and a positron is observed when their energy is greater than 1·02 MeV ($h\nu > 2m_0c^2$, where m_0 is the mass common to the electron and to the positron). This is an inelastic collision involving four particles: it occurs only when the photon meets a nucleus of an atom capable of absorbing the excess momentum (together the two particles, electron and positron, cannot have a momentum greater than that of the photon).

In these three phenomena, the total energy of the projectile (photon or electron) is usually very much smaller than the rest energy of the target atom; and whatever the details of the conservation equations, it can be shown that the target atom takes hardly any kinetic energy. Let

w be the total energy of the incident projectile

m_0c^2 be the rest energy of the target $\boxed{w \ll m_0c^2}$

W be the energy } of the target after the collision

p be the momentum } of the target after the collision

Table 2.4 Table of various types of collisions

	$w \ll m_0 c^2$ Total energy of the projectile $\ll$ rest energy of the target (target takes up hardly any kinetic energy)		$w \gtrsim m_0 c^2$ Total energy of the projectile $\gtrsim$ rest energy of the target (target takes up a significant kinetic energy)
Elastic collisions (conservation of kinetic energy, inclusive of energy of the photons)	(there is hardly any transfer of kinetic energy between the particles) —slow electrons on gas atoms (below the Franck and Hertz resonance potential)		(sizeable transfer of kinetic energy between the particles)
	—deflection of α-particles by nuclei (Rutherford's experiment, see chapter 6)	$\leftrightarrow$	α-particles on hydrogen nuclei (see chapter 6)
	—Thomson scattering	$\leftrightarrow$	Compton scattering at short wavelengths
	—Bremsstrahlung (see chapter 7)		
Inelastic collisions (kinetic energy can be transformed into another form of energy)	(the initial kinetic energy may be completely transformed into another form of energy) —sufficiently fast electrons on gas atoms (above the Franck and Hertz resonance potential) —other collision processes between atoms and electrons (see volume 2, chapter 7)		(a large part of the kinetic energy must remain in the form of kinetic energy)
	—optical resonance and Mössbauer effect	$\leftrightarrow$	absorption and emission of non-resonant γ-photons
	—photoelectric effect (photoionisation)		

The projectile has energy w and possesses a momentum whose modulus is less than or equal to w/c (equality for a photon, inequality for all material particles). During the collision, the target takes up a momentum p whose order of magnitude is equal to the momentum of the projectile, that is to say, p must be smaller or of the same order of magnitude as w/c: $p \lesssim w/c \ll m_0 c$.

The condition $p \ll m_0 c$ shows us that the target retains the behaviour of a non-relativistic particle. Its energy can be calculated by the use of relativistic invariance

$$W = \sqrt{(m_0^2 c^4 + p^2 c^2)} = m_0 c^2 \sqrt{(1 + p^2/m_0^2 c^2)} \simeq m_0 c^2(1 + p^2/2\, m_0^2 c^2)$$

and from this we find a kinetic energy approximating to its non-relativistic value

$$W - m_0 c^2 \simeq p^2/2m_0$$

Taking into account the order of magnitude of p we finally obtain

$$W - m_0 c^2 \lesssim w^2/2m_0 c^2 = w(w/m_0 c^2) \ll w$$

in other words, the kinetic energy taken up by the target is only a very small fraction of the energy w of the projectile. The heavy target absorbs momentum without taking up kinetic energy.

Thus we understand the basic reason for the very general result so often met in the course of collision calculations (see section 1.4.2 and sections 2.3 and 2.4). The comparison between the energy w of the incident projectile and the energy $m_0 c^2$ of the target provides an important criterion for categorising collision phenomena.

In the course of these studies of collisions, we have also come across another important criterion resting on the distinction between kinetic and other forms of energy, it being understood that the energy of the photons is considered entirely as kinetic energy (see section 2.2.2). Thus, from the above theory, where a very heavy target ($w \ll m_0 c^2$) absorbs momentum without taking up kinetic energy, we have drawn two different conclusions: (i) if the collision is elastic, so that the total kinetic energy is conserved, there is practically no transfer of kinetic energy between the two incident particles; (ii) if the collision is inelastic, so that there is a transformation of kinetic energy into another form of energy, this transformation of kinetic energy can be total (justifying the simplifications made in chapter 1).

A table (table 2.4) of all the collision phenomena studied in this book, classified by the two criteria we have discussed, is given on p. 63.

3

Radiative Transition Probabilities

We have seen that exchanges of energy between radiation and matter occur discontinuously in a variety of situations broadly described in the previous chapter as collisions between particles (emission or resonant absorption, the photoelectric effect, the Compton effect, and so on). Each one of these collisions can be considered as an instantaneous event, but what can be said about the moment of collision? Experiment shows that these individual collisions take place at uncertain and unpredictable times. The only kind of prediction that can be made is a statistical one, expressed in terms of probabilities. In this chapter we shall be concerned with the meaning and experimental measurement of these probabilities.

Our discussion in section 3.1 on the absorption of photons can be applied to all the phenomena previously described. However, in this chapter we shall be interested primarily in optical resonance, that is to say, in transitions between two energy levels of an atom E_2 and E_1 in the presence of photons of energy $h\nu_{12} = E_2 - E_1$ (see section 1.3.2). The first section describes the absorption of photons by atoms that are in a low-energy state E_1; the sections following describe the emission of photons by atoms in an excited state of higher energy E_2.

3.1 The Absorption of Photons

3.1.1 The Hard-spheres Model and the Concept of a Cross-Section

For the purposes of a general description of collision experiments it is useful to discuss first a particularly simple model. The target particles are hard spheres of radius r_1, stationary and distributed in space; the projectiles are also hard spheres of radius r_2, moving with parallel velocities in the same direction Oz. Under these conditions the projectiles that happen to hit a particular target particle of centre O are all those whose centre is situated within a cylinder having axis Oz and radius $r_1 + r_2$, and whose cross-sectional area is $\sigma = \pi(r_1 + r_2)^2$ (see figure 3.1). Let N_{proj} be the total number of incident

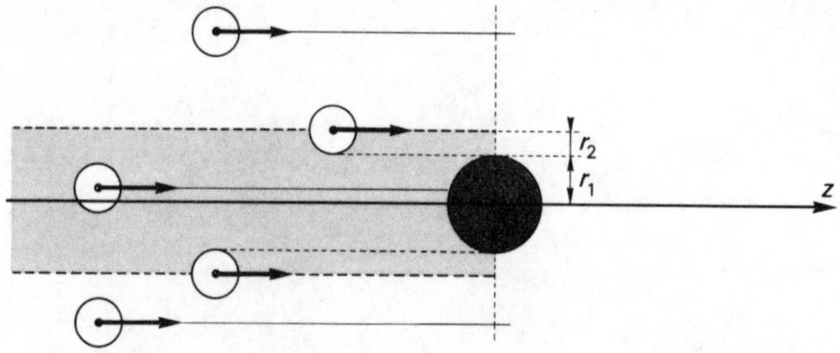

Figure 3.1 Collisions of hard speres—case of a single target

projectiles during an experiment. These projectiles form a cylindrical beam parallel to Oz and of cross-sectional area S; this cylindrical beam sweeps out a certain volume in the target medium containing a total number of target particles equal to N_{target}.

Each of these targets stops projectiles whose centres are contained within a cylinder of revolution of cross-section σ. It could happen that two of these cylinders overlapped partially if two targets were situated approximately on the same line parallel to Oz so that one masked the other (see figure 3.2), but these are exceptional cases and can be disregarded as long as there are not too many targets.

Let us assume that the incident projectiles have *on average* a uniform distribution over the beam, so that the number which undergo a collision N_{coll} is proportional to the volume of N_{target} cylinders of cross-section σ. To be more exact, the proportion of projectiles that undergo a collision $N_{\text{coll}}/N_{\text{proj}}$ is equal to the ratio of the respective volumes of the N_{target} cylinders and of the whole beam, in other words, to the ratio of their cross-sectional areas

$$\frac{N_{\text{coll}}}{N_{\text{proj}}} = \frac{\sigma N_{\text{target}}}{S}$$

giving

$$\boxed{N_{\text{coll}} = \frac{\sigma}{S} N_{\text{proj}} \times N_{\text{target}}}$$

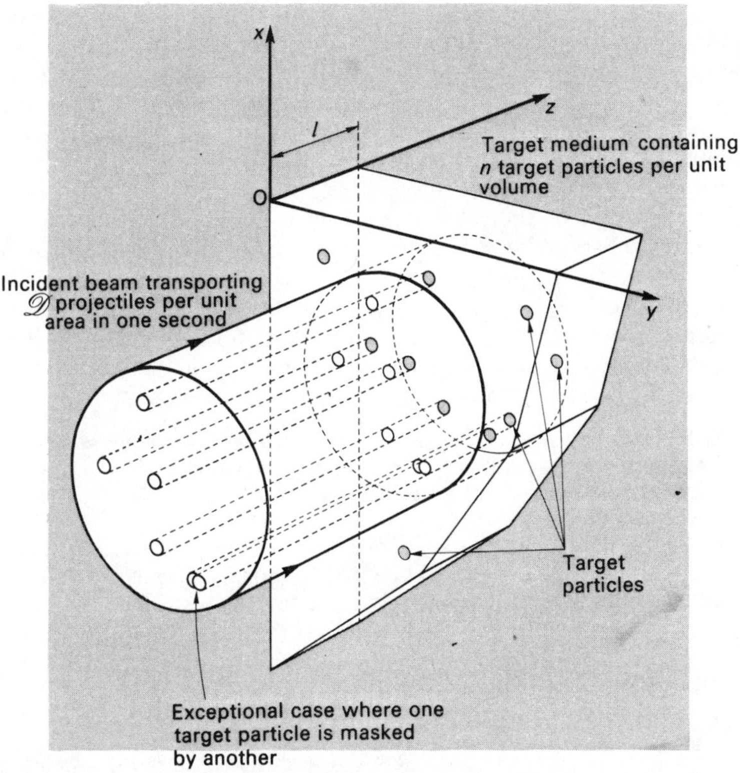

Figure 3.2 Concept of a cross-section

For a more detailed description of the experiment, the following notation can be introduced:

either

n: the number of targets per unit volume

l: the length of the target medium parallel to Oz

so that $N_{target} = nlS$ and *the probability of a collision for one projectile is*

$$\boxed{\frac{N_{coll}}{N_{proj}} = \sigma n l}$$

or

$\mathscr{D}$: the intensity of projectiles in the beam per unit area, in other words the number of projectiles passing through unit cross-sectional area per second

t: the duration of the experiment

so that $N_{proj} = \mathscr{D} t S$ and *the probability of a collision for one projectile is*

$$\boxed{\frac{N_{coll}}{N_{target}} = \sigma \mathscr{D} t}$$

This argument is based on the idea of hard spheres, which does not correspond to any real physical situation, whether the projectiles and targets be photons, electrons or atoms. Nevertheless, it is possible to generalise the formulae for all types of collisions between particles of any kind. These formulae express proportionality relationships between collision probabilities and the numbers of particles per unit area $N_{\text{target}}/S = nl$ or $N_{\text{proj}}/S = \mathscr{D}t$. These proportionality relationships always hold true and the corresponding coefficient of proportionality is called the cross-section σ, having the dimensions of area. (The probabilities are dimensionless and the products nl or $\mathscr{D}t$ both have the dimensions of an inverse area.)

Generally speaking then, one of the above three framed formulae is chosen to define the cross-section σ, and the two others follow directly. The hard-spheres model provides a useful intuitive representation and gives a concrete meaning to the cross section.

Comment I For a collision between two atoms, each of mean radius $\approx 10^{-10}$ m the hard-spheres model gives a collision cross-section of $\approx 10^{-19}$ m^2. This is often called the gas kinetic cross-section and is used as a standard against which actual cross-sections are compared.

Comment II We have defined here the total cross-section of a collision process. In chapter 5, the more elaborate idea of a differential cross-section is introduced with regard to Rutherford's experiment (see also volume 2, section 7.5).

3.1.2 The Absorption Coefficient and Experimental Measurement of Cross-Sections

In most collision processes, the projectile which has undergone a collision is deflected or destroyed, so that it is removed from the incident beam. Experimentally, the simplest observation to make is the loss of intensity of a beam of projectiles when they pass through the target medium.

For a beam of light, the luminous intensity represents the energy transported per unit time or alternatively, the power P. If a monochromatic beam of frequency ν and cross-section S has an intensity of $\mathscr{D}$ photons per unit area, then the power transported $P = \mathscr{D}Sh\nu$. It is convenient to use another quantity in theoretical calculations: the energy density u within the beam of light incident upon the target atoms. By noting that in one second a cylinder of length c (the speed of light) passes through a cross-section, one finds that $P = Scu$. The comparison of these two expressions for the luminous intensity P gives a relationship between the energy density u and the intensity of photons $\mathscr{D}$ per unit area

$$\mathscr{D} = uc/h\nu = u\lambda/h$$

Thus we have a means of relating proportionally the intensity of photons $\mathscr{D}$ to the energy quantities u and P. If a beam of light propagates through an absorbing medium parallel to the Oz axis, these three quantities $\mathscr{D}$, u and P are decreasing functions of z; this is the function we wish to determine.

The calculation of the collision probability $N_{\text{coll}}/N_{\text{proj}}$ carried out using the hard-spheres model is of value only, as we have seen, in a measurement where the target particles do not mask one another. This requires that the cross-section σ and the number of target particles be sufficiently small for the collision probability of the projectiles to be very low: $N_{\text{coll}}/N_{\text{proj}} = \sigma n l \ll 1$.

This condition is realised in most experiments in nuclear physics due to the use of very thin targets (l can be of the order of a micron) and to the small cross-sections (σ is of the order of several barns, 1 barn $= 10^{-28}$ m^2). However, for the interaction between photons and atoms, cross-sections are much larger (σ can be of the order of 10^{-18} m^2) and if the light passes through an

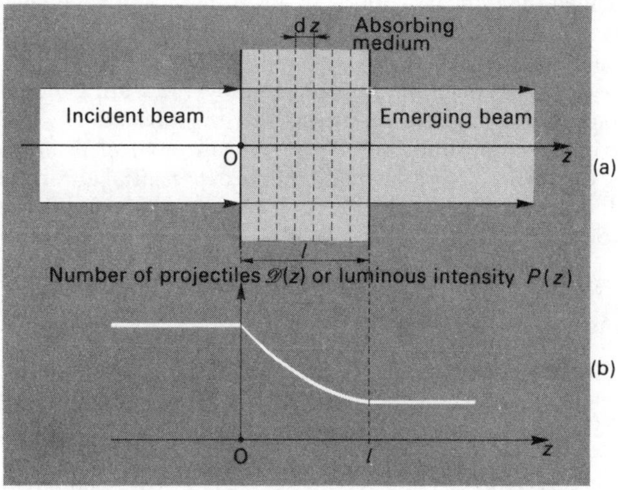

Figure 3.3 Passage through an absorbing medium

absorbing medium of sufficient depth l, the above inequality does not hold. To determine the effect produced in this case, it is necessary to imagine the absorbing medium as divided up into a succession of thin slices, each of thickness dz small enough to allow the above inequality to be satisfied (see figure 3.3(a)). Each collision occurring within the slice between z and $z + \mathrm{d}z$ removes a projectile from the incident beam and causes a decrease in the quantity $\mathscr{D}(z)$. In this way we find the change d$\mathscr{D}$ of the quantity $\mathscr{D}$ passing through a slice of thickness dz

$$\frac{\mathrm{d}\mathscr{D}}{\mathscr{D}(z)} = -\sigma n \,\mathrm{d}z$$

By integrating this differential equation with respect to the variable z we obtain

$$\mathscr{D}(z) = \mathscr{D}_0 \,\mathrm{e}^{-\sigma n z}$$

The intensity $\mathscr{D}(z)$ of the beam of projectiles decreases exponentially as shown in figure 3.3(b).

The intensity $P(z) = Sh\nu$ is proportional to the number of photons and obeys the same exponential law. The intensity remaining after crossing the absorbing medium of length l is

$$\boxed{P(l) = P_0 \mathrm{e}^{-\sigma n l} = P_0 \mathrm{e}^{-Kl}}$$

The parameter $K = \sigma n$ is called the absorption coefficient; it has the dimensions of an inverse length and is usually measured in m^{-1}. It can be interpreted as an absorption probability per unit length. The dimensionless quantity Kl is also called the optical depth of an absorbing medium. When the optical depth is very small ($\sigma nl = Kl \ll 1$), it is unnecessary to use the exponential law; it reduces to a linear law $P(l) = P_0(1 - Kl)$.

It is very easy to determine experimentally the optical depth Kl of an absorbing medium by measuring successively the intensities P_0 in the absence of an absorbing medium and $P(l)$ after passing through an absorbing medium. By varying the thickness l of the medium, the exponential absorption law may be verified. From this the absorption coefficient K may be found and hence the cross-section σ which is a measure of the collision probability of the photons.

3.1.3 Transition Probability per Unit Time

We have just determined what happens to a beam of projectiles when it passes through an absorbing medium; now we are interested in what happens to the targets. The completely symmetric roles played by the targets and the projectiles in the formulae defining the cross-sections have been noted (see section 3.1.1). Because of this symmetry, there corresponds to the length l over which the target atoms are distributed, a time t during which the projectile photons are incident. So the probability of collision with an atom per unit time $\sigma\mathscr{D} = \varpi$ corresponds to the probability of absorption of a photon per unit length $\sigma n = K$.

If *we now confine ourselves to optical resonance*, we may discuss the probability of excitation or transition probability of an atom per second $\sigma\mathscr{D} = \varpi_{12}$. The subscripts 1 and 2 serve to remind us that the atom makes a transition from a lower energy state E_1 to a higher energy state E_2. The photons $h\nu_{12} = E_2 - E_1$ can only be absorbed by atoms in their ground state E_1. More precisely, we must calculate the number of atoms n_1 per unit volume in this state (n_1 is often called the population of the energy level E_1). In most experiments n_1 remains very nearly equal to the total number n, which accounts for the confusion existing until now between n_1 and n, but it can be slightly smaller and strictly we should replace n by n_1.

Each time an atom is excited by absorption of a photon, it leaves the state E_1 and goes into a new state E_2, decreasing the number n_1 by one unit. In an experiment of very short duration $\mathrm{d}t$, the number n_1 changes by an amount $\mathrm{d}n_1$ equal and opposite to the number of atoms excited

$$\frac{\mathrm{d}n_1}{n_1} = -\sigma\mathscr{D}\mathrm{d}t = -\varpi_{12}\,\mathrm{d}t$$

or

$$\frac{1}{n_1}\frac{dn_1}{dt} = -\sigma\mathscr{D} = -\varpi_{12}$$

This relation can be considered as another way of defining ϖ_{12}, the transition probability per unit time. Using the relations given in the last section between the cross-section σ and the absorption coefficient K on the one hand, and on the other hand between the intensity of photons $\mathscr{D}$ and the energy density u, we can express this *probability of excitation per unit time* in various ways

$$\boxed{\varpi_{12} = -\frac{1}{n_1} \times \frac{dn_1}{dt} = \sigma\mathscr{D} = \frac{K}{n_1}\mathscr{D} = \frac{K}{n_1} \times \frac{\lambda}{h}u}$$

This last formula allows a determination of the probability ϖ_{12} from experimental measurements (K, λ, u and $n_1 \simeq n$). It will be noted that this probability per unit time has a dimension: it is measured in s^{-1}. Thus it is distinguished from the dimensionless ratios usually defined as probabilities.

3.1.4 Effects due to Frequency Distribution

Until now we have assumed that optical resonance occurs only with photons having the exact frequency $\nu_{12} = (E_2 - E_1)/h$. In fact the phenomenon also occurs with photons having frequencies ν nearly equal to, but different from, ν_{12}; effects can be observed for a distribution of frequencies over a certain range around the value ν_{12} (see Doppler broadening, section 1.3.3).

If a narrow band of frequencies betwen ν and $\nu + d\nu$ is isolated with a monochromator the coefficient of absorption $K(\nu)$ can be investigated experimentally. Figure 3.4(a) shows the shape of such a curve: $K(\nu)$ passes through

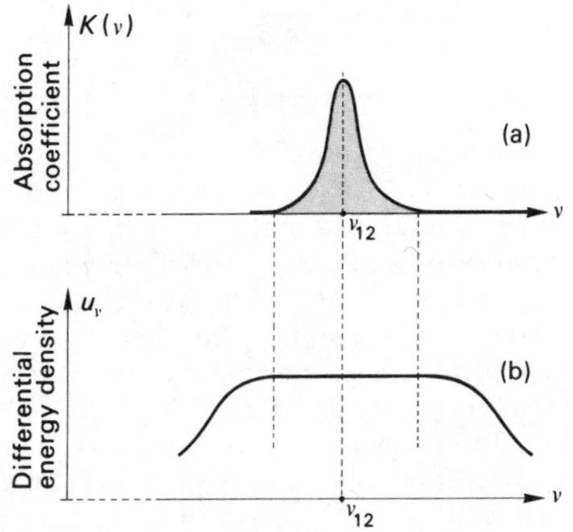

Figure 3.4 Lineshapes: (a) variation of the absorption coefficient with frequency; (b) 'broad line' hypothesis of excitation

a maximum at exactly the frequency $\nu_{12} = (E_2 - E_1)/h$ and vanishes when ν is far removed from ν_{12}.

The luminous intensity isolated by the monochromator clearly depends on the width $d\nu$ of the corresponding band of frequencies and if $d\nu$ is small the luminous intensity is proportional to $d\nu$. The energy density in the light beam thus isolated can be written in the form $u_\nu d\nu$; the coefficient of proportionality u_ν is called the differential energy density. The total energy density in the light beam integrated over frequency is $u = \int u_\nu d\nu$.

The arguments given in the preceding sections may be applied separately to each frequency band of width $d\nu$, and the rate of change in the number of atoms n_1 in the ground state for a single band of frequencies $d\nu$ is

$$\frac{dn_1}{dt} = -\frac{\lambda}{h} K(\nu)\, u_\nu\, d\nu \qquad \text{(contribution from the band } d\nu\text{)}$$

The overall rate of change of the number of atoms is obtained by summing the contributions of each frequency band $d\nu$

$$\frac{dn_1}{dt} = -\frac{\lambda}{h} \int K(\nu)\, u_\nu\, d\nu \qquad \text{(for all frequencies)}$$

From this, the total transition probability may be found

$$\varpi_{12} = -\frac{1}{n_1}\frac{dn_1}{dt} = \frac{\lambda}{h}\frac{1}{n_1} \int K(\nu)\, u_\nu\, d\nu$$

This formula may be simplified by *assuming that the differential energy density* u_ν *is constant over the whole range of frequency where* $K(\nu)$ *is different from zero* (this is called 'broad line' excitation, see figure 3.4(b)). u_ν can then be taken out of the integral, and one obtains

$$\boxed{\varpi_{12} = B_{12} u_\nu \quad \text{where } B_{12} = \frac{\lambda}{h}\frac{1}{n_1} \int K(\nu)\, d\nu}$$

The transition probability of an atom per second ϖ_{12} is proportional to the differential energy density u_ν. The coefficient of proportionality B_{12} can be determined from experimental data since the integral over $K(\nu)$ is equal to the area under the curve in figure 3.4(a). It will be noted that the parameter characterising the interaction between a photon of frequency ν and an atom is the cross-section $\sigma(\nu)$. The absorption coefficient $K(\nu) = \sigma(\nu) n_1$ may be measured experimentally and is proportional to the number of atoms n_1 per unit volume, but the coefficient B_{12} thus calculated is independent of the number of atoms n_1. The coefficient B_{12} is characteristic of the type of atom and of the particular transition studied ($E_1 \to E_2$); it is completely independent of the experimental conditions, and it plays an important role in the quantum theory of electromagnetic radiation from which a theoretical expression for B_{12} may be derived.

Comment I B_{12} can also be derived from classical theory, since the elastically bound electron model allows the integral $\int K(\nu)\,d\nu$ to be calculated (see appendix 2). These classical calculations give rise to results which are often in disagreement with experiment. This disagreement may be expressed by means of a dimensionless ratio called the 'oscillator strength' (see volume 2, section 7.4).

Comment II Self-absorption of a resonance line. We have shown in section 1.3.3 that the broadening of spectral lines in light sources is usually caused by the Doppler effect; furthermore, we have described the corresponding line shape (curve (a) of figure 3.5). However, in the case of resonance lines, the light emitted by atoms at the centre of the discharge can be reabsorbed by atoms situated at the edge of the discharge, near the walls of the lamp.

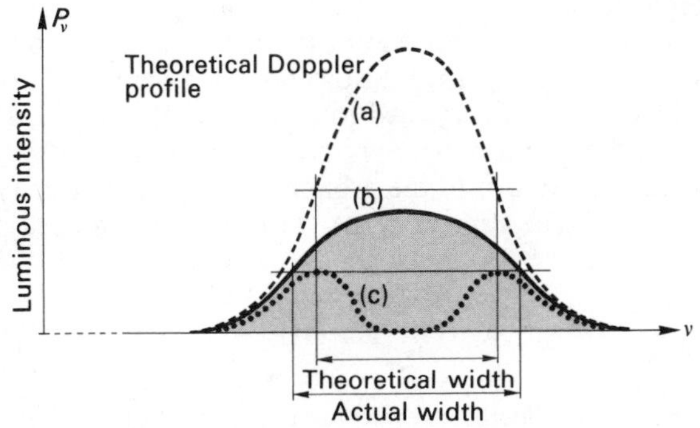

Figure 3.5 Self-absorption of a resonance line

This self-absorption effect is maximum at the centre frequency ν_{12} because the coefficient of absorption $K(\nu)$ is maximum at this frequency. On the other hand, it is unimportant in the wings of the emission line, because the absorbing atoms close to the wall are at a much lower temperature than the emitting atoms at the centre of the discharge. The curve of $K(\nu)$ is therefore narrower than the curve representing the emitted power P_ν. This has the effect of distorting the profile of the curve of intensity as a function of frequency of the light actually leaving the lamp, and gives rise to a considerable broadening (curve (b) of figure 3.5), which is seen as a total reversal of the line (curve (c) of the same figure) if the vapour pressure is too high.

3.2 Spontaneous Emission of Photons

3.2.1 The Probability of Spontaneous Emission and the Lifetime of an Excited State

Spectroscopic and electron bombardment experiments (described in chapter 1) have given rise to the idea of atomic states corresponding to stationary and discrete values of energy. Furthermore, they have taught us that most of these stationary states are not stable; optical resonance and electron bombardment experiments show us that (a) the atom is usually found in the state of minimum energy, called the ground state; (b) in order to put an atom into a state of higher energy, called an excited state, it must be given energy through an external excitation process; (c) the atom does not remain in the excited state, but returns spontaneously to the state of minimum energy or ground state by giving up its excess energy through the emission of light. This phenomenon is called spontaneous emission.

However, a problem arises: is it possible to predict the time at which an excited atom returns spontaneously to the ground state? There is no answer to this question for a particular atom, but an answer is possible in statistical terms applicable to a large number of atoms. All experiments directed to answering this question support the hypothesis of a spontaneous emission probability per unit time, defined analogously to the excitation probability ϖ_{12} in the preceding section.

Let n_2 be the number of atoms per unit volume in the excited energy state E_2 at a given moment of time (n_2 is also called the population of the level E_2).

When the atoms are not subjected to any external influence (therefore the only phenomena that can occur are spontaneous transitions whereby atoms in the excited state E_2 return to the ground state E_1) then it may be assumed that the relative loss in the number n_2 in a short space of time $\mathrm{d}t$ is proportional to $\mathrm{d}t$

$$\frac{\mathrm{d}n_2}{n_2} = -A_{21}\,\mathrm{d}t$$

The probability of spontaneous emission per unit time is denoted by the constant

$$\boxed{A_{21} = -\frac{1}{n_2}\frac{\mathrm{d}n_2}{\mathrm{d}t}}$$

By integrating the above differential equation, the law governing the variation in the number n_2 of excited atoms as a function of time may be found. Assuming that n_0 atoms are in the excited state at the time $t = 0$ whereupon the source of excitation is cut off, one obtains for positive t the exponential law (see figure 3.6)

$$n_2 = n_0\,\mathrm{e}^{-A_{21}t}$$

From this, the mean time that the atoms remain in the excited state can be deduced

$$\tau = \frac{1}{n_0}\sum_{i=1}^{n_0} t_i \qquad (t_i \text{ is the time for atom } i)$$

We know how to find the change of the number n_2 in a short time $\mathrm{d}t$

$$\mathrm{d}n_2 = -A_{21}\,n_0\,\mathrm{e}^{-A_{21}t}\,\mathrm{d}t$$

The absolute value $|\mathrm{d}n_2|$ is the number of atoms whose total time t_i in the excited state lies between t and $t + \mathrm{d}t$, and from this we derive a very simple

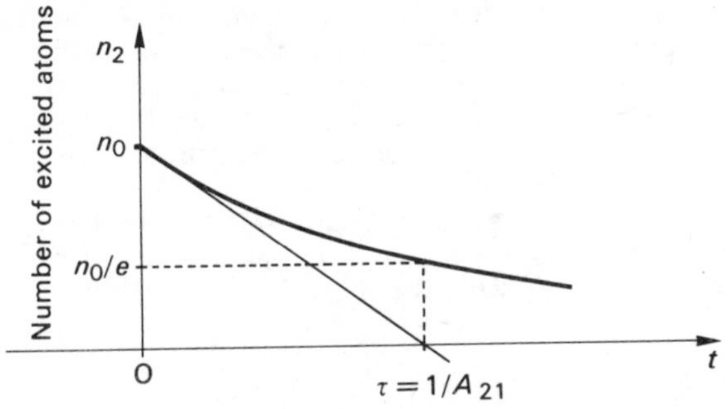

Figure 3.6 Exponential decay

expression for the mean time

$$\tau = \frac{1}{n_0}\int_{t=0}^{\infty} t|\mathrm{d}n_2| = \int_0^{\infty} t\mathrm{e}^{-A_{21}t} A_{21}\,\mathrm{d}t$$

$$= \frac{1}{A_{21}} \qquad \text{(the result of a trivial integration by parts)}$$

This mean time is called the lifetime of the excited state

$$\boxed{\tau = 1/A_{21}}$$

This is also the time after which the number of excited atoms has been reduced by the factor 1/e (e = 2·718).

The assumptions we have made are justified by experiments which measure lifetimes, as described in the next section.

Comment I The above theory assumes that the source of excitation is cut off at time $t = 0$, and the number of atoms at that time is n_0; but the exact times at which these n_0 atoms were excited do not enter into the theory at all. It is possible, as was assumed above, for all the n_0 atoms to be excited simultaneously at time $t = 0$, but equally it is possible for them to be excited at different times in the past (t negative): τ then represents the mean lifetime of the n_0 atoms remaining in the excited state, whatever their past histories.

Comment II The assumptions made to describe the spontaneous emission of photons are completely analogous to those made in describing the phenomenon of radioactive decay. However, care must be taken because nuclear physicists define a radioactive half life $\tau_{1/2}$ as the time during which the number of radioactive atoms falls by $\frac{1}{2}$ (and not by 1/e). The half-life may be derived as

$$\tau_{1/2} = \frac{\ln 2}{A_{21}} = \frac{0{\cdot}693}{A_{21}}$$

Comment III If the particular level being studied is above other excited levels, it is known that the return to the ground state can occur by way of one or more intermediate levels with the successive emission of photons in cascade. Therefore there are different ways of returning indirectly to the ground state. We direct our attention here to the first stage of each of these decay routes: the atoms can make spontaneous transitions from the same particular initial excited level E_p to several levels E_i lower than E_p. The probability of spontaneous emission A_{pi} corresponding to each of these transitions can be defined.

The total number of atoms leaving the excited state E_p in a time dt is obtained by summing over all the atoms that take part in one or other of these spontaneous transitions, and thus one can write the change in the number n_p of atoms in the state E_p

$$dn_p = \sum_i - A_{pi} n_p \, dt = - n_p \, dt \left(\sum_i A_{pi} \right)$$

Here we meet the well-known rule for the addition of probabilities of independent events: the total probability of leaving the excited state E_p is equal to the sum of the probabilities of the various competing processes giving rise to this event. Hence the lifetime τ_p of the state E_p may be found

$$\boxed{-\frac{1}{n_p}\frac{dn_p}{dt} = \sum_i A_{pi} = \frac{1}{\tau_{pi}}}$$

3.2.2 Experimental Measurement of Lifetimes

The assumptions made in section 3.1 are confirmed in various experiments leading to measurements of lifetimes.

(*a*) *The sudden interruption of the excitation process* of atoms does not cause the immediate cessation of light emission; instead a gradual decrease of light intensity may be measured after the interruption of excitation. This is a direct consequence of the evolution law for the number of atoms n_2 in the excited state as a function of time, described in the last section (see figure 3.6). The number of photons emitted per unit time is equal to the number of atoms that leave the excited state in the same time dn_2/dt. A fraction $\Omega/4\pi$ of the photons emitted at random in all directions of space are collected within a solid angle Ω, and so the light power measured is

$$P = h\nu \frac{\Omega}{4\pi} \left| \frac{dn_2}{dt} \right| = h\nu \frac{\Omega}{4\pi} n_0 A_{21} e^{-A_{21}t} = P_0 e^{-t/\tau}$$

The emitted light intensity P varies with the same time constant τ as the number of atoms n_2; by recording this variation of P as a function of time, the lifetime τ may be measured.

The lifetimes of excited states usually observed in spectroscopy range between a nanosecond and a microsecond (these are not necessarily extreme values), and therefore very fast recording methods must be used.

The main difficulty in these experiments, however, is the need to interrupt the excitation very suddenly, in a time which is short compared with the lifetime τ. If the excitation is produced by a beam of primary light (optical resonance experiment), a Kerr cell placed between two crossed polarisers can be used as a shutter: the light is transmitted when an electric field makes the transparent medium in the Kerr cell birefringent. Advances in electronics now allow this voltage to be removed in a time less than a nanosecond, down to a tenth of a nanosecond. If the excitation is produced by electron bombardment, a negative voltage must be applied suddenly to one electrode so as to prevent the passage of electrons; the time constant for applying this negative voltage can be of the same order of magnitude.

In practice, the experiments are not always as simple because it is often difficult to excite just a single level and effects due to cascading occur, requiring careful interpretation (see volume 2, chapter 7).

(*b*) *Alternating excitation* can be used to overcome the difficulties inherent in suddenly interrupting the excitation. Using this method, an alternating voltage V of period T is applied to an electrode controlling the flow of electrons. Under these conditions, the light emitted by the atoms is also alternating with the same period.

But, since the atoms remain in the excited state for a certain time, the light intensity $P(t)$ is re-emitted with a certain delay in relation to the voltage $V(t)$; the alternating function $P(t)$, measured with a photocell, is out of phase with the voltage $V(t)$. If the period T is long compared with the lifetime τ, the delay is of the order of τ and therefore the dephasing ϕ is of the order of $2\pi\tau/T$; the dephasing is small and may be unobservable. However, if the period T is shortened so as to bring it close to τ, a large dephasing may be measured. In the limit when the period T is much shorter than τ, the light intensity $P(t)$ remains practically constant. The changes in amplitude and phase of the alternating part of $P(t)$ can be calculated as a function of the parameter τ/T. By comparing them with the changes observed experimentally as a function of T, the value of τ may be deduced. This second method of measurement is less direct than the first, but has the advantage of requiring only alternating voltages, which are much simpler to produce and use than discontinuous changes of voltage.

(*c*) *Sudden excitation in an atomic beam* enables the transformation of the temporal variation of the light intensity emitted by the atoms into a spatial variation. An atomic beam can be produced (see figure 3.7(a)) if a vessel containing a vapour and a highly evacuated enclosure are connected by a small orifice. The residual pressure is less than 10^{-5} torr so that the mean free path of the atoms is greater than the dimensions of the enclosure; in other words, the atoms no longer collide with one another, and retain rectilinear trajectories. A diaphragm pierced by a second orifice selects atoms having velocities parallel to the Oz axis, passing through the two orifices; only these atoms are able to continue in straight-line paths into the second half of the evacuated enclosure, and there they form a rectilinear beam of atoms (their weight is negligible).

The atoms are excited at a precise point in the beam by bombardment with a thin pencil of electrons as if from a gun, similar to the operation of cathode ray oscilloscopes. The light emitted by the excited atoms can then be observed, but because of the lifetime of the excited state, some atoms have time to move farther along the Oz axis before they emit a photon. The atomic beam appears luminous over a certain length, starting very nearly at the point of impact O of the pencil of electrons.

It is possible to isolate the light emitted by a small fraction of the atomic beam at a distance z from the point of impact O, to measure its intensity

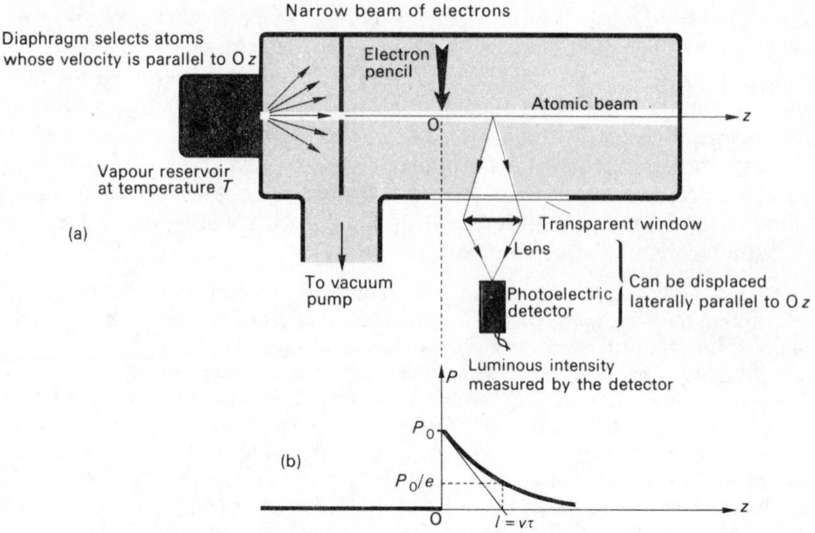

Figure 3.7 Measurement of lifetimes in an atomic beam

$P(z)$ and, by varying the distance z, to draw a curve of the variation of P as a function of z (see figure 3.7(b)). If all the atoms had the same velocity v, this curve could be interpreted from the variation of P as a function of time (see beginning of section 3.2) by simply changing to the variable $z = vt$, thus

$$P(t) = P_0 e^{-t/\tau} \quad \text{gives} \quad P(z) = P_0 e^{-z/v\tau}$$

The light intensity falls by a factor of 1/e at a distance $l = v\tau$ from the point of excitation by the pencil of electrons.

An exact calculation must take into account the distribution of speeds in the atomic beam. This may be calculated, from the kinetic theory of gases, as a function of the temperature T of the vapour whence the beam originates (see section 10.7). To a first approximation, the experiment is well described by the above formula if v is taken as a mean velocity, related to the mean-square velocity of the atoms in the vapour $\overline{C^2} = 3kT/m$. In other words, v is generally of the order of km/s; a relatively long lifetime $\tau = 1$ μs, gives rise to a distance l of the order of a millimetre, which is so small as to make measurements difficult. Therefore this method of measurement is applicable only in exceptional cases of particularly long lifetimes (for instance Brück's experiments in 1942 on the resonance level 4^3P_1 of zinc with a lifetime of 30 μs).

However, the method has been generalised and used extensively for several years in ion spectroscopy. It is quite simple to give ions kinetic energies of several MeV with small modern accelerators: thus monoenergetic ions having speeds greater than 1000 km/s may be obtained, providing entirely favourable conditions for applying the method (see the technique of beam foil spectroscopy described in volume 2, chapter 7).

(*d*) *Other indirect methods* also allow lifetimes to be measured. These will be discussed in volume 2, section 7.4.

Comment *Imprisonment time of resonance radiation.* When a lifetime measurement makes use of an optical resonance line, it is always necessary to operate with a sufficiently dilute medium (a vapour at very low pressure) so that its optical depth Kl is much smaller than unity, l being the mean dimension of the experimental cell (see section 3.1.2). In the opposite situation where $Kl > 1$, a large proportion of the photons emitted at the centre of the experimental cell are reabsorbed by other atoms before leaving the cell. The atoms which are thus indirectly excited at second hand, in turn re-emit photons in all directions, some of which leave the cell while others are absorbed again.

The same sequence can repeat itself many times and the overall phenomenon is called multiple diffusion. Consequently, only a small fraction of the photons received by a detector have been emitted directly by atoms excited by external means. On the contrary, many of the photons received have been absorbed successively by 2, 3, 4 or even more atoms. When the external excitation source is suddenly removed, these photons are delayed by their successive interactions with several atoms and arrive at the detector much later. Overall a much slower decrease of the light intensity is observed, and the experimentally measured time constant can be much longer than the lifetime of the excited state. This is the imprisonment time of resonance radiation.

3.2.3 Comparison with the Classical Theory of Radiation

The classical theory of atomic radiation uses a model of an elastically bound electron (we have already mentioned this in section 1.3.2, Comment II). Within atoms one assumes the existence of electrons which are restored to their equilibrium position by an attractive force proportional to distance. They therefore undergo free oscillation with a natural frequency ω_0 determined by the attractive force: $z = z_0 \cos \omega_0 t$ (see appendix 2).

Such an electron, having a non-uniform motion, generates an electromagnetic wave which carries away energy *ad infinitum*. To ensure conservation of total energy, the energy $W = \frac{1}{2}m\omega_0^2 z_0^2$ stored by the oscillating electron must decrease steadily, as must the amplitude z_0 of its oscillatory motion.

The theory presented in appendix 2 results in an exponential law for the energy W stored by the oscillator: $W = W_0 e^{-t/\tau}$.

The amplitude of the oscillation z_0 varies as $\sqrt{W}$, and so changes exponentially as $e^{-t/2\tau}$. The electric field created by the electron at a point in space also has a decreasing amplitude E_0 proportional to z_0, following an exponential law of time constant 2τ. The electron in free, or spontaneous oscillation emits a damped wave train as illustrated in figure 3.8.

This classical description gives a result in agreement with the quantum description and with experiment: the excitation energy of the medium studied decreases progressively according to an exponential law of time constant 2τ. There is an equation in appendix 2 that enables the calculation of a classical 'lifetime' which is of the correct order of magnitude

$$\tau = \frac{6\pi\varepsilon_0 mc^3}{q^2 \omega_0^2} = \frac{3\varepsilon_0 mc}{2\pi q^2}\lambda_0^2 \simeq 10^{-8} \text{ seconds for a wavelength } \lambda_0 = 500 \text{ nm}$$

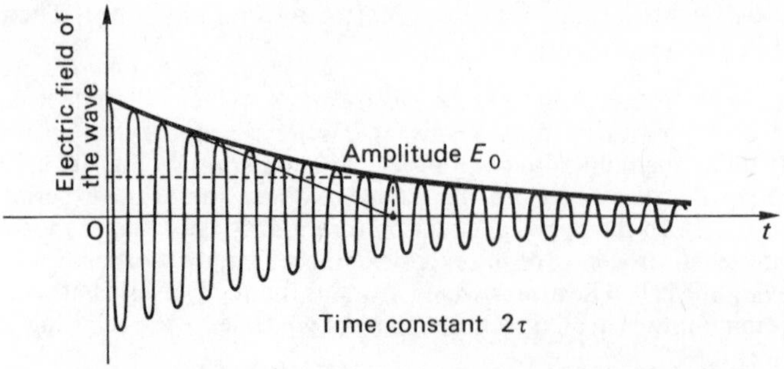

Figure 3.8 The damped wave train of classical emission

However, the classical theory attributes this property to each isolated atom, whereas in the quantum description this property is only valid statistically, when applied to a very large number of atoms. The main failure of the classical theory is to predict the same lifetime τ for all excited states corresponding to spectral lines of the same frequency $\nu_0 = \omega_0/2\pi$, or nearby frequencies; this is shown to be false by experiment.

3.3 Induced Emission and Einstein's Theory of Radiation

3.3.1 The Concept of Induced Emission

We would have an incomplete description of the interaction between atoms and radiation, if we had let the matter rest after the two preceding sections, and too hasty a generalisation would have misled us.

Until now we have confined our study to two phenomena that in normal circumstances play the most important roles, and that may be demonstrated by straightforward experiments: in section 3.1, transitions in which the atom is 'raised' from a lower level E_1 to a higher level E_2 by absorption of an electromagnetic wave of frequency ν_{12}; in section 3.2, transitions in which the atom 'falls' from a higher level E_2 to a lower level E_1 by spontaneous emission in the absence of any external influence. However, our discussion in section 3.1 concerning the effect of an electromagnetic wave is not complete. The spontaneous emission process described in section 3.2 is not the only type of transition that can occur: the action of an electromagnetic wave can also cause emission by transitions in which the atom 'falls' from a higher level to a lower level. These transitions are induced by the electromagnetic wave and the phenomenon is called induced emission or stimulated emission. Without taking this into account, it would be impossible to explain certain experiments or to build up a coherent theory of atom–radiation interaction.

Historically it was by an analogy, and by reliance on the classical theory of radiation, that the concept of induced emission evolved (Einstein, 1917), and the analogy remains the simplest way to introduce this concept. It makes use of the model of an elastically bound electron.

In section 3.2.3 we discussed the similarity between the process of spontaneous emission and the phenomenon of natural oscillation of an electron, damped by the loss of energy due to radiation. We are concerned now with the effect of an electromagnetic wave on an electron of charge q, that is to say in the phenomenon of forced oscillation which arises from the action of a sinusoidal force $qE_0 \cos \omega t$, of frequency ω close to the natural frequency ω_0 of the electron. We shall ignore the difference between ω and ω_0 but assume that initially the electron is in free oscillation and closely concern ourselves with *transient* phenomena obtained by the superposition of the forced oscillation and the free oscillation before the latter is completely damped (not to be compared with appendix 2, where we have dealt only with the steady-state regime after the free oscillation has died away).

Depending on the phase difference ϕ between the free motion of the electron

$$z = z_0 \cos(\omega_0 t + \phi)$$

and the electromagnetic force $qE_0 \cos \omega t$, the latter behaves either as an accelerating force or as a retarding force (one passes from one to the other by adding or subtracting π to the value of ϕ). In other words, depending on the value of ϕ, the amplitude z_0 of the oscillatory motion will increase or decrease, and the energy stored by the oscillating electron $W = \frac{1}{2} m \omega_0^2 z_0^2$ will vary in the same way as the amplitude z_0. Finally, as the phase changes, the oscillating electron within the atom can either (a) increase its energy at the expense of the electromagnetic wave (absorption of energy) or (b) lose energy due to the action of the same wave. This second case is a phenomenon of energy emission under the action of a wave, and it provides us with a classical representation of induced emission.

This classical model brings to light the complete symmetry between the two phenomena of absorption and induced emission. They take place with equal probability since the occurrence of one or the other depends simply on the phase difference ϕ (clearly, induced emission can only occur for electrons that already have stored energy in a free oscillation; electrons initially without energy can give rise only to absorption).

Transposing into quantum language, one assumes that there is an induced emission probability with a definition entirely analogous to that of the absorption probability (see section 3.1.3). Induced emission transitions can be brought about only by atoms in an excited state E_2; as a result, the number n_2 decreases in the same way as absorption transitions result in a decrease in the number n_1 in the ground state. We may define the probability of induced emission ϖ_{21} (the subscripts remind us that the atoms pass from the state E_2 to the state E_1) by the formula

$$\varpi_{21} = -\frac{1}{n_2} \frac{\mathrm{d}n_2}{\mathrm{d}t}$$

For induced emission, as for absorption, we assume also that this transition probability is proportional to the differential energy density u_ν of the electromagnetic wave (see section 3.1.4). Finally, we may write *the probability of induced emission* (*or stimulated emission*) *per unit time*

$$\boxed{\varpi_{21} = -\frac{1}{n_2}\frac{dn_2}{dt} = B_{21}\, u_\nu}$$

where the coefficient B_{21} is characteristic of the particular transition $E_2 \to E_1$ of the atom, but independent of experimental conditions.

At this stage of our study, the equalities framed above may be considered as an hypothesis. Confirmation of this hypothesis rests on the consistency of the theories consequent upon it (see section 3.3.3) and on the striking effects observable experimentally with lasers and masers (see the following chapter). The quantum theory of electromagnetic fields also confirms this hypothesis, and it enables a theoretical calculation of the coefficient B_{21}.

3.3.2 Equations for the Three Types of Transitions in an Optical Resonance Experiment

To define the probabilities per unit time corresponding to each of the three types of transition, ϖ_{12}, A_{21}, and ϖ_{21}, we have argued as if only the one type of transition occurred at a time; but in an actual experiment, transitions of the three types occur simultaneously and their effects must be added to calculate exactly the evolution of the numbers of atoms n_1 and n_2, also called populations of the energy levels E_1 and E_2.

Furthermore, it is necessary to take into account the fact that the atoms make transitions between the two levels E_1 and E_2 alone (we confine ourselves to this simple case), and effects which cause a certain variation dn_1 of one of the two populations bring about simultaneously an opposite variation $dn_2 = -dn_1$ of the other population. Bearing this in mind, the contributions towards the evolution of the populations n_1 and n_2 of each of the three phenomena are given below

$$\begin{cases} \text{absorption:} & \dfrac{dn_2}{dt} = -\dfrac{dn_1}{dt} = \varpi_{12}\, n_1 & \text{(see section 3.1.3)} \\ \text{spontaneous emission:} & \dfrac{dn_2}{dt} = -\dfrac{dn_1}{dt} = -A_{21}\, n_2 & \text{(see section 3.2.1)} \\ \text{induced emission:} & \dfrac{dn_2}{dt} = -\dfrac{dn_1}{dt} = -\varpi_{21}\, n_2 & \text{(see section 3.3.1)} \end{cases}$$

Finally, by adding the three contributions, we obtain

$$\frac{dn_2}{dt} = -\frac{dn_1}{dt} = \varpi_{12}\, n_1 - A_{21}\, n_2 - \varpi_{21}\, n_2 = (B_{12}\, n_1 - B_{21}\, n_2)\, u_\nu - A_{21}\, n_2$$

In most experiments, a state of dynamic equilibrium is established rapidly between the incident wave and medium under study, such that the populations n_1 and n_2 are stationary in spite of continuous exchanges of atoms between the two levels E_1 and E_2. In other words $dn_2/dt = dn_1/dt = 0$, and we find that

$$\boxed{\left(B_{12}\frac{n_1}{n_2} - B_{21}\right)u_\nu = A_{21}}$$

This formula enables, for example, the number n_2 of excited atoms existing in a steady-state optical resonance experiment to be calculated, once the intensity of the incident beam of light is known. The calculation may usually be simplified when $n_1/n_2 \gg 1$ and when the induced emission term B_{21} can be neglected.

Let us state the conditions of validity of this calculation: it assumes that the atoms are subjected only to the influence of an electromagnetic wave and that there occur only transitions of the three types described above, called radiative transitions. The calculation is no longer valid if other causes of excitation exist (electron bombardment for example) or other causes of de-excitation. The return of excited atoms to the ground state can occur, for example, by non-radiative transitions when the atoms under study are mixed with a foreign gas, or 'buffer gas', and collide with gas molecules; the excitation energy of such an atom is then transformed into rotational or vibrational energy of the foreign gas molecule. This phenomenon is known as 'quenching'.

3.3.3 Relations between the Radiative Transition Probabilities

The dynamic equilibrium discussed above is not a thermal equilibrium in the usual sense of the term, for two reasons which have a bearing on recent experiments: (a) the atoms exchange their excitation energy only with the electromagnetic wave and not with the thermal bath constituting the surrounding medium; (b) the electromagnetic radiation emitted by a spectral lamp bears no relation to the thermal equilibrium radiation described by Planck's law.

Although scarcely corresponding to conditions realisable in the laboratory, nevertheless we can imagine a situation where an effective thermal equilibrium is realised (this can happen in a star). We assume that the atoms are enclosed in a container at an absolute temperature T and wait for the time required to establish thermal equilibrium, so that (i) the differential energy density u_ν of the electromagnetic radiation inside the container is the thermal equilibrium density given by Planck's law; (ii) the populations n_1 and n_2 of the two energy levels E_1 and E_2 are stationary, and obey Boltzmann's statistical law

$$\frac{n_1}{n_2} = \frac{G_1}{G_2}\,e^{(E_2-E_1)/kT} \qquad (n_1 > n_2 \text{ for } E_2 > E_1)$$

where G_1 (or G_2) is the statistical weight of the energy level E_1 (or E_2), that is to say the number of distinct quantum states having the same energy E_1 (or E_2). This is also called the order of degeneracy of this level.

We assume that only radiative transitions occur between these two energy levels; since their populations are stationary, we may rewrite the framed relationship of the preceding section

$$A_{21} = \left(B_{12}\frac{n_1}{n_2} - B_{21}\right) u_\nu = \left[B_{12}\frac{G_1}{G_2} e^{(E_2-E_1)/kT} - B_{21}\right] u_\nu$$

The electromagnetic energy within the container is distributed over all frequencies, but here we are interested only in those corresponding to transitions between E_1 and E_2, so we use a value of u_ν at a particular frequency ν such that $h\nu = E_2 - E_1$.

This relation holds whatever the temperature T of the container. In particular, it holds when T becomes infinite

$$\text{As } T \to \infty \begin{cases} e^{(E_2-E_1)/kT} \to 1\text{; that is, the expression in the second bracket tends to } [B_{12}(G_1/G_2) - B_{21}] \\ \text{from Planck's law } u_\nu \to \infty \end{cases}$$

Since the first term of the equation is a finite constant, the product in the second term must also remain finite, so the bracket must tend to zero. A first important relation is then found

$$\boxed{G_1\, B_{12} = G_2\, B_{21}}$$

The probabilities of absorption $\varpi_{12} = B_{12} u_\nu$ and of induced emission $\varpi_{21} = B_{21} u_\nu$ are in the ratio of the statistical weights G_2 and G_1 of the two levels concerned. This relation shows that the two probabilities are always of the same order of magnitude, and is a good illustration of the symmetric character of the two phenomena. Their experimentally asymmetric behaviour arises from the difference of populations n_1 and n_2.

Comment The reasoning leading to the relation between B_{12} and B_{21} can be given a firmer basis: when the energy density becomes very large, the transitions caused by the electromagnetic wave outnumber the spontaneous transitions, which remain constant in number. To maintain stationary populations, the transitions caused by the wave (absorption and induced emission) must come to equilibrium amongst themselves.

With this new relation, we can transform the equation that we started with

$$A_{21} = B_{21}[e^{(E_2-E_1)/kT} - 1]\, u_\nu = B_{21}[e^{h\nu/kT} - 1]\, u_\nu$$

so we find that

$$u_\nu = \frac{A_{21}}{B_{21}} \times \frac{1}{e^{h\nu/kT} - 1}$$

and u_ν also obeys Planck's law because of the assumption of thermal equilibrium

$$u_\nu = \frac{8\pi h\nu^3}{c^3} \times \frac{1}{e^{h\nu/kT} - 1} = \frac{8\pi h}{\lambda^3} \times \frac{1}{e^{h\nu/kT} - 1}$$

The coefficients A_{21} and B_{21} are independent of experimental conditions and in particular of temperature; thus we find that the law of variation of the energy density u_ν as a function of the temperature T conforms to Planck's law. This justifies the assumptions we have made. In particular it will be noted that if induced emission is omitted the -1 in the denominator of u_ν disappears, contradicting Planck's law: *the hypothesis of induced emission is necessary to support the quantum description of the atom–radiation interaction.*

Continuing the comparison with Planck's formula, we obtain a second relation between the coefficients which we have introduced to express the transition probabilities

$$\boxed{\frac{A_{21}}{B_{21}} = \frac{8\pi h}{\lambda^3}}$$

These relations have been derived by assuming very special experimental conditions (thermal equilibrium), but the coefficients B_{12}, B_{21} and A_{21} depend only on the atom and on the particular transition $E_1 \rightarrow E_2$ studied; they do not depend in any way on the experimental conditions. The relations obtained are thus generally applicable. Therefore an extremely precise relationship exists between the probabilities for spontaneous emission and induced emission. This relationship is characterised by a $1/\lambda^3$ law: in passing from optical transitions ($\lambda \approx 1$ μm) to radiofrequency transitions ($\lambda \approx 1$ m), the wavelength increases by a factor of 10^6 and the ratio A_{21}/B_{21} then decreases by a factor of 10^{18}; that is to say the probability of spontaneous emission becomes very small compared with the probability of induced emission (clearly it is necessary to assume a reasonable, not infinitely small, energy density u_ν). Experiments show that normally the phenomenon of spontaneous emission can be disregarded when studying radiofrequency transitions, but on the other hand, induced emission can often be disregarded when studying optical transitions.

To summarise, we have confirmed the hypothesis of induced emission and have obtained two fundamental relations between the three coefficients B_{12}, B_{21} and A_{21}. These relations may also be derived from the quantum theory of electromagnetic radiation, but it should be noted that they were obtained by Einstein in 1917, before the basis of quantum mechanics had been established. We have not finished discussing induced emission; the radiation produced through induced emission has very special and important properties which we shall examine in the next chapter. There we shall describe experiments with masers and lasers, which are themselves the best experimental confirmation of the phenomenon of induced emission.

Comment The relationship between A_{21} and B_{21} may be expressed in another form, in terms of the lifetime τ of the excited state and of the coefficient B_{12} corresponding to absorption, which we know how to measure experimentally from the absorption coefficient $K(\nu)$ (see section 3.1.4)

$$\frac{1}{\tau} = \frac{8\pi h}{\lambda^3} \times \frac{G_1}{G_2} \times B_{12} = \frac{8\pi}{\lambda^3} \times \frac{G_1}{G_2} \times \frac{1}{n_1} \int K(\nu)\,\mathrm{d}\nu$$

Thus it may be seen that absorption measurements allow lifetimes to be calculated (see section 3.2.2), since the integrated absorption coefficient of a resonance line is inversely proportional to the lifetime of the corresponding excited state.

4

Wave–Corpuscle Duality

4.1 The Relationship between Photons and the Amplitude of a Classical Field

4.1.1 Energy Density and Number of Photons

First we shall review some familiar ideas. We have calculated several times the number of photons in a beam of light by dividing the energy of the light by the energy $h\nu$ of a photon, particularly in section 3.1 dealing with the absorption cross section of photons. Here we retain the same notation: let $\mathcal{D}$ be the intensity of photons per unit area, in other words the number of photons passing through a unit cross-sectional area per second. The number of photons passing through an area S in an experiment of duration t is

$$\boxed{N = \mathcal{D}tS} \quad \text{(from the definition of } \mathcal{D}\text{)}$$

We have also established the relation between this intensity of photons $\mathcal{D}$ and the energy density u in the beam of light, using the power P transported by the photons as an intermediate step

$$P = Suc = S\mathcal{D}h\nu \rightarrow \mathcal{D} = \frac{uc}{h\nu} = \frac{\lambda}{h}u$$

Elsewhere, in section 2.1, we have discussed the classical relation between the energy density and the two fields E and B of the electromagnetic wave. Here we write the relationship in terms of instantaneous values

$$u_{\text{inst}} = \frac{\varepsilon_0}{2} E^2 + \frac{1}{2\mu_0} B^2 = \varepsilon_0 E^2$$

using the relation between the two fields in a plane wave

$$B = E\sqrt{(\varepsilon_0 \mu_0)}$$

By comparing these two results, a linear relationship is found between the intensity of photons and the square of the electric field

$$\mathscr{D}_{\text{inst}} = \frac{\lambda}{h} \varepsilon_0 E^2$$

The electric field E at a point having co-ordinates x, y, z varies sinusoidally with time; we let E_0 be the amplitude of this sinusoidal function

$$E = E_0(x, y, z, t) \cos[\omega t - \phi(x, y, z, t)]$$

(The amplitude E_0 and phase ϕ of the field depend upon space co-ordinates, but they may also depend *slowly* on time.) We shall also use the complex function $\mathscr{E}$, the real part of which is the field E

$$\mathscr{E} = \mathscr{E}_0(x, y, z, t)\, e^{i[\omega t - \phi'(x,y,z)]}$$

The amplitude E_0 of the actual field is equal to the modulus of the complex amplitude $\mathscr{E}_0$.† Generally, time-averaged values are used (the average calculated over one period)

$$\boxed{\mathscr{D} = \frac{\lambda}{h} \varepsilon_0 \bar{E}^2 = \frac{\lambda}{h} \frac{\varepsilon_0}{2} E_0^2 = \frac{\lambda}{h} \frac{\varepsilon_0}{2} \mathscr{E}\mathscr{E}^*}$$

($\mathscr{E}^*$ is the complex conjugate of $\mathscr{E}$). Thus we have obtained a direct relationship between the number of photons and the amplitude of the classical field.

However, it will be easier to appreciate the practical significance of these formulae after calculating orders of magnitude for the number of photons in two typical cases.

(*a*) *The reception of broadcast waves by a modern transistor set.* Let us suppose that we are at a distance $r = 100$ km from an isotropic radiator of power $P = 1$ kW and frequency $\nu = 1$ MHz ($\lambda = 300$ m). This power is distributed uniformly over the surface of a sphere of radius r and we may calculate the

† *Translators' note:* The time-dependent part of $\phi(x, y, z, t)$ is contained in the complex part of $\mathscr{E}_0$.

power received per unit area

$$P/S = P/4\pi r^2 \approx 10^{-8}\ \mathrm{W\,m^{-2}}$$

Hence we deduce the intensity of photons per unit area

$$\mathscr{D} = P/Sh\nu \approx 10^{19}\ \mathrm{m^{-2}\,s^{-1}}$$

By assuming that the ferrite rod which acts as an aerial collects the wave over an area $S = 1\ \mathrm{cm}^2$, we receive in one period of the wave $T = 10^{-6}$ s, a number of photons $N = \mathscr{D}TS \sim 10^9$. Moreover, even in a thousandth of a period we receive an average of a million photons. We can thus ascertain the numbers of photons received in each thousandth of a period, and so the instantaneous value of the photon intensity $\mathscr{D}_{\text{inst}}$ can be used, thereby giving a definite meaning to the instantaneous value E of the electric field.

Under the same experimental conditions we are able to detect, by use of a cathode ray oscilloscope, the voltage $V = V_0 \cos(\omega t - \phi)$ received by the aerial, and thus the sinusoidal curve representing the values at each moment of the instantaneous electric field E may be observed on the oscilloscope screen.

(*b*) *The visibility of a weak light source.* We look at a lighted candle by night at a distance $r = 100$ m. The luminous power radiated by a candle may be calculated if we restrict ourselves to the frequency where the eye is most sensitive (over an interval $\delta\lambda \approx 20$ nm around $\lambda = 600$ nm, or $\nu = 5 \times 10^{14}$ Hz); P is found to be of the order of 1 mW. Hence we deduce the power per unit area at the distance r where we are situated

$$P/S = P/4\pi r^2 \approx 10^{-8}\ \mathrm{W m^{-2}}$$

This is the same value as in the previous example, but the photons are now 10^9 times higher in frequency and thus the intensity of photons is much lower

$$\mathscr{D} = P/Sh\nu = 3 \times 10^{10}\ \mathrm{m^{-2}\,s^{-1}}$$

Estimating the area of the pupil as $S = \frac{1}{3}\ \mathrm{cm}^2$, we find that the number of photons received by the eye in one period $T = 2 \times 10^{-15}$ s of the light wave, is

$$N = \mathscr{D}TS = 2 \times 10^{-9}$$

This time the number N which we have calculated does not have any meaning; it indicates that we must wait a thousand million periods to receive just one photon. (However, the very low light flux we have calculated is about 10 000 times greater than the limit of sensitivity of the eye; an eye accustomed to the dark is sensitive to a flux of about 100 photons per second.) Only the average intensity of photons $\mathscr{D}$, calculated over times much greater than the period of the wave, can now have any meaning. Therefore it appears that the instantaneous value E of the classical field is meaningless, and it may be asked whether in conditions of equally weak luminosity the idea of a classical wave $E = E_0 \cos \omega t$ is a useful description of the radiation. We return to this question in section 4.3.

4.1.2 Intensities of Weak Light Sources and Photoelectron Counting

In this section we are concerned with the detection of very low light fluxes. Formerly the only method of detection, other than by eye, was by photographic recording with exposure times from several hours to several days. (The phenomenon being studied had to be sufficiently still.) Photoelectric cells have an even lower sensitivity than the eye. For example, in the experiments described in the previous section with a power per unit area $P/S = 10^{-8}\ \mathrm{W\ m^{-2}}$,

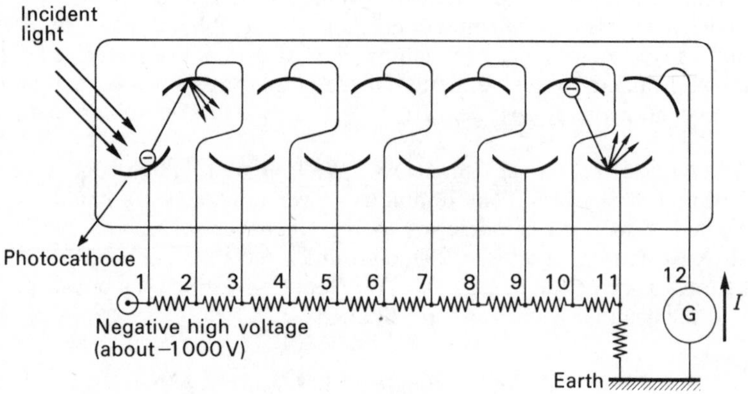

Figure 4.1 Diagram showing the principle of a photomultiplier tube

a photocathode of 1 $\mathrm{cm^2}$ would receive a power of 10^{-12} W, and assuming an average sensitivity of 10 mA/W (see section 1.2.3), the current is found to be $I = 10^{-14}$ A, a value normally too small to be measured.

However, the sensitivity of photoelectric detection can be increased considerably by an arrangement for multiplying the electrons. Figure 4.1 is a diagram of a photomultiplier tube which uses this method. The evacuated tube contains 12 evenly spaced electrodes, at steadily increasing potentials (about 100 V between two consecutive electrodes); the first electrode, at the lowest voltage, is a photocathode emitting photoelectrons under the action of a beam of light. They are attracted by the second electrode and strike it with a large kinetic energy. Part of this energy is given to some of the free electrons in the metal and provides them with the escape energy W_S necessary for detachment from the metal. Each of these incident photoelectrons thus brings about the ejection of three or four other electrons; this is called *secondary emission*. The electrons extracted from the second electrode are then attracted by the third where they in turn cause secondary emission. The number of electrons is thus multiplied gradually, and the current I at the last electrode is amplified considerably compared with the current measured directly at the photocathode. The gain of this device can be as great as 10^6; that is to say, a single photoelectron causes the arrival of a million electrons at the last electrode.

The gain is large enough to allow individual photoelectrons to be detected. A million electrons form an effective electric charge of

$$Q = -10^6 e = -1{\cdot}6 \times 10^{-13}\ \mathrm{C}$$

whose electrostatic effect, although weak, is certainly measurable. Since they all arrive at the same time, they momentarily charge up the condenser of capacity C formed between the last electrode and earth, and this electrode reaches a potential $V_0 = Q/C$ instantaneously. If care is taken to avoid additional

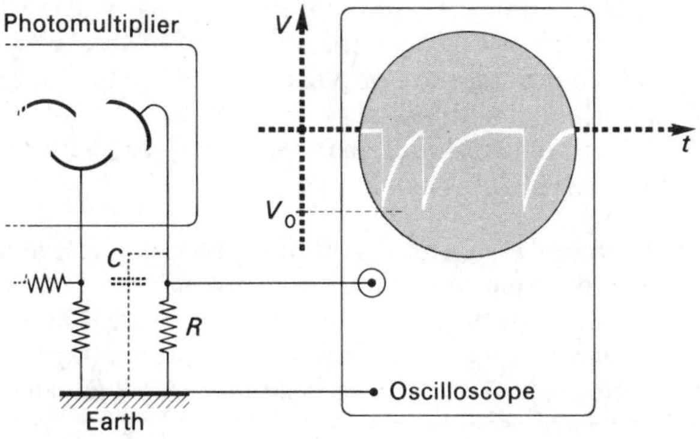

Figure 4.2 Detection of individual photoelectrons

parasitic capacitances arising from connections external to the tube, a capacitance $C \approx 10\ \mathrm{pF} = 10^{-11}\ \mathrm{F}$ can be achieved, giving $|V_0| \approx 10\ \mathrm{mV}$. The condenser discharges very rapidly through the resistance R connecting the last electrode to earth; the discharge is exponential with time constant $\tau = RC$. A short voltage pulse of maximum amplitude V_0 occurs at the last electrode each time a photoelectron is liberated from the photocathode, and this can be very easily observed on an oscilloscope (see figure 4.2). These pulses can also be recorded by using techniques common in nuclear physics, and in this way the number of photoelectron pulses produced in a time t may be counted. In other words, after multiplying by $1/\eta$ (the inverse of the quantum efficiency η of the photocathode) the number of incident photons ($N = \mathscr{D}tS$) in time t may be counted.

These techniques for detecting individual photoelectrons form the basis of the following two ideas.

(*a*) *The possibility of carrying out experiments with very few photons.* Once photons can be counted one by one (or five by five taking into account the quantum efficiency), it would seem that there was no longer any limit to the intensity of the photon flux and, if necessary, an experiment could be carried out with a single photon. This ignores the inevitable parasitic effects arising from 'thermal electrons' which spontaneously detach themselves from the various electrodes due to their kinetic energy, as in thermionic emission; this is a small effect since none of the electrodes are heated. However, because of this, a dark current $I \approx 10^{-9}$ A is received at the last electrode of a modern photomultiplier. Assuming that this current is entirely due to thermal electrons from the photocathode, the number of thermal electrons may be calculated

$$I/|Q| = 10^{-6}\,I/e$$

as many thousands per second. This number can be reduced by cooling the photomultiplier, but it should not be cooled too much lest the quantum efficiency of the photocathode be reduced as well. By taking special precautions in the construction of the photomultiplier (in particular by minimising the useful area of the photocathode), the number of thermal electrons may be reduced to about ten per second. For pulse counting to have any meaning, it is necessary to work in conditions where the number of photoelectrons produced is greater than ten per second; in other words, the photocathode should receive more than 100 photons per second.

(*b*) *The random arrival of photons.* Although the photocathode receives a constant luminous flux, photons do not arrive regularly. The times between successive pulses observed on an oscilloscope (see figure 4.2) vary randomly around a particular average value. By using the counting technique, the number of photoelectrons $N(t)$ counted in a time t also fluctuates and this can be explained in terms of probability theory. These fluctuations are small if the time t is long, but their magnitude increases as the time t shortens. *It is impossible to predict the number of photons that will arrive during a very short time t; only the probability of arrival of photons may be defined.*

Comment This counting technique is suitable only for sufficiently low photon fluxes. If two photoelectrons are liberated very nearly simultaneously (within a time interval $t < \tau = RC$) the two pulses they produce are superimposed, and the electronics will count a single pulse instead of two. Therefore the number of pulses counted is meaningful only if the average time interval between the arrival of two successive photons is sufficiently long compared with the time constant $\tau = RC$, so that the probability of superposition of two pulses remains very small.

4.1.3 Interference Between Photons and the Probabilistic Interpretation of a Wave

The importance of the field in a classical wave is well illustrated by the simple experiment we shall now describe. The same phenomena can be described with any two-beam interferometer. We have chosen the simplest, which is Young's slits.

The diagram of the experiment is shown in figure 4.3. The light from a single slit F (perpendicular to the plane of the diagram) passes through two neighbouring slits S_1 and S_2, parallel to F and at equal distances from it. Both have the same very small width, such that the light passing through them is diffracted to both sides of the geometrical paths (which are extensions of the planes FS_1 and FS_2 beyond the slits S_1 and S_2 towards the right-hand side of figure 4.3). Thus the central region, symmetric about the plane Fz, is crossed simultaneously by two waves coming from two coherent sources S_1 and S_2, and the addition of these two waves gives rise to interference fringes. We have indicated them in figure 4.3 by a sinusoidal curve (dotted line) which represents

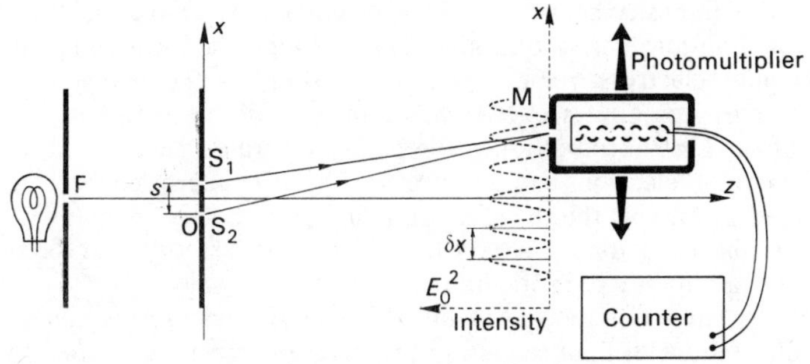

Figure 4.3 Interference, photon by photon

the variation of luminous intensity $E_0{}^2$ as a function of x in a plane z = constant. We can show that the calculated fringe separation $\delta x = \lambda(z/s)$ agrees with the value measured experimentally either by photographic recording or by moving, parallel to the Ox axis, a slit M having a width much smaller than the fringe separation, in conjunction with a photomultiplier.

The resultant wave at a point M, situated at a distance z from the plane S_1S_2 and at a distance x from the plane of symmetry Fz, may be obtained by calculating the path difference between the optical paths of the two waves (assuming $z \gg x$).

$$\delta = S_2M - S_1M = s\frac{x}{z}$$

Hence the phase difference between the two waves may be found, $\phi = 2\pi\delta/\lambda$, and consequently the resultant wave

$$E = A\cos\omega t + A\cos(\omega t + \phi) = 2A\cos\frac{\phi}{2}\cos\left(\omega t + \frac{\phi}{2}\right) = E_0(x, z)\cos\left(\omega t + \frac{\phi}{2}\right)$$

We can calculate the variation of light intensity as a function of x at a fixed distance z from S_1S_2

$$E_0{}^2 = 4A^2\cos^2\frac{\phi}{2} = 2A^2(1 + \cos\phi) = 2A^2\left(1 + \cos 2\pi\frac{s}{\lambda}\frac{x}{z}\right)$$

This is the equation of the curve drawn in dotted lines in figure 4.3 and from it the fringe separation δx may be found.

Why this review of a classic experiment of wave optics? Simply to emphasise the following experimental fact:

The luminous intensity from the first slit F can be reduced in any proportion without any essential change occurring in the experiment. However weak the incident light intensity, the same interference fringes are always observed.

If the observation were carried out by photographic recording, lengthy exposure times would certainly be required. If the fringe system is explored with a measuring slit M in conjunction with a photomultiplier (see figure 4.3) the luminous intensity cannot be reduced indefinitely: a photon flux of the order of a thousand per second must pass through the measuring slit M so that the photoelectrons remain greatly in excess of the parasitic thermal electrons (see preceding section). Now, instead of measuring the average current I from the photomultiplier, the clearly separated pulses resulting from each of the photoelectrons can be counted. Thus the interference fringes may be observed by moving the slit M, since many more pulses are counted at the site of a bright fringe than at a dark fringe (the pulses recorded at the site of a dark fringe are mainly parasitic due to thermal electrons).

So we may carry out an experiment which demonstrates *simultaneously* the corpuscular nature of light (counting the random arrival of photons) and its wave nature (the existence of interference fringes).

Comment To indicate orders of magnitude, we give a numerical example. Two slits S_1 and S_2 of height 15 mm and width $a = 30$ μm have a total area $S \approx 10^{-6}\,m^2$.

Assuming that the light from the slit F in the plane $S_1 S_2$ is reduced in intensity to

$$P/S = 10^{-7}\,W/m^2$$

(equivalent to the light from a candle at 30 m; see example (b) in section 4.1.1), we find that the total luminous power passing through the slits $S_1 S_2$ is $P = 10^{-13}$ W. Photons of wavelength $\lambda = 600$ nm ($\nu = 5 \times 10^{14}$ Hz) have $h\nu = 3 \times 10^{-19}$ J, and so the total number of photons passing through the two slits $S_1 S_2$ is

$$N/t = P/h\nu = 3 \times 10^5 \text{ photons per second}$$

These photons are nearly all distributed within the central diffraction spot of the slits, that is to say within a cone of angle $\lambda/a = 0{\cdot}02$ radian whose width, parallel to the Ox axis, is 40 mm at a distance $z = 2$ m from the slits $S_1 S_2$.

Let us assume that the separation of the two slits is $s = 5a = 150$ μm. We may calculate the distance between fringes $\delta x = \lambda z/s = 8$ mm for $z = 2$ m, and thus we find that only five interference fringes are visible. To investigate these five fringes a measuring slit M of width 1 mm can be used, which will receive on average 1/40 of the total number of photons, in other words several thousand photons per second.

By enabling the arrivals of individual photons to be counted, the method of photoelectric observation merely confirms our calculations for converting a luminous intensity into a number of photons per second, but it also makes the following discussion for the case of a very low luminous intensity easier to understand. We use the numerical values of the previous example, which are of a reasonable order of magnitude without in any way being a limiting case;

it is possible to observe fringes even if the total power passing through the two slits $S_1 S_2$ is as low as $P = 10^{-13}$ W which, for an optical frequency $\nu = 5 \times 10^{14}$ Hz, corresponds to a number of photons $N/t = P/h\nu = 3 \times 10^5$ photons per second. It should be emphasised that this is the total number of photons passing through the slits S_1 and S_2. The average time elapsing between the passage of two successive photons is therefore

$$\tau = t/N \sim 3 \times 10^{-6} \text{ s}$$

We can compare this for example with the transit time of each photon through the apparatus of length z; this transit time is $z/c \approx 10^{-8}$ s, so that $\tau \sim 300\, z/c$. The average time interval between two photons is about 300 times longer than the transit time of each photon; in other words, the apparatus contains a photon for only 1/300 of the time, and for most of the time it is completely empty. It may be asserted that the photons pass through the apparatus one by one.

Finally, we can express the statement emphasised on page 94 more precisely: *interference fringes remain observable when the photons pass through the apparatus one by one.*

In these conditions, the propagation of each photon is unrelated to other photons, justifying consideration of the passage of a single photon. We cannot predict at which bright fringe a particular photon will arrive, nor whether it will arrive at the centre or on the edge of a bright fringe, although we do know that it will not arrive at the centre of a dark fringe. However, we can predict the result obtained over several seconds, after millions of *identical and independent* experiments have been carried out: the greatest number of photons will be recorded at the centres of the bright fringes and, in general, the number of photons counted at each point will be proportional to $E_0{}^2$, the square of the amplitude of the electric field of the wave at that point.

To summarise:

(a) the propagation of a photon can be described by a classical wave in every case,

(b) the classical wave has a probabilistic interpretation: the probability of finding a photon at a point is proportional to the square of the amplitude of the field at that point $E_0{}^2 = \mathscr{E}\mathscr{E}^*$.

By analogy with quantum mechanics, we can say that the complex function $\mathscr{E}$ associated with the electric field is the wave function of the photon.

When applied to the case of a single photon, this probabilistic description of the propagation of photons gives rise to a great degree of uncertainty. We are forced to admit the impossibility of further enquiry; we have reached the limit of our knowledge. In Young's slits experiment in particular, we cannot know whether the photon has passed through the slit S_1 or the slit S_2. All we can say is that the wave which accompanies and determines the propagation of the photon passes simultaneously through the two slits $S_1 S_2$. The question whether the photon has passed through S_1 or through S_2 is meaningless, because we cannot devise an experiment enabling simultaneous observation of the interference fringes and the passage of photons through S_1 or S_2.

Any experiment attempting to obtain this information would interfere with the propagation of the photons so that the interference fringes could no longer be observed. The photon manifests itself only by exchanges of energy and momentum with matter, and even if the photon is not completely destroyed (see the Compton effect), these exchanges alter the frequency or direction of the wave considerably. The description of a photon as an individual localised particle is justified only at the moments when these exchanges with matter occur, and we cannot generalise this description beyond those moments. This will be made clearer in the next section.

4.2 Characteristics of Photons in Relation to the Spatiotemporal Function of the Classical Field—the Uncertainty Principle

4.2.1 Infinitely Long Plane Wave

In a great many problems, electromagnetic waves can be represented to a good approximation by an infinitely long plane wave, whose amplitude E_0 is a constant independent of both position and time, this being the simplest solution of Maxwell's equations. If the plane of the wave is parallel to xOy, the electric field is independent of x and y and is given by

$$E(z, t) = E_0 \cos \omega(t - z/c)$$

Let $\boldsymbol{N}$ be the unit vector normal to the plane of the wave. The plane of the wave passing through the origin and the plane passing through the point of radius vector $\boldsymbol{r}$ are separated by a distance $\boldsymbol{N}\cdot\boldsymbol{r}$. The field E then depends on the three components xyz of the radius vector $\boldsymbol{r}$

$$E(\boldsymbol{r}, t) = E_0 \cos \omega \left(t - \frac{\boldsymbol{N}\cdot\boldsymbol{r}}{c}\right) = E_0 \cos(\omega t - \boldsymbol{k}\cdot\boldsymbol{r}) = E_0 \cos(\omega t - k_x x - k_y y - k_z z)$$

after introducing the wave vector $\boldsymbol{k}$

$$\boxed{\boldsymbol{k} = \frac{\omega}{c}\boldsymbol{N} = \frac{2\pi}{\lambda}\boldsymbol{N}}$$

By analogy with the angular frequency ω, the components k_x, k_y, k_z of the wave vector behave as spatial angular frequencies. In other words, the quantities

$$k_x/2\pi, k_y/2\pi, k_z/2\pi$$

represent spatial frequencies. They are reciprocal lengths parallel to the three axes after which the field E repeats itself (at a fixed time).

In previous chapters, we have assumed an infinite plane wave approximation, associated with photons having

$$\begin{cases} \text{Energy } W = h\nu = \hbar\omega \\ \text{Momentum} \begin{cases} \boldsymbol{p}, \text{ parallel to the direction of propagation } \boldsymbol{N} \\ \text{of magnitude } |\boldsymbol{p}| = h\nu/c = h/\lambda = \hbar(\omega/c) = \hbar(2\pi/\lambda) = \hbar|\boldsymbol{k}| \end{cases} \end{cases}$$

Hence

$$\boxed{\boldsymbol{p} = \hbar \boldsymbol{k} \text{ just as } W = \hbar\omega}$$

A similar relationship exists between on the one hand the frequency ω and the energy W and on the other hand the spatial frequencies k_x, k_y, k_z and the components of momentum p_x, p_y, p_z. This means the electric field of the wave can be expressed in terms of the quantities characteristic of the photon

$$E = E_0 \cos(\omega t - \boldsymbol{k}\cdot\boldsymbol{r}) = E_0 \cos \frac{Wt - \boldsymbol{p}\cdot\boldsymbol{r}}{\hbar}$$

or alternatively, by introducing the complex function $\mathscr{E}$ whose real part is the electric field E, and which is the wave function of the photon

$$\boxed{\mathscr{E} = E_0\, \mathrm{e}^{\mathrm{i}(\omega t - \boldsymbol{k}\cdot\boldsymbol{r})} = E_0\, \mathrm{e}^{\mathrm{i}(Wt - \boldsymbol{p}\cdot\boldsymbol{r})/\hbar}}$$

Thus we obtain an exact relationship between the infinite plane wave and the characteristic quantities W, p_x, p_y, p_z of the associated photons. However, an infinitely long plane wave is merely a simple and exact solution of Maxwell's equation with no bearing on any real physical situation. In actual problems we always have to deal with waves of finite spatial and temporal extent, and we must therefore examine the consequences of such a limitation.

4.2.2 A Wave Limited in Time to a Rectangular Pulse

We start with a simple example in which we consider only the temporal function of the wave, leaving aside its spatial function; that is to say, we take up a fixed position and observe the electromagnetic wave at this point alone.

We consider the case where the amplitude of the electric field is a rectangular function of time

$$E_0(t) = \text{constant} = a \quad \text{for } -\tau < t < +\tau$$

$$E_0(t) = 0 \text{ for } t < -\tau \text{ and for } t > +\tau \text{ (see figure 4.4(a))}$$

This type of wave is frequently used in experimental physics, for example in experiments measuring the speed of light by interrupting a beam of light with an indented wheel rotated at high speed.

The corresponding wave function $\mathscr{E}(t) = E_0(t)\mathrm{e}^{\mathrm{i}\omega_0 t}$ is not truly harmonic since its amplitude depends on time. Because of this we cannot entirely predict its behaviour when it is passed through an optical component or an interfero-

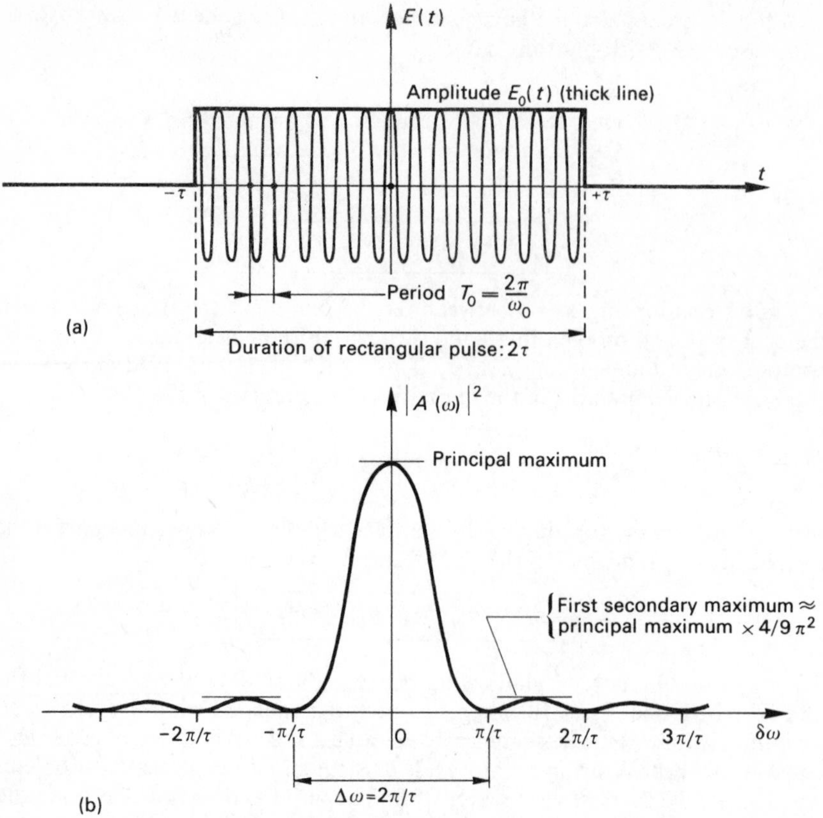

Figure 4.4 (a) Rectangular wave form (at one fixed position); (b) spectral decomposition of the rectangular wave (square of the modulus of the Fourier transform)

meter. General solutions to these propagation problems have been calculated by decomposition into harmonic functions of constant amplitude, because they alone lead to easily calculable straightforward solutions. The same difficulty occurs in dealing with the problem of group velocity, and it is resolved in the same way by using Fourier transforms, which enable any function to be expressed as a sum of harmonic functions of different frequencies ω. The method involves a continuous summation, in other words an integration

$$\mathscr{E}(t) = \int_{-\infty}^{+\infty} A(\omega)\,\mathrm{e}^{\mathrm{i}\omega t}\,\mathrm{d}\omega$$

the differential complex amplitude $A(\omega)$ being obtained from the equation

$$A(\omega) = \frac{1}{2\pi}\int_{-\infty}^{+\infty} \mathscr{E}(t)\,\mathrm{e}^{-\mathrm{i}\omega t}\,\mathrm{d}t$$

For our particular problem we may calculate the differential amplitude $A(\omega)$ as follows

$$A(\omega)=\frac{1}{2\pi}\int_{-\infty}^{+\infty}E_0(t)\,e^{i(\omega_0-\omega)t}\,dt=\frac{a}{2\pi}\int_{-\tau}^{+\tau}e^{i(\omega_0-\omega)t}\,dt$$

$$=\frac{a}{2\pi}\left(\frac{e^{i(\omega_0-\omega)\tau}-e^{-i(\omega_0-\omega)\tau}}{i(\omega_0-\omega)}\right)$$

or

$$A(\omega)=\frac{a}{\pi}\left(\frac{\sin(\omega_0-\omega)\tau}{\omega_0-\omega}\right)$$

(in this particular example $A(\omega)$ is therefore a real function).

For each component of frequency ω there is in classical language a corresponding power or, in quantum language, a probability of finding a photon, both of which are proportional to the square of the real amplitude, or alternatively to the square of the modulus of the complex amplitude

$$P_\omega \approx |A(\omega)|^2=\frac{a^2}{\pi^2}\frac{\sin^2(\omega-\omega_0)\tau}{(\omega-\omega_0)^2}=\frac{a^2}{\pi^2}\frac{\sin^2\delta\omega\tau}{\delta\omega^2}$$

(where $\delta\omega$ is the frequency difference $\omega-\omega_0$).

Comment The proportional relationship obtained for the power can be transformed into an equality by normalisation of the above formula. The differential power P_ω is defined so that the total power is

$$P_{\text{total}}=\int_{-\infty}^{+\infty}P_\omega\,d\omega$$

In this way the differential power is found to be

$$P_\omega=P_{\text{total}}\frac{1}{\pi\tau}\times\frac{\sin^2(\omega-\omega_0)\tau}{(\omega-\omega_0)^2}$$

Figure 4.4(b) shows the variation of $|A|^2$ as a function of $\delta\omega$. The probability $|A|^2$ is zero for $\delta\omega=k(\pi/\tau)$ (k is an integer), except for $\delta\omega=0$ where, in contrast, it passes through a maximum (using the fact that $(\sin x)/x \to 1$ as $x\to 0$). The other maxima of $|A|^2$ are much smaller than this central maximum: the first secondary maximum for $\delta\omega \simeq 3\pi/2$ is already 1/20 the size of the central maximum, and the others decrease as $1/(\delta\omega)^2$. Thus the rectangular wave train has been spectrally decomposed; it is equivalent to the sum of a number of harmonic functions whose frequencies ω are distributed continuously over a wide range, most of which have negligible intensity. In practice the rectangular wave can be accurately represented without including frequencies whose intensities are very small, in other words, by limiting ourselves to frequencies within an interval $\Delta\omega = 2\pi/\tau$. The frequency ω of this rectangular wave is therefore determined with an uncertainty of the order of $\Delta\omega$.

Returning to the photons whose behaviour is described by the wave function $\mathscr{E}(t)$, their energy $W = \hbar\omega$ is determined with the same uncertainty as the frequency ω. The curve in figure 4.4(b) showing $|A|^2$ as a function of $\delta\omega$ can be interpreted, after multiplying the scale of the abscissae by $\hbar$, as representing $|A|^2$ as a function of $\delta W = W - \hbar\omega_0$. It indicates the probability of the energy W of the photon being greater or less than the median energy $W_0 = \hbar\omega_0$. With this change of scale, the first two minima are separated by

$$\Delta W = 2\pi\hbar/\tau = h/\tau$$

this being the degree of uncertainty in the energy W of the photon.

We have dealt with a problem where the temporal extent of the wave is limited to a duration 2τ. This means that the photon must have passed the point $\boldsymbol{r}$ where we are situated within this period of time; therefore the time when the photon passes us is known to within an uncertainty of τ. As a result there is an uncertainty in the photon energy of the order of $\Delta W = h/\tau$ varying inversely as the uncertainty in the time of its passage. If we could predict the time of passage of the photon with greater certainty (τ very short) its energy W would be more uncertain. Conversely, if the energy W were defined exactly, the duration 2τ of the wave train would become infinite, so that the wave would be infinitely extended in time (as in the preceding section); but then the passage of the photon is completely unpredictable. This property has been derived for a particular case, but we shall see how it may be generalised.

4.2.3 The Damped Wave Train—Natural Width

Continuing our study of the temporal variation of the wave, we now consider a second example of great importance since it deals with the phenomenon of spontaneous emission. In the preceding chapter we derived the damped wave train resulting from the laws of classical electromagnetic radiation (see section 3.2.3 and figure 3.8), and we found its amplitude E_0 which decreases exponentially with a time constant 2τ

$$E = E_0(t)\cos\omega_0 t = a\mathrm{e}^{-t/2\tau}\cos\omega_0 t$$

We examined previously the consequences of a spontaneous emission probability $A_{21} = 1/\tau$ (sections 3.2.1 and 3.2.2) and hence the number of photons emitted per unit time, at time t, when n_0 atoms were excited at time zero

$$\left|\frac{\mathrm{d}n_2}{\mathrm{d}t}\right| = n_0 A_{21}\mathrm{e}^{-A_{21}t} = n_0 A_{21}\mathrm{e}^{-t/\tau}$$

This result can be applied to a single atom by using probability language. If we know that an atom is excited at time zero, the probability of observing the emission of a photon at time t varies as $e^{-t/\tau}$. This probability is then proportional to the square of the amplitude of the classical wave train

$$E_0{}^2 = a^2 e^{-t/\tau}$$

With the aid of the probabilistic interpretation of the wave introduced in section 4.1.3, the *damped wave train of classical electromagnetism correctly describes the features of spontaneous emission of a photon.* We may associate this photon with the wave function

$$\mathscr{E}(t) = \mathscr{E}_0(t)\cdot e^{i\omega_0 t} \text{ with amplitude } \begin{cases} \mathscr{E}_0(t) \text{ for } t < 0 \\ \mathscr{E}_0(t) = a e^{-t/2\tau} \text{ for } t > 0 \end{cases}$$

This function $\mathscr{E}(t)$ is not a true harmonic function because its amplitude depends on time. As in the previous section, we must break it down into a spectrum of harmonic functions by means of a Fourier transformation, and thus find the complex amplitude $A(\omega)$ for this particular problem

$$A(\omega) = \frac{1}{2\pi}\int_{-\infty}^{+\infty} \mathscr{E}_0(t)\, e^{i(\omega_0-\omega)t}\, dt = \frac{1}{2\pi}\int_0^{\infty} a e^{[i(\omega_0-\omega)-(1/2\tau)]t}\, dt$$

$$= \frac{a}{2\pi}\,\frac{-1}{i(\omega_0-\omega)-(1/2\tau)}$$

The probability of finding photons is proportional to the square of the modulus of the complex amplitude

$$|A(\omega)|^2 = A(\omega)\,A^*(\omega) = \frac{a^2}{4\pi^2}\,\frac{1}{(\omega-\omega_0)^2+(1/4\tau^2)} = \frac{a^2}{4\pi^2}\,\frac{1}{[\delta\omega^2+(1/4\tau^2)]}$$

The variation of $|A|^2$ as a function of the difference $\delta\omega = \omega - \omega_0$ is represented by a curve usually described as a lorentzian (figure 4.5). $|A|^2$ is a maximum

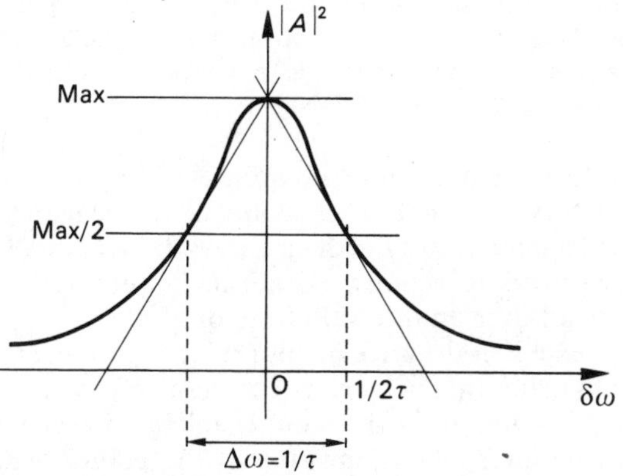

Figure 4.5 Spectral decomposition of a damped wave train (Lorentz curve)

when $\delta\omega = 0$, and is half its maximum value when the difference $\delta\omega = 1/2\tau$. The quantity $\Delta\omega = 1/\tau$, which is twice this difference, is the full width at half height of the curve. The difference itself is sometimes called the half width at half height.

Comment I These terms, however, should be used with caution. Some authors use the term 'half width' as an abbreviation for the width at half height.

Comment II In order to normalise the probability, it should be remembered that

$$\int_{-\infty}^{+\infty} \frac{d\omega}{(\omega - \omega_0)^2 + (1/4\tau^2)} = 2\pi\tau$$

Comment III It will be noted that the two tangents at the half height points meet at the peak, a characteristic property of the lorentzian curve.

This calculation results once again in an uncertainty in the frequency ω and from this two different conclusions may be drawn.

(*a*) *Concerning spectroscopic measurements.* Because of the damping of the wave train, the spectral frequency resulting from spontaneous emission of atoms is marked by a certain imprecision. This is the reason for the broadening of spectral lines, independent of experimental conditions. The width at half height $\Delta\omega$ of the lorentzian curve (figure 4.5) is also called the *natural width* of the corresponding spectral line. However, in most spectroscopic experiments this natural width is negligible compared with the Doppler width resulting from thermal agitation of the atoms (see section 1.3.3) and does not hinder the accuracy of measurements. The lifetimes of energy levels usually observed are rarely less than a nanosecond: $\tau \gtrsim 10^{-9}$ s.

We may therefore deduce a natural width $\Delta\omega \lesssim 10^9\ \text{s}^{-1}$ and $\Delta\nu = \Delta\omega/2\pi \lesssim 10^8\ \text{Hz} = 100$ MHz, whereas Doppler widths are usually of the order of 1000 MHz.

The natural width can be the predominant cause of spectral line broadening both in the exceptional case of a very short lifetime and when there is no Doppler effect. An example of the latter situation was given in chapter 2 with the recoilless emission of γ-photons (section 2.4.3, Mössbauer effect), but the absence of the Doppler effect is met most often in radiofrequency transitions (see volume 2, chapter 7).

(*b*) *Concerning the properties of photons.* As in the previous section, we can interpret the curve of figure 4.5 as signifying the probability $|A|^2$ of a photon being emitted spontaneously with an energy $W = W_0 + \hbar\delta\omega$. The conclusions drawn in the preceding section are confirmed by noting that the possible values of W are spread over an interval of the order of $\Delta W = \hbar\Delta\omega = \hbar\tau$. Although the temporal extent of the wave in this problem is in theory infinite, in practice it is limited to a duration of a few τ. This results from our assumed knowledge of the time of excitation of the atom enabling an accurate determination of the moment of spontaneous emission of the photon within a time interval δt of several τ (for example, the probability of spontaneous emission of a photon occurring after a time of 10τ is $e^{-10} \simeq 1/20\,000$ and may be completely disregarded).

Here, as in the previous example, *the uncertainty δW in the energy varies inversely as the uncertainty δt in the determination of the moment of the passage of a photon*

$$\boxed{\delta W \delta t \approx \hbar}$$

This result is completely general and does not depend on the particular form of the wave train studied. This may be explained by a qualitative argument: by Fourier transformation, the wave $\mathscr{E}(t) = \mathscr{E}_0(t)e^{i\omega_0 t}$ can be expressed in terms of harmonic functions $A(\omega)e^{i\omega t}$ the sum of which reconstructs an amplitude $\mathscr{E}_0(t)$, large in an interval of time δt but very small, however, outside this interval. This means that the phase relations between these harmonic functions give rise to constructive interference at a time t_0 centred on the time interval δt; but for times different from t_0, the harmonic functions of different frequencies become out of phase with one another and after a while this results in destructive interference giving a very small amplitude $\mathscr{E}_0$. If the frequency difference between two of these harmonic functions is $\delta\omega$, they will be out of phase by an angle $\delta\omega \cdot \delta t$ after a time δt. For the interference to be constructive within a time δt, this dephasing angle must not be too large: $\delta\omega\,\delta t \approx 1$ radian, so, after multiplying by $\hbar$, $\delta W \delta t \approx \hbar$.

Comment A concrete example may clarify the meaning of the fundamental law expressed mathematically by the Fourier transformation. The simplest method of measuring the frequency ν of a sinusoidal quantity is to measure the number of periods N in a time t and thereby find that $\nu = N/t$. Moreover this is the method presently used in the laboratory for all voltages having frequencies less than about 50 MHz; electronic counters register the number of positive pulses in a predetermined time t. If fractions of a period are not counted, the number N is determined with an uncertainty equal to a fraction less than one, but which can be close to one. In other words the whole number N does not correspond exactly to the time t and has an error $\delta N < 1$. Therefore the error in the frequency $\delta\nu < 1/t$ varies inversely as the time taken for its measurement. If any sinusoidal phenomenon is limited to a duration δt, its oscillations can be counted only for a maximum time δt and the error in the measurement of its frequency will then be at least $\delta\nu < 1/\delta t$.

4.2.4 Generalisation to the Spatial Function of the Wave

Until now we have considered only the temporal function of the wave, but all the properties that have been studied are easily transposed to the spatial function. We shall now, conversely, choose a particular instant and observe the spatial variation of the wave at this instant alone. Furthermore we shall reduce the problem to one dimension by choosing fixed y and z co-ordinates and studying the wave only along a straight line parallel to the Ox axis. In this case the wave can be expressed in the general form

$$\mathscr{E}(x) = \mathscr{E}_0(x)\,e^{-ik_{0x}x}$$

where k_{0x} is the component of the wave vector $\boldsymbol{k}_0$ in the Ox direction parallel to the direction of propagation, and of modulus $|\boldsymbol{k}_0| = \omega_0/c$ (see section 4.2.1).

If we are dealing with an infinite plane wave, the amplitude $\mathscr{E}_0(x)$ is constant and the wave function $\mathscr{E}(x)$ is a harmonic function of the co-ordinate x with frequency k_{0x}. When the amplitude depends on x, the wave $\mathscr{E}(x)$ must be decomposed as previously into harmonic functions of different frequencies k_x by means of a Fourier transformation

$$\mathscr{E}(x) = \int_{-\infty}^{+\infty} A(k_x)\,\mathrm{d}k_x\,\mathrm{e}^{-\mathrm{i}k_x x}$$

where

$$A(k_x) = \frac{1}{2\pi}\int_{-\infty}^{+\infty} \mathscr{E}(x)\,\mathrm{e}^{+\mathrm{i}k_x x}\,\mathrm{d}x$$

(k_x plays the same role as ω previously. The only change is i for –i in order to conform to the opposite signs of the terms ωt and $\boldsymbol{k}\cdot\boldsymbol{r}$ in the plane wave.) Everything we derived from the Fourier transformation in the two previous sections can then be generalised without recalculation.

Let us assume that the amplitude $\mathscr{E}_0(x)$ is only significant over an interval of length δx (see figure 4.6(a)), that is to say we should be able to predict the position of the photon with a high probability over this segment of length δx. In these conditions, there would be an uncertainty in the wave vector $\delta k_x \approx 1/\delta x$ (see figure 4.6(b)) and multiplying by $\hbar$, we find that the component $p_x = \hbar k_x$ of the momentum vector of the photon has an uncertainty $\delta p_x = \hbar/\delta x$.

The uncertainty of δp_x in the momentum of a photon varies inversely as the uncertainty δx with which the position of the photon can be determined.

The same argument can be applied to the two other co-ordinates y and z, and analogous relations may be found. In addition to the relation already obtained between time and energy, we therefore have the following uncertainty relations

$$\delta W \delta t \approx \hbar; \quad \delta p_x \delta x \approx \hbar; \quad \delta p_y \delta y \approx \hbar; \quad \delta p_z \delta z \approx \hbar$$

The uncertainty relations are a very general consequence of the probabilistic interpretation of the wave, in other words of wave–corpuscle duality.

Comment Let us return to the experiment of a photon interfering with a photon described in section 4.1.3 (see figure 4.3). In order to observe the interference fringes the two slits S_1 and S_2 must act as coherent sources so that the waves received from F by S_1 and S_2 are in phase. It is not necessary for them to have exactly the same phase, but their phase difference must be constant. In other words, the normal of the plane wave arriving at S_1S_2 need not be exactly perpendicular to the plane S_1S_2; the wave vector $\boldsymbol{k}$ may have a small component k_x in the Ox direction so that there may be a phase difference $\phi_{12} = k_x s$ between the two slits separated by s. This phase difference ϕ_{12} must, however, be constant in time; to be precise, let us suppose that the fluctuations of ϕ_{12} must be less than a radian: $\delta\phi_{12} < 1$. This imposes

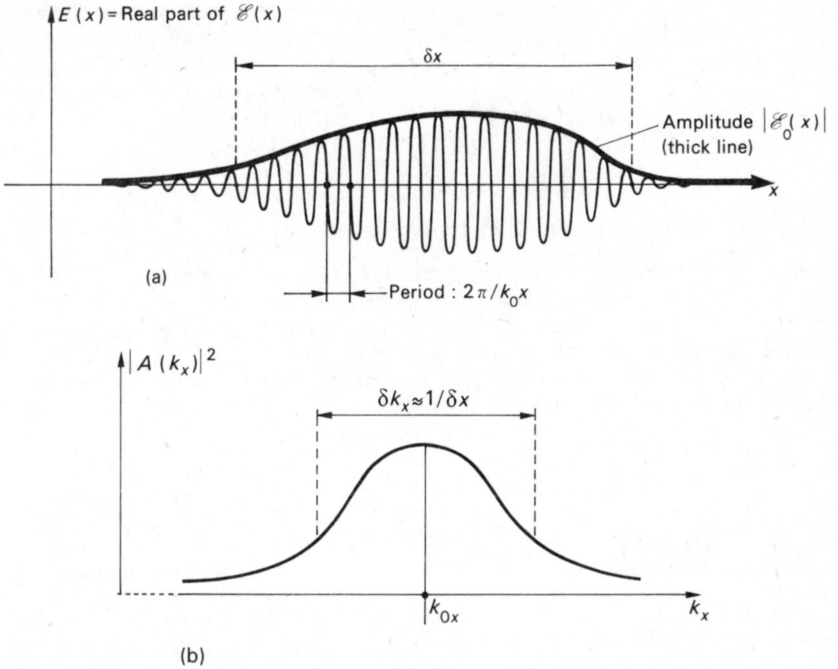

Figure 4.6 (a) Spatial variation of the wave at a fixed instant; (b) spectral decomposition of the spatial function

a similar limit on the fluctuations of k_x

$$\delta k_x = \delta\phi_{12}/s < 1/s$$

The component k_x of the wave vector $\boldsymbol{k}$ must be determined therefore with an uncertainty less than $1/s$.

From the uncertainty principle, the spatial extent δx of the wave is related to the uncertainty in the wave vector by the relation

$$\delta x \sim 1/\delta k_x > s$$

Thus we find that the spatial extent of the wave received by S_1 and S_2 is greater than the separation s between the two sources. Observation of fringes requires that the amplitude of the incident wave should be of a similar order of magnitude at the two slits S_1 and S_2, so that the probabilities that a photon will pass through either slit are also of a similar order of magnitude. Conversely, if we knew the particular slit through which the photon passed, the wave associated with this photon would have an amplitude of zero at the other slit, and the spatial extent of such a wave would be less than the separation between the slits $\delta x < s$. The uncertainty in the wave vector would then be too great to allow observation of interference fringes. The uncertainty principle forces us to choose: either we observe the fringes and do not know through which of the slits each photon passes, or we know through which slit they pass but can then no longer observe interference fringes.

In another connection in section 4.1.3 we compared the mean interval between two successive photons $\tau = t/N \sim 3$ μs with the transit time of each photon through the apparatus $z/c \sim 10^{-8}$ s; but the really significant comparison is between the mean interval and the duration δt of the wave train associated with each photon. We are unable to predict the time at which the photon passes within this duration δt: δt is a sort of 'transit time' of the photon at a fixed point.

4.2.5 Precise Form of the Uncertainty Relations

Precise statements of the uncertainty relations between the uncertainties δW, δt, δp_x, δx and so on, are set out in quantum mechanics texts (see, for example, Messiah p. 129). Use is made of the mean-square deviations of the corresponding quantities, such as they are usually defined in probability theory.

The probabilistic interpretation of the wave enables the calculation of, for example, the average time t_0 for the passage of a photon

$$t_0 = \frac{\int t|\mathscr{E}(t)|^2\,\mathrm{d}t}{\int |\mathscr{E}(t)|^2\,\mathrm{d}t}$$

Next, the mean-square deviation between the time t and its average value t_0 can be found

$$\overline{\delta t^2} = \frac{\int (t-t_0)^2\,|\mathscr{E}(t)|^2\,\mathrm{d}t}{\int |\mathscr{E}(t)|^2\,\mathrm{d}t}$$

Similarly, when we have decomposed a wave into a Fourier sum, the square of the modulus of the complex amplitude $|A(\omega)|^2$ is proportional to the probability that the photon has a frequency ω and the mean-square deviation between the frequency ω and its average value ω_0 can be calculated

$$\overline{\delta\omega^2} = \frac{\int (\omega-\omega_0)^2\,|A(\omega)|^2\,\mathrm{d}\omega}{\int |A(\omega)|^2\,\mathrm{d}\omega}$$

From this the root-mean-square deviation of the energy may be found

$$\sqrt{(\overline{\delta W^2})} = \hbar\sqrt{(\overline{\delta\omega^2})}$$

With these definitions, an exact form of the time–energy uncertainty principle may be written

$$\sqrt{\overline{\delta W^2}}) \times \sqrt{(\overline{\delta t^2})} \geqslant \hbar/2$$

However, this form is not of very great value because the root-mean-square deviation $\sqrt{(\overline{\delta W^2})}$ or $\sqrt{(\overline{\delta t^2})}$ *is not always the best parameter with which to characterise a statistical distribution.* In the two examples studied in sections 4.2.2 and 4.2.3, it was seen that the root-mean-square deviation of the frequency, and therefore of the energy, is infinite (but nevertheless in both cases, the probability of the frequency falling outside the interval between $\omega_0 - \Delta\omega/2$ and $\omega_0 + \Delta\omega/2$ is extremely small).

All the same, it will be recalled that for gaussian functions (e^{-x^2}) the root-mean-square deviation is a meaningful quantity, and the uncertainty relation reduces to the limit of an equality. It is known that the Fourier transform of a gaussian is a gaussian; more generally, if the modulus of a complex function is a gaussian, its Fourier transform is also a gaussian. A complete calculation allows an association between

$$\begin{cases} \text{the wave} & \mathscr{E}(t) = a e^{-\alpha^2(t-t_0)^2} e^{-i\omega_0 t} \\ \text{and the complex amplitude} & A(\omega) = \dfrac{a}{2\alpha\sqrt{\pi}} e^{-(\omega-\omega_0)^2/4\alpha^2} e^{+i(\omega_0-\omega)t_0} \end{cases}$$

(the equations may be simplified by choosing the origin of time at the instant t_0).

For the photons, we derive the probabilities of

$$\begin{cases} \text{observation at time } t & \approx |\mathscr{E}(t)|^2 = a^2 e^{-2\alpha^2(t-t_0)^2} \\ \text{having frequency } \omega & \approx |A(\omega)|^2 = \dfrac{a^2}{4\pi\alpha^2} e^{-(\omega-\omega_0)^2/2\alpha^2} \end{cases}$$

Using a general result from the laws of gaussian statistics and the probability function for ω, we may show that the root-mean-square deviation is exactly equal to the parameter α

$$\sqrt{(\overline{\delta\omega^2})} = \alpha$$

Similarly, from the probability function for t, we may derive its root-mean-square deviation

$$\sqrt{(\overline{\delta t^2})} = \frac{1}{2\alpha}$$

Therefore, in this particular example

$$\sqrt{(\overline{\delta\omega^2})} \times \sqrt{(\overline{\delta t^2})} = 1/2 \quad \text{and} \quad \sqrt{(\overline{\delta W^2})} \times \sqrt{(\overline{\delta t^2})} = \hbar/2$$

This result has an important application: the width of a spectral line is usually governed by the Doppler effect (see section 1.3.3), and the spectral distribution of frequencies is therefore a gaussian curve. It may then be shown that the amplitude of the wave associated with the photons is a gaussian function of time. The amplitude of this wave is large only over a time δt of a few standard deviations whose order of magnitude is therefore equal to the inverse of the Doppler width $\Delta\omega$ (remembering that the width at half height $\Delta\omega$ is related to the root-mean-square deviation α by the equation $\Delta\omega = 2\alpha\sqrt{(2 \ln 2)} \simeq 2{\cdot}4\ \alpha$).

When this spectral line is used as a luminous source in a two beam interferometer, such as Michelson's interferometer, interference fringes are observable only when the delay between the two waves is smaller than, or of the same magnitude as, this time interval δt, which is called the coherence time of the wave; in other words the path difference l between the two arms of the interferometer must be sufficiently short: $l < c\delta t \sim c/\Delta\omega$.

4.3 Properties of Induced Radiation—the Laser and Maser

4.3.1 Characteristics of the Induced Wave

In chapter 3 we have already discussed induced or stimulated emission of photons caused by an interaction between an incident wave at the resonance frequency $\nu_{12} = (E_2 - E_1)/h$ and atoms in an excited state E_2. However, we have studied this phenomenon only partially, in relation to the transition probabilities. Now that we have investigated the idea of a wave associated with a photon we are better equipped to understand the properties of induced radiation as well as their applications.

In section 4.2.3, we were led into associating the damped wave train of classical theory with a photon produced by spontaneous emission. What type of wave should now be associated with a photon produced by induced emission?

For several reasons physicists *have attributed to the induced photon a wave completely identical to that of the incident photon which brought about the emission*:

$$\boldsymbol{E}(\boldsymbol{r}, t) = E_0(\boldsymbol{r}, t) \cos(\omega t - \boldsymbol{k} \cdot \boldsymbol{r} - \phi)$$

with

(i) the same frequency ω,

(ii) the same wave vector $\boldsymbol{k}$, so that the direction of the induced photon is exactly determined (in contrast with a spontaneously emitted photon which can leave randomly in all directions),

(iii) the same spatial and temporal function, and thus the electric field E has the same phase ϕ, and

(iv) the same polarisation, that is to say, the electric field vector $\boldsymbol{E}$ has the same direction as the incident photon.

In other words, the induced photon reinforces the incident wave and increases its energy.

In the following paragraphs we shall discuss the experimentally observable consequences of this hypothesis, and explain the reasoning which led to its formulation well before the quantum theory of electromagnetic radiation was able to explain it.

(*a*) *First we present an argument by analogy with the classical theory of radiation*, in other words, we return to the model of the elastically bound electron which was used in section 3.3.1 to introduce the idea of induced emission.

This model is refined by the introduction of slightly more detail:

(i) the plane electromagnetic wave propagates along the Oz axis with a wave vector $\boldsymbol{k}$, and the component of its electric field in the Ox direction is

$$E_x(t, z) = E_0 \cos(\omega t - kz)$$

(ii) the electron is bound to a fixed point of co-ordinates (x, y, z) but is initially driven into free oscillation about this point. This oscillatory motion is parallel to the Ox axis: the co-ordinate of the electron is therefore $x + \xi$ where $\xi = \xi_0 \cos(\omega t - kz + \phi)$.

As in chapter 3, ϕ is the phase difference between the initial free oscillation of the electron and the electric field which is applied to it.

In chapter 3 we showed that the action of the wave depends essentially on the phase difference ϕ. We saw that for a certain value ϕ_a of the phase, the energy of the oscillation W stored by the electron increases (absorption) whereas for the value $\phi_a + \pi$ on the other hand, the energy W of the electron decreases (induced emission). In both cases, the change δW of this energy is equal to the work done by the force qE_x applied to the electron of charge q: if the phase is ϕ_a, positive work is done by the electric force and the electromagnetic wave which does this work loses energy; if, on the other hand, the phase is $\phi_a + \pi$, negative work is done by the electric force, that is resistive work, and the wave on which this work is done increases its energy.

The effect produced on the electromagnetic wave can be studied in greater detail by assuming that a large number of electron oscillators are situated at points (x, y, z) distributed at random within a certain volume. A particular category of electrons may be singled out, all having the same phase difference ϕ, and the wave radiated by the electrons in this category may be calculated. The calculation consists of adding the assembly of electric fields radiated by each of the electrons in this category ϕ, taking into account the phase differences introduced between them by propagation from separate points whose co-ordinates (x, y, z) are random. In all directions other than Oz, these phase differences are themselves random, giving rise to destructive interference, and the wave radiated by the assembly in any direction other than Oz has negligible amplitude. On the other hand, in the Oz direction, an exact compensation occurs between the dephasing introduced by the propagation of the wave radiated by the electron, and the dephasing–kz of the incident wave applied to this electron, a dephasing accounted for in our theory in writing the oscillatory motion ξ of the electron and by defining the phase ϕ. Consequently, the electric fields radiated in the Oz direction by the electrons in category ϕ are all in phase and their constructive interference gives rise to a resultant wave of large amplitude.

To conclude this study, we shall distinguish two cases according to the phase difference ϕ between the oscillatory motion of each electron and the incident electric field which it experiences.

(1) For the value ϕ_a, it is found that the field E_x' radiated in the Oz direction is out of phase with the incident field E_x. From this it is found that the resultant field $E_x + E_x'$ after passing through a medium containing electrons, has an amplitude smaller than the incident field E_x. This is described by a loss of energy by the wave after passing through the medium, in other words an absorption phenomenon.

(2) If on the other hand the phase ϕ has the value $\phi_a + \pi$ (a situation where the energy of electrons decreases, due to induced emission) the field E_x' radiated in the Oz direction also has a phase change of π compared with the previous case; it is $E_x + E_x'$ after passing through the medium, has a larger amplitude than the incident wave E_x. The phenomenon of induced emission has amplified the incident wave.
E_x. The phenomenon of induced emission has amplified the incident wave.

(*b*) *The conservation of momentum exchanged* between radiation and atoms was studied by Einstein in his famous paper of 1917. In section 2.4 we showed that the exchanges of momentum between atoms and photons in the optical region are too small to be experimentally observable; but theoretically they must obey the general law of conservation of momentum. Einstein made a detailed analysis of these exchanges in the case of a gas in thermal equilibrium and showed that the conservation law is obeyed only if a momentum vector $\boldsymbol{p}$, identical to that of the incident photons which caused their emission, is attributed to the induced emission photons.

4.3.2 Condition for Amplification of the Incident Wave

When it increases the energy of the incident wave, induced emission plays a role diametrically opposed to the phenomenon of absorption; but induced emission occurs only from excited atoms (energy level E_2) whereas absorption occurs from ground state atoms (energy level E_1). Under normal conditions, the number n_2 of excited atoms per unit volume (or the population of the level E_2) is much smaller than the number n_1 of atoms in the ground state (or the population of level E_1). This is why we disregarded induced emission at the beginning of chapter 3, when studying the change of intensity of a beam of light passing through a material medium (see figure 3.3) and when defining the coefficient of absorption. We shall now consider the modifications necessary to the results given in chapter 3 to account for the energy increase of the wave by induced emission when the number n_2 of excited atoms is not negligible.

Accordingly we calculate the change δP in the power P of a light beam of cross-sectional area S when it passes through a medium of length δz parallel to the direction of propagation Oz. We obtain δP by counting the number of photons absorbed, or created by induced emission, in one second within a volume $\mathscr{V} = S\delta z$

$$\text{Number of photons absorbed } \mathscr{V}\left|\frac{\mathrm{d}n_1}{\mathrm{d}t}\right| = \mathscr{V}\varpi_{12}n_1 = \mathscr{V}B_{12}u_\nu n_1$$

$$\text{Number of photons induced } \mathscr{V}\left|\frac{\mathrm{d}n_2}{\mathrm{d}t}\right| = \mathscr{V}\varpi_{12}n_2 = \mathscr{V}B_{12}u_\nu n_2$$

We calculate the change in power by subtracting the number of photons absorbed from the number of photons emitted

$$\boxed{\delta P = h\nu\left(\mathscr{V}\left|\frac{\mathrm{d}n_2}{\mathrm{d}t}\right| - \mathscr{V}\left|\frac{\mathrm{d}n_1}{\mathrm{d}t}\right|\right) = \mathscr{V}h\nu(B_{21}n_2 - B_{12}\mathrm{n}_1)\,u_\nu}$$

Comment The expression above suffices for the following discussion, but it is often necessary to express it in another form by introducing the absorption coefficient $K(\nu)$ measured experimentally at the frequency ν. This coefficient is defined by isolating a band of frequencies of width $\mathrm{d}\nu$ within which the beam has a power $P_\nu\mathrm{d}\nu$. The differential power P_ν is defined analogously to the differential energy density u_ν

$$P = \int P_\nu\,\mathrm{d}\nu \quad \text{just as } u = \int u_\nu\,\mathrm{d}\nu$$

In a similar way to P and u, P_ν and u_ν are related by the equation $P_\nu = Scu_\nu$. From this we find

$$u_\nu = S\delta z u_\nu = (1/c)P_\nu\,\delta z$$

We can thus write the definition of $K(\nu)$ as

$$K(\nu) = -\frac{1}{P_\nu \, d\nu} \frac{\delta(P_\nu \, d\nu)}{\delta z} = -\frac{1}{P_\nu} \frac{\delta P_\nu}{\delta z}$$

The frequency differential $d\nu$ should not be confused with the increment δz of the distance z travelled; the operations of differentiation or integration over frequency ν are independent of the operations of incrementing a function of z. The order of these operations can be changed around

$$\int K(\nu) \, d\nu = -\frac{1}{P_\nu} \frac{\int \delta P_\nu \, d\nu}{\delta z} = \frac{1}{P_\nu} \frac{\delta P}{\delta z} = -\frac{h\nu}{c} (B_{21} n_2 - B_{12} n_1)$$

so finally

$$\boxed{\int K(\nu) \, d\nu = -\frac{1}{P_\nu} \frac{\delta P}{\delta z} = \frac{h}{\lambda} (B_{12} n_1 - B_{21} n_2)}$$

This equation may be compared with the one framed at the end of section 3.1.4; it differs from the latter only by the additional term $-B_{21} n_2$.

In these formulae, the term $B_{12} n_1$ represents absorption and the term $B_{21} n_2$ induced emission. Under normal experimental conditions n_2 is smaller than n_1 so the variation of power δP is negative; thus absorption exceeds stimulated emission. (The experimentally measured absorption coefficient $K(\nu)$ may be slightly reduced if n_2 is not too small—Landenburg's experiments around 1930.) For the opposite to occur, so that induced emission exceeds absorption, allowing amplification of the wave and thereby producing a positive variation of power δP, the following condition would have to be satisfied

$$B_{21} n_2 - B_{12} n_1 > 0$$

or alternatively

$$\boxed{\frac{n_2}{n_1} > \frac{B_{12}}{B_{21}} = \frac{G_2}{G_1}}$$

(G_1 and G_2 are the statistical weights, or orders of degeneracy, of the two energy levels E_1 and E_2 (see section 3.3.3)). However, Boltzmann's law gives the ratio of the populations n_1 and n_2 of the two energy levels under conditions of thermal equilibrium at an absolute temperature T

$$\frac{n_2}{n_1} = \frac{G_2}{G_1} e^{(E_1 - E_2)/kT} < \frac{G_2}{G_1}$$

since $E_1 - E_2 < 0$.

The framed condition above can never be satisfied under conditions of thermal equilibrium, whatever the temperature.

This condition can, nevertheless, be realised by means of various techniques that destroy thermal equilibrium; we shall discuss these in the next section. This is described by saying that there is a *population inversion* compared to conditions of thermal equilibrium, or that a fictitious *negative temperature* has been achieved since whenever the temperature T is negative the ratio n_2/n_1 from Boltzmann's formula is greater than G_2/G_1. The concept of negative temperature is often used nowadays by physicists, but it must be used with care. In particular it should be noted that when the ratio n_2/n_1 becomes greater than G_2/G_1, the temperature T changes discontinuously from $+\infty$ to $-\infty$; there is a continuous change from a system at positive temperature to one at negative temperature through an intermediate state of infinite temperature.

Spontaneous emission has not yet entered into this argument. However, it plays a part indirectly since it always tends to decrease n_2 and therefore prevent attainment of a population inversion. In chapter 3 we saw that the probability of spontaneous emission A_{21} varies as ν^3 and that it becomes negligible at low frequencies where the transition frequency ν_{12} occurs in the radiofrequency region.

In 1954 the first experimental device enabling observation of the amplification of a wave passing through a medium due to a population inversion was operated in the radiofrequency region. Its inventor, the American physicist Townes, gave it the name maser, the initials of 'Microwave Amplification by Stimulated Emission of Radiation' (microwaves fall in that part of radiofrequency region where waveguide techniques are used). The analogous device in the optical region is called the laser, obtained by substituting the word 'light' for 'microwave', since it is an amplifier of light.

4.3.3 Achievement of a Population Inversion

In this section we describe the various techniques used to disturb thermal equilibrium and produce a population inversion. Clearly it is impossible to be exhaustive.

(*a*) *Separation in an atomic or molecular beam.* This was the method used in the first maser constructed by Townes in 1954 and we explain it in this context. The ground state of the molecule in gaseous ammonia NH_3 is split into two closely spaced levels, the transition between them occurring at a frequency in the waveguide region $\nu_{12} = 23\,870$ MHz ($\lambda_{12} \approx 1{\cdot}25$ cm). At ordinary temperatures, $T = 300$ K, the ratio $h\nu_{12}/kT = (E_2 - E_1)/kT$ is about 1/250, and the populations n_1 and n_2 of these two energy levels are very nearly equal. Molecules in these two levels differ in their electric dipole moment, and Townes used this property to separate them.

A reservoir of ammonia gas is connected to an evacuated enclosure by a small opening; molecules emerging through it without encountering an obstacle have straight line paths. A second aperture selects molecules whose trajectories have a certain direction. The molecular beam thus formed crosses a region where there is a very strong electric field gradient. Under the influence of this field gradient, the electric dipole moment is subjected to a translation force which deflects the molecular trajectories. Since molecules in the two levels E_1 and E_2 have different dipole moments, they also have different trajectories, and are thus separated. Those in the lower level E_1 are stopped and those in the higher level E_2 pass on across a waveguide (see figure 4.7).

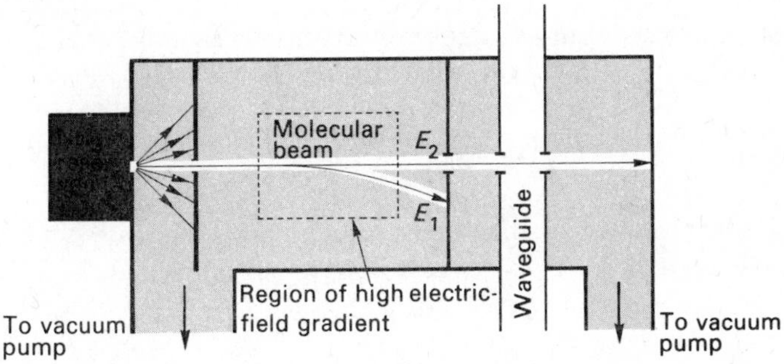

Figure 4.7 Diagram showing the principle of Townes's maser

Comment This separation technique in a molecular beam is merely an extension to the case of electric dipoles of the technique used in 1920 by Stern and Gerlach with magnetic dipoles (see chapter 10).

(*b*) *Pumping with an electromagnetic wave on another transition.* This method makes use of at least three energy levels of an atom, which we call E_1, E_2 and E_3 in order of increasing energy. By causing a great many transitions between the levels E_1 and E_3 with a high intensity wave of frequency $\nu_{13} = (E_3 - E_1)/h$, the populations n_1 and n_3 of the levels E_1 and E_3 can be altered so as to obtain a population inversion between E_1 and E_2 or between E_2 and E_3. Amplification of the frequencies ν_{12} or ν_{23} will be observed, the frequency ν_{13} being called the pumping frequency. We shall give three typical examples.

(i) *Pumping with a radiofrequency wave.* This is the three-level maser proposed independently by Basov and Prokhorov in 1955 and by Bloembergen in 1956. Here, the levels E_1, E_2, E_3 are all very near the ground state (which could be E_1, but need not be) and the frequencies of transitions between them fall in the radiofrequency region. Thus (1) spontaneous transitions are negligible; (2) the ratios $(E_i - E_j)/kT$ are small compared with unity, even at low temperatures, and the populations of the three levels in thermal equilibrium are very nearly equal (see method (a): separation in a molecular beam).

To simplify the argument, let us assume that the statistical weights are equal: $G_1 = G_2 = G_3$. An intensely powerful electromagnetic wave at frequency ν_{13} causes many transitions by absorption and stimulated emission whose overall effect is to equalise the populations n_1 and n_3 of the levels E_1 and E_3: the transition ν_{13} is said to be 'saturated'. Before applying the pumping wave, the populations are determined by the Boltzmann law

$$\frac{n_3}{n_2} = e^{-h\nu_{23}/kT} \simeq 1 - \frac{h\nu_{23}}{kT} \quad \text{or} \quad n_3 \simeq n_2\left(1 - \frac{h\nu_{23}}{kT}\right)$$

$$\frac{n_1}{n_2} = e^{+h\nu_{12}/kT} \simeq 1 + \frac{h\nu_{12}}{kT} \quad \text{or} \quad n_1 \simeq n_2\left(1 + \frac{h\nu_{12}}{kT}\right)$$

To a first approximation, irradiation by the pumping wave changes neither n_2 nor the sum $n_1 + n_3$; from this we find n_1 and n_3 after irradiation

$$n_1 = n_3 \simeq n_2\left[1 + \frac{h(\nu_{12} - \nu_{23})}{2kT}\right] \quad \begin{cases} > n_2 \text{ if } \nu_{12} > \nu_{23} \\ < n_2 \text{ if } \nu_{12} < \nu_{23} \end{cases}$$

Population inversion occurs between n_3 and n_2 in the first case (figure 4.8(a)) or between n_1 and n_2 in the second case (figure 4.8(b)).

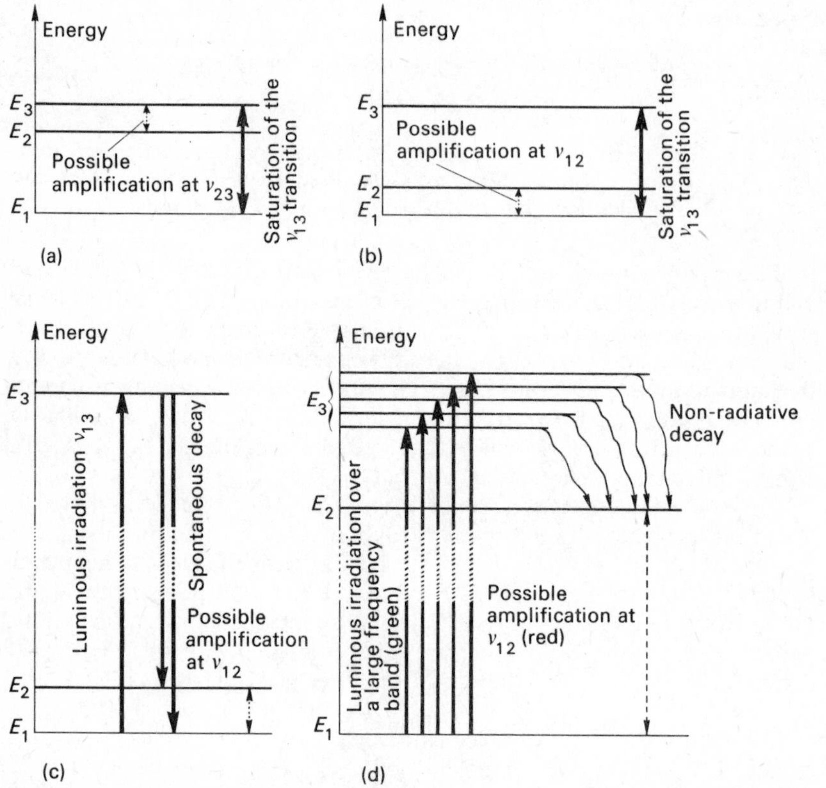

Figure 4.8 Pumping methods: (a) 3-level maser ($\nu_{12} > \nu_{23}$); (b) 3-level maser ($\nu_{12} < \nu_{23}$); (c) optical pumping of a radiofrequency transition; (d) particular case of ruby (optical pumping of an optical transition)

(ii) *Optical pumping of a radiofrequency transition.* Here the transition frequency between the two lower levels E_1 and E_2 is always in the radio-frequency region, but the pumping frequency ν_{13} is in the optical region (see figure 4.8(c)). Conditions are chosen such that luminous irradiation can produce transitions only between E_1 and E_3 but not between E_2 and E_3 (either because the energy difference between E_2 and E_1 is much greater than the width of a spectral line, or because of the polarisation of the light). Atoms which are excited to the state E_3 decay by spontaneous emission; this decay may proceed via several alternative transitions, and some of the atoms end up in the level E_2.

Some of the atoms in the level E_1 are thus transferred to the level E_2. To create a population inversion between E_2 and E_1, the pumping process must be moıe rapid than the random interaction processes which tend to re-establish thermal equilibrium. (In chapter 11 we shall describe in greater detail the use of polarised light in optical pumping.)

(iii) *Optical pumping of an optical transition.* This includes the somewhat exceptional case of ruby with which Maiman produced the first laser in 1960. Ruby is a crystal of alumina Al_2O_3 doped with Cr^{3+} ions, which are responsible for all its optical properties in the visible region. Figure 4.8(d) shows representatively some of the energy levels of the ion Cr^{3+} when embedded in the crystal lattice of Al_2O_3: a large number of transitions very close in frequency allow excitation of the Cr^{3+} ions from a ground level E_1 to a series of excited levels E_3 (the lines are too close together to be distinguishable and form a wide absorption band in the green; it is because it absorbs green light that the ruby crystal appears red). However, spontaneous emission is not observed from these levels E_3; before there is time for this to take place, extremely fast non-radiative transitions occur in which surplus energy is transferred to the crystal lattice in the form of energy of thermal agitation. All these non-radiative transitions bring the Cr^{3+} ions back to the same energy level E_2 which happens to have an exceptionally long lifetime τ (of the order of a millisecond). By illuminating a ruby crystal with the light from an extremely powerful photographic flashlamp, a population inversion between E_2 and E_1 is realised, resulting in laser action on the corresponding red line ν_{12}.

(*c*) *Electric discharge in a gas.* The case of the Cr^{3+} ion in ruby that we have described is exceptional (many equivalent access paths to the same level E_2 on the one hand; small probability of spontaneous emission $A_{21} = 1/\tau$ from the level E_2 on the other hand). Usually it is impossible to create a population inversion between the ground state of an atom and one of its excited states in the optical region. However, it is very easy to produce population inversions between two excited levels both of which are empty under normal conditions. The establishment of a discharge in a gas at low pressure may give rise to discharge conditions in which atoms are lifted in great numbers to certain excited states. This is the principle of operation of the modern gas laser.

(*d*) *Interaction with other atoms, ions or molecules.* The equilibrium populations n_1 and n_2 of an atom A in a sample material being studied can be attained through a wide variety of complex microscopic mechanisms. These mechanisms may involve other atoms B in the same sample. If any one of the preceding methods is used to change the populations of the energy levels of atom B without however causing an inversion, these interaction mechanisms result in changes of the populations n_1 and n_2 of atom A. In certain cases a population inversion in A can be obtained.

We shall give only one example, that of the first gas laser, operated by Javan in 1960, shortly after the ruby laser: an electric discharge is created in a mixture of the two noble gases helium and neon. The discharge raises a certain number of helium atoms (which here play the role of B) into an excited metastable state, that is to say, an excited state from which radiative transitions are highly improbable, and having therefore a very long lifetime. The energy of this metastable level of helium practically coincides with the energy of several closely separated excited states of neon; this coincidence allows transfer of energy from a metastable helium atom to an atom of neon when they interact in a collision. These few excited levels of neon are thus preferentially populated, to the exclusion of the others.

Different mechanisms give rise to other examples, in both liquids and solids.

(*e*) *Chemical reaction producing a molecule directly in an excited state.* If the chemical agents producing a reaction are continually renewed, and the products of this reaction are removed simultaneously, a constant excess of molecules in an excited state can be maintained in an experimental cell.

4.3.4 Amplification and Oscillation—the Function of a Resonant Cavity

In section 4.2, we calculated the power gained by an electromagnetic wave when it passes through a medium of volume $\mathscr{V}$ in which a population inversion exists

$$\delta P = \mathscr{V} \hbar\omega(B_{21} n_2 - B_{12} n_1) u_\nu$$

The increase in power is proportional to the differential energy density u_ν of the incident wave, and the amplification of the wave will therefore be greater if, for a given incident power P_i, the energy density u_ν can be made larger. This can easily be arranged in the microwave region by making the waveguide open into a resonant cavity; this is a rectangular metal box whose dimensions have been accurately determined for a particular frequency such that the waves reflected by the walls add up in phase and form a system of *stationary waves* of large amplitude. By passing across the same region many times, the wave created there has an energy density u_ν much greater than in a simple travelling wave; in other words the cavity builds up in a limited volume a large energy W which had been widely spread in the form of a travelling wave before entering it.

The process of accumulation and storage in itself could not increase the total energy transported by the wave and is certainly not an amplification process. However, by increasing the energy density u_ν corresponding to a given incident power P_i, it adds to the power amplification δP produced by the medium. It increases the gain of the amplifier, consisting of a medium in which there is a population inversion.

The improved gain resulting from the use of a resonant cavity can be explained in another way. Instead of considering the whole stationary wave present in the cavity, we can analyse the internal action of this cavity in terms of a progressive wave reflected many times by the walls. The wave amplified by the first pass through the medium is amplified again on the second pass, and then on the third pass and so on.

The phenomenon thus described is totally analogous to the phenomenon observed in electronics where an amplifier is used and a certain fraction α of the alternating voltage from the output of the amplifier is fed back to the input of the amplifier. This fraction α is called the feedback ratio, and feedback is sometimes used to increase the gain of an amplifier artificially.

Comment Texts on electronics show that if g is the normal gain of the amplifier for a single passage of the signal the resultant gain G obtained in the presence of feedback is $G = g/(1 - \alpha g)$. If we apply this formula to the maser or laser under normal operating conditions, α and g are both very close to unity. For example it might be that $\alpha = 0{\cdot}99$ and $g = 1{\cdot}01$, giving $G = 10^4$.

If there is too much feedback, the resultant gain G becomes infinite, that is to say, the amplifier produces an alternating voltage at its output, without any external source. The device goes into oscillation of its own accord and becomes an alternating voltage generator. Anyone who has ever tinkered with radios knows that unless great care is taken to avoid parasitic feedback, a high-gain amplifier is very likely to 'hoot' and go into free oscillation. Exactly the same occurs in the resonant cavity of a maser. If the cavity is of good quality and if the population inversion is large enough, free oscillation is obtained; a wave is generated within the cavity.

Thus masers may operate as amplifiers or oscillators. Either way they are characterised by a very precise frequency since this is defined by the atomic transition (the latter is not Doppler broadened in the microwave region, see volume 2, chapter 7). This very well-defined frequency helps to give maser amplifiers the very low noise levels which provide their main points of interest. Maser oscillators can serve as frequency standards, but their use for this purpose is not presently unrivalled; equivalent performance may be obtained (with a relative frequency uncertainty less than 10^{-10}) with the more traditional quartz crystal-controlled oscillator. (The frequency ν from a quartz oscillator is locked to an atomic transition; if it moves off the central value ν_{12} of this transition, a change in the signal level is observed. The voltage which measures this effect is used to change the operating conditions of the quartz crystal slightly and thus automatically correct its frequency ν.)

All that has been said about masers can be applied equally well in the optical region to lasers. Following a suggestion by Townes and Schawlow in 1958, a Fabry–Perot interferometer is used as a resonant cavity. This explains the diagram of a gas laser shown in figure 4.9. The amplifying medium is a gas or a mixture of gases at low pressure, through which an electric discharge is

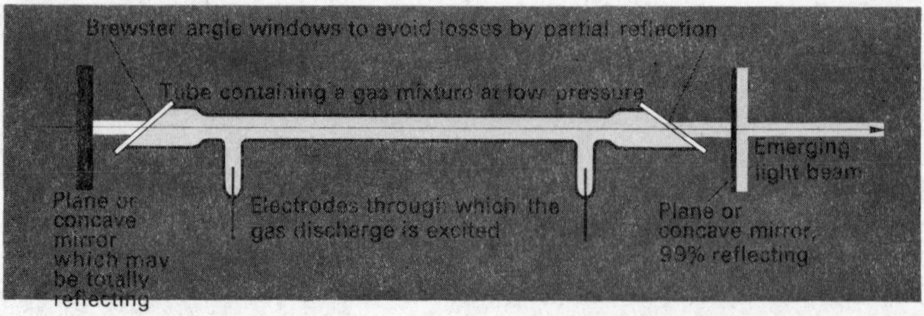

Figure 4.9 Diagram of a typical gas laser

passed (see preceding section). The tube containing the gas is placed between the two mirrors of an interferometer in such a way that the wave passes through it after each partial reflection. The distance between the two mirrors (of the order of a metre) is very much greater than in conventional interferometers; this extension of the cavity allows a great increase in the qualify factor Q (see following section), as well as the volume of the amplifying medium.

Comment The cavity losses are often reduced by substituting spherical mirrors for the plane mirrors normally used in interferometers.

In certain lasers, these Fabry–Perot mirrors define the volume occupied by the gas, but this arrangement complicates the construction and carries the risk of deterioration of the mirrors. It is much easier to enclose the gas in the tube independently. To avoid reduction in the intensity of the wave by partial reflection at the two end windows of the tube, they are inclined at Brewster's angle ($\tan i = n$, the refractive index). The wave whose electric vector $\boldsymbol{E}$ is in the plane of incidence is then transmitted entirely without reflection, as may be shown from the theory of electromagnetism. The wave generated by the laser is thus polarised (the other polarisation is attenuated by partial reflection and cannot be generated).

In contrast to masers, lasers are nearly always used as oscillators. They have revolutionised optics, by giving physicists sources of coherent light for the first time, similar to those commonly existing at radiofrequencies or microwave frequencies. In section 4.6 we shall study in detail the properties of the radiation that they emit. (The use of a laser as an amplifier is not generally of interest because there are no other light sources giving a coherent wave useful for amplification; in certain exceptional circumstances it enables an increase of the power produced in a short pulse from another laser.)

Before concluding this section, we must emphasise *the secondary nature of the role played by the resonant cavity*. It increases the gain of the device and is of use in the attainment of the oscillation threshold. Nevertheless, it is the pumping that is essential for the population inversion to be achieved. The best demonstration of this is in the existence of *optical superradiant sources*,†

† *Translator's note:* The term superradiance was first used to describe a completely different phenomenon that involved only spontaneous emission. Here it refers to sources that involve stimulated emission and is sometimes called 'amplified spontaneous emission' to avoid confusion.

which are oscillators using induced emission without a cavity: if the length z of the amplifying medium is sufficient and if the gain per unit length $\mathrm{d}G/\mathrm{d}z$ is high enough, a large intensity can be obtained in a single pass, starting from a single initial photon.

4.3.5 Calculation of the Gain and the Condition for Oscillation

Our aim is to express quantitatively the ideas presented in the preceding section.

(*a*) *The main properties of a resonant cavity.* These may be summarised by its *coefficient of quality* which may be defined exactly as for a resonant circuit; it is equal to the product of the wave frequency ω and the time constant τ for the decay of the energy W stored in the cavity when the incident wave is prevented from leaving it: $Q=\omega\tau$. We review briefly the significance of the time duration τ. The decrease of W arises from two distinct causes:

(i) the cavity is not completely closed—it is connected to a waveguide in which there is an outgoing wave carrying a certain power P_s;
(ii) there are losses in the walls of the cavity which are not perfectly reflecting and because of this, even if the cavity were completely closed, there would be a certain power lost P_p. These two powers P_s and P_p are proportional to the square of the electric field of the wave, in other words to the mean energy density u in the cavity and therefore to the total energy $W=\mathscr{V}u$ stored in its volume $\mathscr{V}$. Letting $1/\tau$ be the coefficient of proportionality

$$-\frac{\mathrm{d}W}{\mathrm{d}t}=P_s+P_p=\frac{1}{\tau}W$$

and by integrating the differential equation in W, an exponential law of time constant τ is obtained. (In other words, it may be stated that $Q=2\pi\nu\tau$ is a measure of the number of periods completed before the wave dies away). Thus we derive the definition of the coefficient of quality in terms of energy

$$Q=\omega W/(P_s+P_p)$$

(*b*) *Calculation of the gain of an amplifier.* We shall evaluate the role of the cavity by means of conservation of energy under steady-state conditions, when the energy W is kept constant. The total power put into the cavity by the incident wave P_i and by induced emission δP must be equal to the total power removed from it by the outgoing wave P_s and through the losses P_p

$$P_\mathrm{i}+\delta P=P_s+P_p=\frac{\omega}{Q}W$$

The cavity can be constructed in such a way that the output power P_s is much greater than the losses P_p, and this optimistic assumption is made in order to calculate the gain in power

$$G=\frac{P_s}{P_\mathrm{i}}\simeq\frac{P_s+P_p}{P_\mathrm{i}}=\frac{(\omega/Q)\,W}{(\omega/Q)\,W-\delta P}=\frac{1}{1-Q\delta P/\omega W}$$

(This result may be compared with the formula $G=g/(1-\alpha g)\simeq 1/(1-\alpha g)$ given in the comment in the previous section.) It is clear that the gain is greater than unity as long as δP is positive; however, the calculation requires some refinement in order to show that the energy density may be eliminated. This requires the introduction of the uncertainty in frequency ν, but without giving a precise definition that would depend on the individual problem being studied. Let $\Delta\nu$ be the width of the frequency band interacting with the atoms. We can then express the stored energy $W = \mathscr{V} u_\nu \Delta\nu$, on the basis that u_ν represents the mean value of the differential energy density within the cavity. For greater accuracy, we should also take into account the fact that the amplifying medium does not usually fill the whole cavity: we shall retain the letter $\mathscr{V}$ as designating the volume of the cavity and write the increase in power due to induced emission as

$$\delta P = \eta \mathscr{V} \hbar\omega (B_{21} n_2 - B_{12} n_1) u_\nu$$

The dimensionless coefficient η, called the filling factor, is always less than one; its evaluation must take into account the spatial distribution of energy within the cavity, not just the actual volume of the amplifying medium. Finally we find that

$$Q \frac{\delta P}{\omega W} = \frac{Q\eta\hbar}{\Delta\nu}(B_{21} n_2 - B_{12} n_1)$$

which allows us to calculate the gain G of the maser; it should be noted that it increases with the quality factor Q of the cavity.

(c) *The condition for free oscillation.* This may be obtained by finding the condition for the gain to become infinite. It may also be found by simply comparing (1) the power supplied by induced emission δP, and (2) the total power removed from the cavity $P_p + P_s = (\omega/Q) W$. Both these quantities are proportional to the energy density u_ν.

If for a given value of u_ν

$$\boxed{\delta P > \frac{\omega}{Q} W}$$

or alternatively

$$\boxed{\eta\hbar(B_{21} n_2 - B_{12} n_1) > \frac{\Delta\nu}{Q}}$$

the cavity receives continuously more energy than it loses, and its energy W increases. However small the initial value of the density u_ν it increases exponentially. A single photon is sufficient to initiate the process; the device then generates a wave of its own accord, and behaves as an oscillator. If the matter rested there, the impression could be given that u_ν increases indefinitely which would present problems. Fortunately this is not so. The number of induced transitions increases proportionally to u_ν, and when the amplitude of the oscillations becomes very large, the population n_2 of the upper level tends to fall because the pumping mechanism is unable to compensate for the losses by induced emission. And so the energy density u_ν and the populations n_1, n_2 stabilise themselves at values such that $\delta P = (\omega/Q) W$.

(d) *The particular case of Fabry–Perot cavities used in lasers.* We confine our discussion to the simple case of plane mirrors. Let r be the reflection coefficient of a mirror.

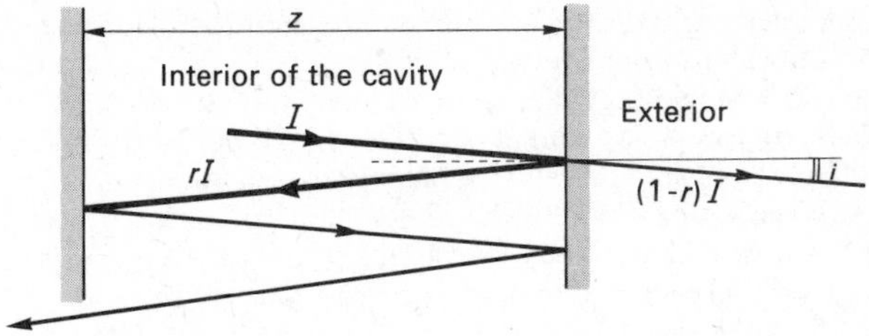

Figure 4.10 Operation of a Fabry–Perot cavity

Each time a beam of light of intensity I within the cavity hits the mirror, a beam of intensity rI is reflected, and a beam of intensity $(1-r)I$ is transmitted to the outside (see figure 4.10; we disregard losses in passing through the mirror). If the phase relations give rise to a system of stationary waves of large amplitude, the intensities of the incident and reflected beams may be added; from this may be found the ratio between the mean energy densities u_{int} and u_{ext} within the cavity and in the external emerging beam respectively

$$\frac{u_{\text{int}}}{u_{\text{ext}}} = \frac{1+r}{1-r} \simeq \frac{2}{1-r} (\approx 100 \text{ if } r \approx 0{\cdot}98)$$

In a single-mode microwave cavity whose dimensions approximate to the wavelength λ, the ratio of the energy densities is of the order of quality factor Q. However, this is not so in the case of a Fabry–Perot cavity whose length z is much greater than λ, as will be shown.

Let us consider an interferometer whose two mirrors have the same reflection coefficient. The power leaving the cavity in the two emerging beams is $P_s = 2\,Scu_{\text{ext}}$ (S is the cross-sectional area of the cavity). The definition of the quality factor in terms of energy allows us to write, neglecting the losses,

$$2Scu_{\text{ext}} = P_s = \omega\frac{W}{Q} = \frac{Szu_{\text{int}}}{Q} = \frac{2\pi c}{\lambda}\frac{Szu_{\text{int}}}{Q}$$

from which we find

$$\boxed{Q = \frac{u_{\text{int}}}{u_{\text{ext}}}\pi\frac{z}{\lambda} = \frac{2\pi}{1-r}\frac{z}{\lambda}}$$

(Choosing for example $z = 50$ cm, then for $\lambda = 0{\cdot}5$ μm and $r = 0{\cdot}98$, one obtains $Q \sim 3 \times 10^8$; twice this value would be obtained if one of the mirrors in the cavity was perfectly reflecting.) The origin of the factor z/λ may readily be understood: it represents the number of periods occurring between two internal reflections of the light beam of the mirrors. That is to say that in z/λ periods, the light beam undergoes a single reflection and therefore each time a corresponding loss of energy.

There is another important consequence of the ratio z/λ. The Fabry–Perot cavity can support a large number of stationary wave modes of frequency or of distinct geometric configurations. In fact the laser can operate only in axial modes corresponding to an angle of incidence normal to the mirrors. If the angle of incidence i is not exactly zero, the rays are diverted from the centre at each reflection. Some of the luminous power is thus lost through the sides of the cavity (see figure 4.10) and the losses are no longer negligible. The ratio u_{int}/u_{ext} and the quality coefficient Q are then less than the values calculated above.

The frequencies of the different axial modes may be calculated from the condition that the length of the cavity contains a whole number of wavelengths ($z=k\lambda$). Hence the difference of frequency $\delta\nu$ between two consecutive axial modes may be found, and is given by $\delta\nu/\nu=\lambda/z$. This should not be confused with the width $\Delta\nu$ of each mode, which is determined by the quality factor: $\Delta\nu/\nu=1/Q$. Further, the ratio between the difference of two neighbouring modes to the width of a mode may be found (see figure 4.11)

$$\frac{\delta\nu}{\Delta\nu}=Q\frac{\lambda}{z}=\frac{2\pi}{1-r}=\pi\frac{u_{int}}{u_{ext}}$$

(Alternatively, these results may be derived from the Airy function, a calculation given in textbooks on wave optics.)

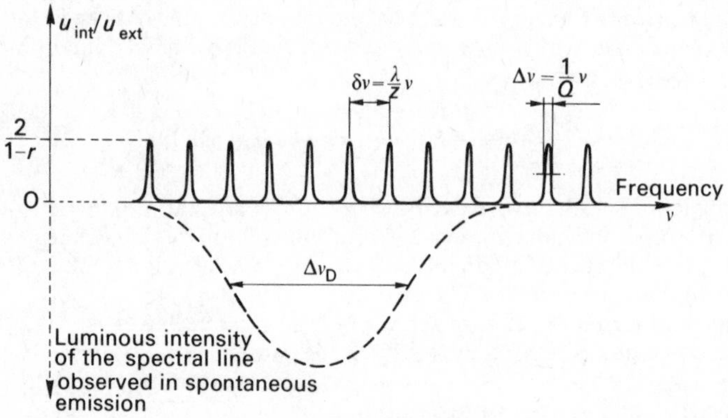

Figure 4.11 Axial modes of a Fabry–Perot cavity (laser)

Generally there are many axial modes whose frequencies lie within the Doppler width $\Delta\nu_D$ of the spectral line used (see figure 4.11). The laser can operate in any one of these modes, and usually operates simultaneously in several of these modes or jumps randomly from one to the other; there are subtle techniques which allow the laser to be locked into one particular mode. In this connection it should be noted that (i) the precision of the oscillation frequency depends on the Fabry–Perot cavity and not on the atomic transition; (ii) when the populations n_1 and n_2 of the energy levels are evaluated, in order to determine the condition for oscillation for example, account need be taken only of the small fraction of atoms whose component of velocity v_z allows them to absorb or emit the exact frequency of that mode (see section 1.3.3).

It is interesting to compare figure 4.11 with figure 4.12 which applies to the operation of a maser. The microwave cavity usually possesses only a single stationary mode having a quality factor Q of the order of 10^4 to 10^5. This single mode therefore has a relative width $\Delta\nu/\nu=1/Q$, much greater than any one of the modes of the Fabry–Perot. On the other hand the atomic transition is not subject to the Doppler effect (see volume 2, chapter 7) and its natural width $\Delta\nu_N$ (see section 4.2.3) is much narrower than the width $\Delta\nu$ of the cavity mode. It should be noted that the atomic

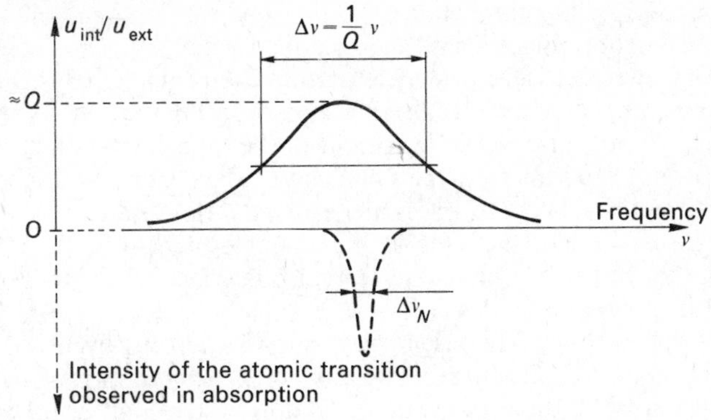

Figure 4.12 Single mode of a microwave cavity (maser)

transition frequency may be slightly modified by interaction with the very strong electromagnetic field inside the cavity.

In both the laser and maser, the wave is generated at the frequency where the amplification is a maximum, in other words at the centre of the narrower of the two lines (instrumental and atomic) shown in figure 4.11 or in figure 4.12. The spectral width of the wave generated (the uncertainty in its frequency) is much less than the width of this narrower line, which means (i) in the case of the laser it is much smaller than the width of a Fabry–Perot mode; (ii) in the case of the maser it is much smaller than the width of the atomic transition.

4.3.6 Properties of Laser Radiation

The special properties of laser radiation arise from the fact that all the photons produced by induced emission from the same initial photon are in the same stationary wave mode

$$\mathscr{E}(\boldsymbol{r},t) = \mathscr{E}_0(\boldsymbol{r},t)\,e^{i(\omega t - \boldsymbol{k}\cdot\boldsymbol{r})} = \mathscr{E}_0(\boldsymbol{r},t)\,e^{i(Wt - \boldsymbol{p}\cdot\boldsymbol{r})/\hbar}$$

On the other hand, there is never more than one photon per mode in the light emitted by an ordinary spectral lamp; each spontaneously emitted photon is represented by a different wave from the others. The existence of a large number of photons in the same mode is characteristic of waves emitted by radio or radar aerials, and the light emitted by lasers bears more resemblance to radiofrequency waves than to light waves emitted by ordinary optical sources. The properties of laser radiation can be classified into two main categories related to two distinct concepts.

(a) Spatial coherence

This term expresses in a general way the breadth of a plane wave, over a surface perpendicular to the direction of propagation, within which the photons are represented by the same wave $\mathscr{E}(\boldsymbol{r},t)$. This surface is called the coherence area.

If the amplifying medium is not homogeneous, light emission occurs instantaneously from only a small region of it (for a ruby laser, observation of very short pulses provides evidence for thin filaments of light), but if the amplifying medium is homogeneous (as in a gas laser) the wave corresponding to the initial photons fills the whole volume of the cavity; the latter is filled by induced photons represented by the same wave. The coherence area occupies the whole width of the beam of light emerging from this laser. In other words, the cross-sectional area of the laser beam is a single-plane wave, described by saying that this beam has complete spatial coherence. This property gives rise to several experimental observations.

(i) *The directionality* of the laser beam may be derived by using the uncertainty principle. Let Oz be the direction of propagation (normal to the Fabry–Perot mirrors); the dimensions δx and δy of the light beam represent the breadth of the wave and the uncertainty principle may be written

$$\delta x\, \delta p_x \approx \hbar \quad \text{and} \quad \delta y\, \delta p_y \approx \hbar$$

As a result of our choice of the Oxyz axes, the momentum of the photons $\boldsymbol{p}$ must on average be parallel to Oz; p_x and p_y are on average zero whereas

$$p_z \simeq p = \hbar k = \hbar \frac{2\pi}{\lambda}$$

The angle α between the z axis and the vector $\boldsymbol{p}$ is also on average zero, and has an uncertainty

$$\delta\alpha \approx \frac{\delta p_x}{p_z} \quad \text{or} \quad \frac{\delta p_y}{p_z}$$

giving either

$$\delta\alpha \approx \frac{\lambda}{\delta x} \quad \text{or} \quad \frac{\lambda}{\delta y}$$

There is another way of expressing the same idea. The precision with which a stationary wave mode may be defined is limited by diffraction due to the finite dimensions δx and δy of the light beam, and when a beam of light is limited by an aperture of dimension δx, the angular aperture of the central diffraction spot is $\delta\alpha \approx \lambda/\delta x$. If the dimensions δx and δy are several millimetres and the wavelength $\lambda \approx 500$ nm, then $\delta\alpha \approx 10^{-4}$ radian.

This directionality can be further increased at the cost of increasing the diameter of the light beam, by letting it pass through a confocal system formed by two lenses (the image focus of the first lens of short focal length f coinciding with the object focus of the second lens of long focal length F; the angular aperture $\delta\alpha$ is then reduced by a factor F/f). This directionality allows light echoes from satellites or from the moon to be observed, and has other telemetric applications.

To obtain a similar directionality with a classical source, a very small circular source would have to be placed at the focus of a long focal length lens, and only a very small fraction of its intensity would be collected. Despite the length of the laser, it behaves as if all its energy came from a virtual point source placed at the focus of a lens.

(ii) *The possibility of focusing virtually to a point* arises directly from the converse of the preceding remark. By passing the laser beam through a lens of focal length f, it forms an extremely small spot of light of dimensions $f\delta\alpha$ at the image focus. If we choose $f = 1$ cm, the total power of the laser is concentrated in a spot having a diameter of the order of one micrometre.

By concentrating all the power of the laser into such a small area, a very high energy density u is obtained, capable even of drilling through a thin sheet of metal (micro-drilling applications). By passing the laser beam into a microscope, in the opposite direction to normal use, cells of any particular kind contained in a microscopic sample can be destroyed one by one.

(iii) *The very high luminous intensity* that can be obtained in a laser beam also results from its directionality: the total energy emitted by a long discharge tube (as in a gas laser) is normally dispersed in all directions, but is now concentrated into a narrow beam within a small solid angle. (With the exception of the small fraction of energy spontaneously emitted in lateral directions.) With a continuously operating gas laser, a luminous power of the order of a watt is often obtained. This is also the order of magnitude of power emitted by a powerful incandescent lamp or electric arc, but in those cases the power is spread out into a large solid angle and distributed throughout the visible spectrum. The frequency of laser light on the other hand is very well defined and therefore an extremely large differential energy density is obtained. Even higher instantaneous powers may be achieved with a pulsed laser. The energy stored in the population inversion is radiated in a single pulse of very short duration, and during the pulse instantaneous luminous powers greater than a megawatt may be obtained.

(iv) *Non-linear optics* has been developed in recent years; it makes use of both the spatial coherence and the high energy density obtained in laser beams. By comparing an example of radiofrequency radiation and an example of optical radiation we have shown (in section 4.1.1) that the classical notion of an electric field has a practical meaning only when the number of photons is sufficiently large; furthermore, all these photons must be in the same mode. Both these conditions are realised in the laser beam, and the electric field $E = \sqrt{(u/\varepsilon_0)}$ can be very high.

The propagation of a wave through a dielectric medium depends on the electric polarisation $\boldsymbol{P} = \chi \boldsymbol{E}$, imposed upon the medium by the electric field of the wave. The dielectric constant may be found from the electric susceptibility

$$\varepsilon_r = 1 + \chi/\varepsilon_0$$

and the refractive index is given by $n = \sqrt{\varepsilon_r}$. However, the linear law $P = \chi E$ is valid only to a first approximation; when the electric field is very high, a saturation effect occurs as in magnetisation and the relation between P and E is no longer linear. Even the ideas of dielectric constant and refractive index become meaningless, and the laws of optics must be reconstructed on a new basis.

One of the more spectacular results of non-linear optics is frequency doubling. By expanding the relationship $P=f(E)$ to a second order of approximation

$$P=\chi_1 E+\chi_2 E^2+\ldots=\chi_1 E_0 \cos \omega t+\chi_2 E_0{}^2 \cos^2 \omega t+\ldots$$

The variation of the electric polarisation as a function of time contains a term in $\cos^2 \omega t=(1+\cos 2\omega t)/2$. This modulation of the polarisation at the frequency 2ω generates a wave of the same frequency. When a dielectric medium is irradiated with a powerful red laser, ultraviolet radiation of twice the frequency may be seen emerging from its midst the dielectric.

(b) Temporal coherence

The duration of the wave train is described under this heading, that is to say the time interval δt over which the photons are represented by the same wave $\boldsymbol{E}(\boldsymbol{r}, t)$. This time interval is called the coherence time. We have explained that the coherence time δt for normal light sources is determined by the Doppler width $\Delta\omega$ of the spectral line (see section 4.2.5)

$$\delta t=1/\Delta\omega \approx 10^{-9}\ \mathrm{s}$$

With a continuously operating laser, a much longer coherence time is obtained. Theoretically it could be almost infinite; this is not attained in practice because of insufficiently stable operating conditions for the laser but nevertheless coherence times of the order of a microsecond may easily be obtained and in favourable conditions, a millisecond may be achieved. This long coherence time is demonstrated experimentally in several ways.

(i) *Observation of interference effects with path differences as large as a hundred metres*, whereas with classical sources interference fringes are blurred once path differences are as large as

$$l=c\,\delta t \approx 1\ \mathrm{m}$$

(ii) *Observation of interference fringes using two independent lasers.* In all the classical interference experiments, two coherent sources are used, which could be either two slits illuminated from the same slit source, or two images of the same slit source; it is usually impossible to observe interference fringes with two independent sources. Such an observation can be made with two independent lasers whose beams meet at a small angle, as long as it is carried out sufficiently quickly.

Over the whole coherence time δt that we assume to be the same for both lasers, the same wave $\mathscr{E}_1(\boldsymbol{r}, t)$ is emitted by the first laser, and the same wave $\mathscr{E}_2(\boldsymbol{r}, t)$ by the second laser. In a region of space where the two beams cross one another, the same resultant wave $\mathscr{E}_1+\mathscr{E}_2$ is obtained during this time δt, the amplitude of which has maxima and minima as a function of the position in space $\boldsymbol{r}$. During the time δt a detector placed at a maximum of $\mathscr{E}_1+\mathscr{E}_2$ receives many photons (a bright fringe) and a detector placed at a minimum of $\mathscr{E}_1+\mathscr{E}_2$

receives very few photons (a dark fringe). An instant later, the waves $\mathscr{E}_1$ and $\mathscr{E}_2$ are replaced by waves $\mathscr{E}_1'$ and $\mathscr{E}_2'$ whose phase difference has changed, so the amplitude of the new resultant wave $\mathscr{E}_1' + \mathscr{E}_2'$ has maxima and minima at different points. This means that the bright fringes and the dark fringes are changing places from one moment to another; in a long observation, equal numbers of photons are recorded at different points over the superposition region of the laser beams and no fringes are observable; but the interference fringes become clearly defined if the measuring time is limited to the coherence time of the two lasers. The experiment was carried out in 1963 by Magyar and Mandel by using an image tube intensifier, similar to that used for taking television pictures in poor lighting conditions; normally the tube is off, but may be turned on momentarily with a very short high voltage pulse.

(iii) *Observation of beats between two laser beams.* This is simply the temporal analogue of spatial interference fringes. When the two waves $\mathscr{E}_1$ and $\mathscr{E}_2$ with slightly different wave vectors $\boldsymbol{k}_1$ and $\boldsymbol{k}_2$ are superposed (at a small angle to one another) the amplitude of the resultant wave $\mathscr{E}_1 + \mathscr{E}_2$ is modulated in space. The spatial periodicity of the interference fringes is given by the vector difference $\boldsymbol{k}_1 - \boldsymbol{k}_2$. In the same way, if two waves $\mathscr{E}_1$ and $\mathscr{E}_2$ with nearly equal frequencies ν_1 and ν_2 are superposed, the amplitude of the resultant wave $\mathscr{E}_1 + \mathscr{E}_2$ is modulated as a function of time at the difference frequency $\nu_1 - \nu_2$; this is the phenomenon of beating.

The beating between two independent lasers is the best method of controlling their frequency stability and measuring their coherence time. The fluctuations in the beat frequency $\nu_1 - \nu_2$ provide us with an order of magnitude of the fluctuations $\delta\nu_1$ and $\delta\nu_2$ in the two frequencies, and from the uncertainty principle the coherence time $\delta t \approx 1/2\pi\delta\nu$ may be deduced. The beat frequencies $\nu_1 - \nu_2$ observed are usually several megahertz, but in well-stabilised conditions beat frequencies of a few kilohertz can be observed. (In order to observe the frequency $\nu_1 - \nu_2$ it should be much greater than the fluctuations $\delta\nu_1$ and $\delta\nu_2$.)

In this context, short-term frequency fluctuations which are related to the coherence time, should be distinguished from long-term frequency drifts which are usually an order of magnitude greater:

(1) Coherence times around a millisecond corresponding to short-term frequency fluctuations of the order of a kilohertz may be measured, implying a relative uncertainty $\delta\nu/\nu \approx 10^{-12}$. This frequency uncertainty represents the spectral width of the radiation emitted by the laser. In highly stabilised conditions, the spectral width of laser radiation therefore can be 100 to 1000 times narrower than the smallest natural widths (usually of the order of 100 kHz to 100 MHz; see section 4.2.3). Until a few years ago, physicists considered the natural width to be the ultimate limit of precision, impossible to exceed; the notion of this so-called limit was destroyed since an assembly of atoms interacting within a cavity collectively emit a wave of much longer duration than that from an isolated atom.

(2) However, the frequency of laser radiation does not remain stable over a long time to this same precision. We have indicated that this frequency is

determined by the Fabry–Perot cavity mode selected (see section 4.3.5); frequency fluctuations arise from fluctuations of the cavity length z: $\delta\nu/\nu = \delta z/z$. (To achieve a frequency stability of 10^{-12} with a cavity 1 m long, the length z must be stabilised so that the fluctuations $\delta z < 1$ pm.)

The method of beats has another practical application: it allows measurement of small frequency changes occurring in a light wave as a result of certain types of scattering; all that is necessary is to beat the scattered light with a small fraction of the incident light. In most scattering experiments, the fraction of incident energy that is scattered laterally is very small, and the incident beam should have a well-defined direction. The intensity and directionality of laser beams have allowed much progress to be made in all kinds of scattering experiments, which would otherwise have been very difficult to perform using classical light sources (Raman scattering, Brillouin scattering and so on).

Comment Interference and beating between two independent lasers is observed even when the light intensity is reduced to such an extent that photons arrive one by one (see section 4.1.3). Let us assume that the total number of photons carried per unit time by the two attentuated laser beams is $N/t \approx 10^6$ to 10^7 s^{-1}, so that the mean time between the arrival of two successive photons $t/N \sim 10^{-7}$ to 10^{-6} s is very much longer than the transit time of the photons through the apparatus $z/c \approx 10^{-8}$ s. In a coherence time $\delta t \approx 10^{-3}$ s, a total number $N \sim 1000$ photons is received, all represented by the same wave $\mathscr{E}_1 + \mathscr{E}_2$ and therefore obeying the same statistical distribution law (a spatial distribution in the case of interference, a temporal distribution in beating). Because of the quantum efficiency (say 10 per cent) of photoelectric detectors only about 100 photons are in fact detected, in practice less, so that investigation of the statistical law becomes more difficult. Nevertheless, it may be verified that the photons are distributed on arrival (at the detector) according to the probability law $|\mathscr{E}_1 + \mathscr{E}_2|^2$ (Pflegor and Mandel, 1967; Radloff, 1968).

Half the luminous energy arises from each of the two lasers. In terms of the emission of photons, we may say that one photon out of two is emitted by each of the two lasers. Since the photons arrive one by one, there is very little chance that a photon emitted by laser No. 1 is coincident with a photon emitted by laser No. 2. However, the photons received in the superposition region of the two laser beams do not 'belong' to either laser No. 1 or laser No. 2, since they are represented by the resultant wave $\mathscr{E}_1 + \mathscr{E}_2$. The photons should be considered as appearing only at the moment of energy exchange processes (emission or absorption), but they vanish while propagating. Only when making a measurement is the wave decomposed into photons (this may be compared with the many experiments described in terms of atomic magnetic moments: the Stern–Gerlach apparatus for example (see section 10.1) imposes an axis of quantisation Oz, and thus allows the component $\mathscr{M}_z$ to be determined).

4.4 Material Particles

Particles whose rest mass is not zero are called material particles (contrast to photon). They include electrons, atoms, nuclei and molecules.

4.4.1 De Broglie Waves

In discussing the operation of lasers, we reviewed the phenomenon of stationary waves that occur within resonant cavities. Although the equations for wave propagation are written in terms of classical physics where all quantities vary continuously, within a cavity solutions of these equations are found that are practically discontinuous. Large amplitudes are obtained only for a particular set of frequencies, corresponding to different stationary modes.

Since wave phenomena could give rise to discontinuous solutions of continuous equations, Louis De Broglie envisaged wave phenomena also explaining the discrete energy levels of atomic states as discontinuous solutions of continuous equations. Accordingly in 1923 he generalised the wave properties of photons to the constituent electrons of the atom. Because waves of wavelength λ had been associated with photons of momentum $p = h/\lambda$ (see chapter 2), he proposed conversely to associate material particles of momentum $p = mv$ with a wave of wavelength $\lambda = h/p$. This idea opened the way two years later to Schrödinger's equation and the development of wave mechanics.

The case of material particles is in fact much more complicated than that of photons because the former more often interact with other particles; Schrödinger's equation involves the law expressing the interaction energy of particles. However, when material particles are isolated, and move in space *without interacting, the analogy with photons can be taken a long way*. The first experimental evidence of this analogy was presented in 1927 by Davisson and Germer when they observed the diffraction of a beam of electrons by a crystal, an effect in every way comparable with X-ray diffraction by the same crystal. It may be shown that non-relativistic electrons accelerated by a potential difference V possess a momentum $p = mv = \sqrt{(2meV)}$, which leads to a wavelength $\lambda = h/p = h/\sqrt{(2meV)}$ of about 0·1 nm when $V = 100$ V. This is about the order of magnitude of X-ray wavelengths and of interatomic distances in a crystal. Other experiments analogous to those of wave optics have been carried out with electron beams: Fresnel diffraction at the edge of a screen (Boersch, 1940), two-beam interferometry (Marton, 1953; the division of amplitude of the wave is obtained by means of a thin crystalline sheet, used analogously to a semi-reflecting mirror). Diffraction experiments have also been carried out with molecules in a molecular beam (Stern, 1932). At the present time, diffraction of neutrons is being used as a method of studying crystal lattices.

4.4.2 Interference between Electrons

We shall describe in detail an experiment the interpretation of which is very simple and the results particularly spectacular. It is an interference experiment by division of a wave front entirely similar to Fresnel's biprism experiment, the diagram of which is shown in figure 4.13: the two very small angle prisms result in the superposition of rays coming from the same source S by bending to the left rays that have been emitted to the right and vice versa.

The corresponding experiment with electrons (Möllenstedt and Düker, 1954; Faget and Fert, 1956) is shown in figure 4.14. A thin metallic wire F is placed in the path of an electron beam which emerges from an electron gun at the point S. Electrons are slightly attracted by the wire; if they travel to the left of the wire, they are deflected to the right, but if they travel to the right of the wire, they are deflected to the left. The electrons that are received at a point

M at a certain distance from the electron gun therefore may have followed either of two different trajectories.

All the electrons have trajectories nearly parallel to the axis Sz of the electron gun, and therefore they all pass very close to the wire F (the dimensions perpendicular to Sz have been greatly exaggerated in figure 4.14). Under these

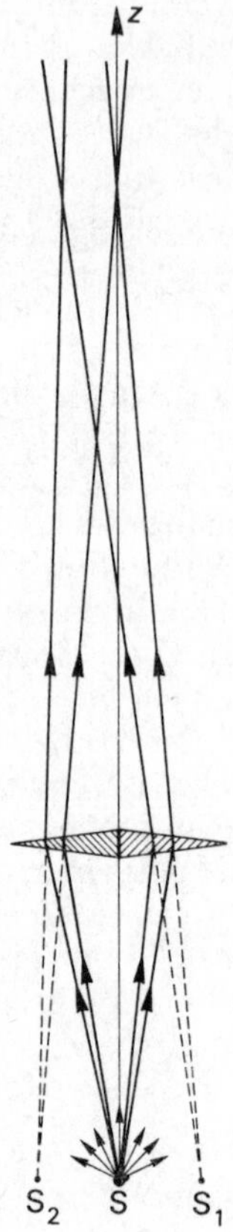

Figure 4.13 Fresnel's biprism

conditions, it may easily be shown that the angle of deviation β is practically the same for all trajectories whatever their distance (small) from the wire F. From this it will be seen that all the trajectories passing to the right of the wire F appear to come from the same points S_1 after deflection such that $SS_1 = FS \times \beta = a\beta$. Similarly the trajectories passing to the left of the wire appear after deflection to come from the point S_2, symmetric to S_1 in relation to S. Thus, everything behaves as if the electrons received at M came from the two images S_1 and S_2 of electron source S, separated by a distance $s = S_1S_2 = 2a\beta$.

To calculate the deviation β, we study the trajectory in polar co-ordinates by taking the wire F as the origin and the axis Fz as a polar axis for measuring the angle θ. The deviation arises from the electric field caused by the wire; this field is radial,

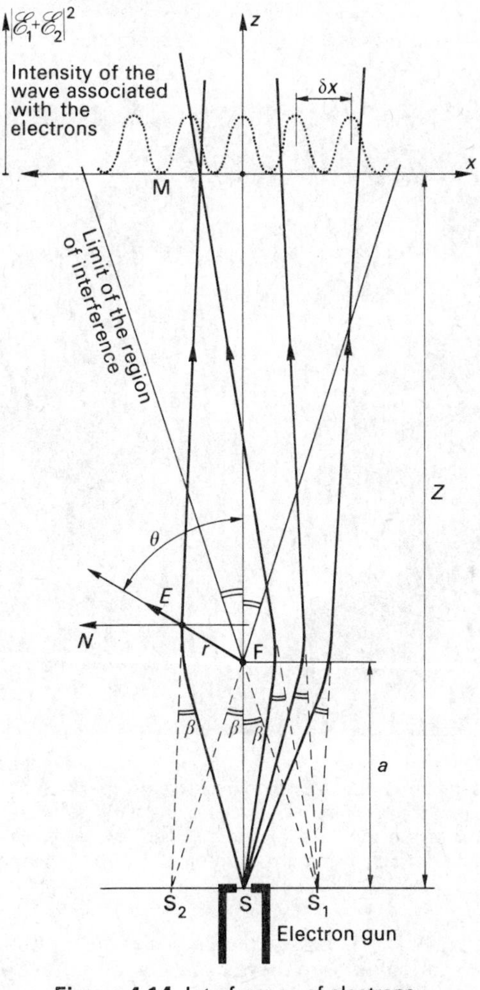

Figure 4.14 Interference of electrons

and because of the cylindrical symmetry around the wire, it depends only on the distance r from the wire. Gauss' theorem enables us to write

$$Er = E_0 r_0 \quad \text{or} \quad E = E_0 r_0 / r$$

(r_0 is the radius of the wire and E_0 the electric field at the surface of the wire). We resolve this field into two components tangential and normal to the trajectory: (a) the sole effect of the tangential component is to change the speed slightly, which we shall ignore (we are dealing with very fast relativistic electrons; the electrons' accelerating potential V within the gun is far greater than the potential U_F of the metallic filament with respect to the outside of the gun): (b) the normal component E_N allows us to calculate the radius of curvature R of the trajectory: $mv^2/R = eE_N$ (where m is the relativistic mass applicable at high velocities). If we assume that the deviation β is very small, the trajectory remains practically parallel to Fz and the

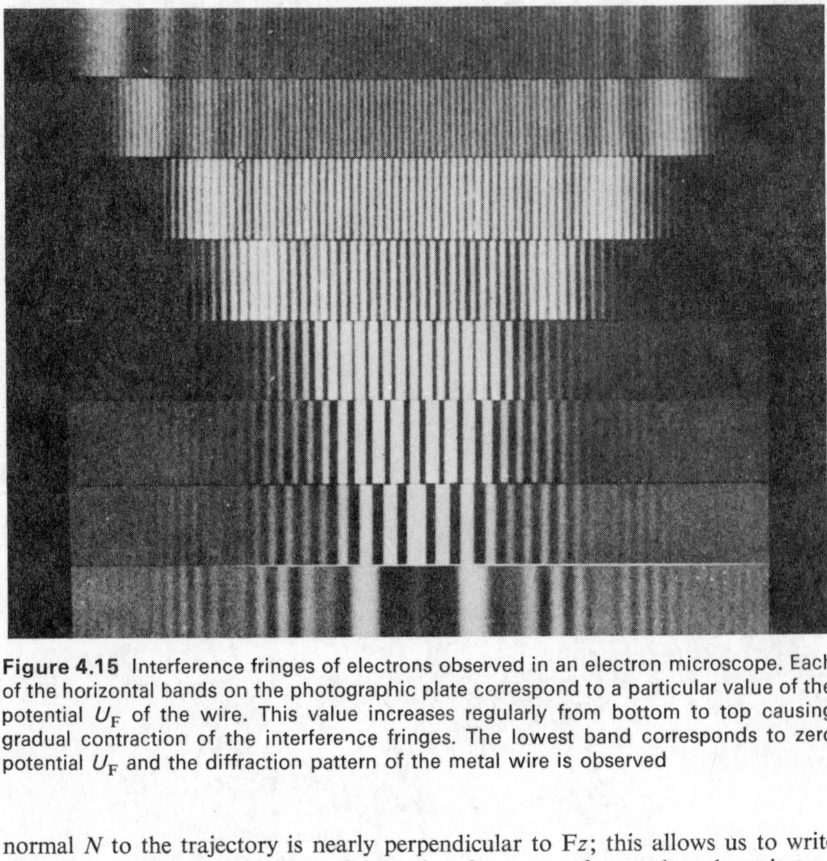

Figure 4.15 Interference fringes of electrons observed in an electron microscope. Each of the horizontal bands on the photographic plate correspond to a particular value of the potential U_F of the wire. This value increases regularly from bottom to top causing gradual contraction of the interference fringes. The lowest band corresponds to zero potential U_F and the diffraction pattern of the metal wire is observed

normal N to the trajectory is nearly perpendicular to Fz; this allows us to write $E_N \simeq E \sin\theta$. From the definition of the radius of curvature, for a path on the trajectory of length dl, the tangent to it turns through an angle

$$d\beta = \frac{dl}{R} = \frac{dl eE_N}{mv^2} = \frac{eE\,dl \sin\theta}{mv^2} = \frac{eE_0 r_0}{mv^2} \times \frac{dl \sin\theta}{r} = \frac{eE_0 r_0}{mv^2} d\theta$$

dθ being the angle through which the radius vector turns in a path dl. The total deviation β is found by integrating this expression from the polar angle $\theta = 0$ to the polar angle $\theta = \pi$, and one obtains $\beta = \pi e E_0 r_0/mv^2$.

β is thus the same for all trajectories within the limits of approximation that we have made.

It we associate a wave with the classical electron trajectory we have to calculate at the point M the superposition of the waves $\mathscr{E}_1(x,y,z,t)$ and $\mathscr{E}_2(x,y,z,t)$ coming from S_1 and S_2 respectively. This problem is identical to that of Young's double slits reviewed in section 4.1.3, and the same solution is obtained: the intensity of the resultant wave $|\mathscr{E}_1+\mathscr{E}_2|^2$ is a sinusoidal function of space co-ordinates. The dotted curve drawn in figure 4.14 represents its variation as a function of x in the plane perpendicular to the axis Sz, at a distance Z from the electron gun. The distance between two consecutive maxima is given by the same formula as in the optical interference experiment

$$\delta x = \lambda \frac{Z}{s} = \lambda \frac{Z}{2a\beta}$$

where $\lambda = h/mv$.

We give the wave associated with the electrons the same probabilistic interpretation as the classical wave associated with photons, and we conclude that the electrons are not distributed equally in space; dark and bright fringes may be recorded on a plate sensitive to electron bombardment (see figure 4.15). When the potential U_F of the wire F is progressively increased, and consequently the angle of deviation β, it may be observed simultaneously that (a) the fringes become narrower since the distance δx becomes smaller; (b) the region occupied by the fringes becomes wider because this is limited by the projection of the lines S_1F and S_2F (see figure 4.14). The wave associated with the electrons is the only way of explaining the formation of these fringes.

Numerical calculations and orders of magnitude

(*a*) *The wavelength λ associated with the electrons.* Once the accelerating potential V within the electron gun is known, the momentum of the electrons may be found

$$p = mv = \frac{1}{c}\sqrt{(W^2 - m_0{}^2c^2)} = m_0 c\sqrt{[(1+eV/m_0c^2)^2-1]}$$

and

$$\lambda = \frac{h}{p} = \frac{\Lambda}{\sqrt{[(1+eV/m_0c^2)^2-1]}}$$

By using the Compton wavelength $\Lambda = h/m_0c = 2{\cdot}4$ pm

$$V = 10^5\ \mathrm{V} \rightarrow eV/m_0c^2 \approx 1/5 \rightarrow \lambda \simeq 3.7\ \mathrm{pm}$$

(*b*) *The angle of deviation of the electron trajectories, $\beta = \pi e E_0 r_0/mv^2$*

$$\frac{v^2}{c^2} = 1 - \left(\frac{m_0}{m}\right)^2; \qquad mv^2 = mc^2 - m_0c^2\frac{m_0}{m} = m_0c^2 + eV - \frac{m_0c^2}{1+(eV/m_0c^2)}$$

so,

$$mv^2 = 2eV\,\frac{1+(eV/2m_0c^2)}{1+(eV/m_0c^2)}$$

On the other hand, a relationship can be established between the electric field E_0 and the potential difference U_F between the wire F and the exit point S of the electron gun, by assuming that the rotational symmetry about F is still valid at S

$$U_F = \int_{r=r_0}^{r=a} E\,dr = \int_{r_0}^{a} E_0 r_0 \frac{dr}{r} = E_0 r_0 \ln \frac{a}{r_0}$$

$$r_0 = 1\ \mu m \quad \text{and} \quad a = 1\ \text{cm} \rightarrow U_F \approx 9 E_0 r_0$$

hence $\beta \approx U_F/V$; $U_F = 10$ V $\rightarrow \beta \approx 10^{-4}$ radian. This value justifies the approximations we have made.

(*c*) *To calculate the distance between fringes,* we choose $Z = 10a$, thus $\delta x = \lambda Z/2\beta a \approx 0{\cdot}2\ \mu m$.

This value is too small for direct observation of the fringes. In practice the arrangement shown in figure 4.14 is placed under an electron microscope which gives a magnified image of a plane perpendicular to the axis S*z* and passing through M. The image is recorded on a sensitive plate and is reproduced in figure 4.15.

The intensity of the electric current *I* carried in a carefully focused beam of electrons can be measured. It is always very small

$$I \approx 10^{-12}\ \text{A}$$

From this we find the number of electrons *N* carried in a time *t*

$$N/t = I/e \approx 10^7\ \text{electrons/s}$$

For an accelerating voltage $V = 10^5$ V, the electrons already have a speed v greater than $c/2$, and consequently their transit time through the apparatus

$$Z/v \approx 2Z/c \approx 10^{-9}\ \text{s}$$

is very short compared with the mean time between the passage of two successive electrons $t/N \approx 10^{-7}$ s.

In electron optics experiments, electrons always arrive one by one (see section 4.1.3): each electron passes through the apparatus in isolation, without interacting with others. It is the wave associated with each electron that divides into two possible paths to the right and to the left of the metallic wire.

This explains why the metallic wire must be extremely thin. Its diameter must be less than the spatial extent of the wave (the coherence area or width of the plane wave) and the latter is very small, because of the large uncertainty in direction of the momentum vector of the electron (see uncertainty principle, section 4.2.4). Diameters of the order of a micron may be obtained with quartz filaments coated with metal by evaporation under a vacuum. The finest filaments can go down to 0·1 μm with the same treatment.

Consequently the entire discussion of photons in this chapter may be generalised to electrons. Just as for photons, it is impossible to know whether the electron has passed to the right or to the left of the metallic wire; all we can say is that the associated wave has passed both sides of the wire.

Comment We shall not, however, try to generalise what has been said about lasers (see section 4.3.6). Interference effects between independent lasers are observable for a very special reason: the existence of a large number of photons in the same mode. This property arises because photons are particles called 'bosons' which obey Bose–Einstein statistics. Electrons on the other hand are 'fermions' obeying Fermi–Dirac statistics: it is impossible for two of them to be in the same state and to be represented by the same wave function.

4.4.3 Application of the Waves Associated with Electrons and Neutrons to the Study of Crystal Structure

The regularly arranged atoms in a crystal form a diffraction grating for an incident wave. All the effects observed result from scattering of the incident wave by different atoms in the crystal, followed by interference between the waves scattered by different atoms. The quantitative study of the interference uses the Bragg law: there will be constructive interference when the direction of scattering makes the same angle θ with a given lattice plane as the direction of the incident wave (figure 4.16). In addition, the angle θ must satisfy the relation

$$2d \sin\theta = k\lambda$$

d being the distance between two lattice planes and k an integer.

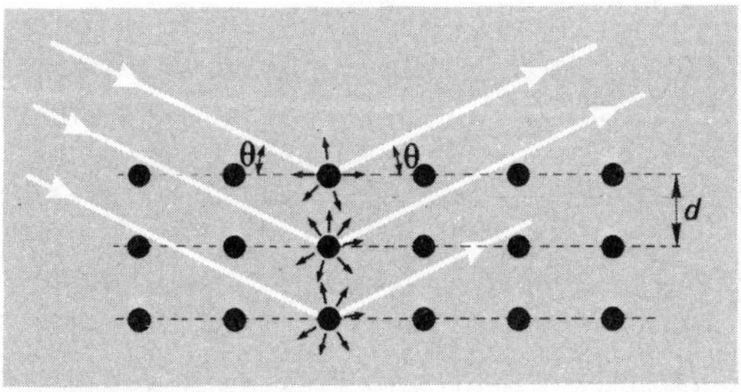

Figure 4.16 Diffraction of waves by a crystal lattice

Kinetic energy of an electron in eV	Wavelength in pm
1	1223
10	387
100	122
1000	38
10 000	12
100 000	3

Comment For an exact calculation, the fact that the waves propagate in a medium with a certain refractive index n must be taken into account; the angle θ and wavelength λ will then have slightly different values from those in a vacuum.

It will be noted that the intensity of interference phenomena is a function of the scattering cross-section of the incident radiation by different atoms. If the crystal contains different types of atoms arranged in a complex pattern, analysis of the interference patterns for different positions of the crystal gives information about the crystal structure. For convenient study of the patterns the wavelength of the incident radiation must, from the Bragg relation, be of the order of d, in other words of the order of 100 pm. The table in figure 4.16 gives the wavelengths associated with electrons of various energies, and we suggest that the reader confirms that a neutron, whose kinetic energy is thermal at the ambient temperature ($\frac{1}{2}mv^2 \approx kT$), has a wavelength of about 100 pm. Such neutrons can be obtained from nuclear reactors (atomic piles) in the form of a relatively intense beam.

The wavelengths associated with beams of thermal neutrons or electrons, as well as the wavelengths of X-rays, may have very similar values; but important differences arise between them in the scattering of these radiations by atoms, and in their penetration of matter. This gives rise to widely different applications in the study of the solid state.

(*a*) *Electron diffraction.* Electrons have little penetrating ability. Two main aspects can be isolated:

(i) Electrons with an energy of 50 keV penetrate about 100 nm into matter. The information that can be obtained differs little from results obtained using X-rays. Except for special cases where information is required concerning the surface layer of a specimen this cumbersome experimental technique is used relatively little. It is now mainly associated with electron microscope equipment allowing physicists to make a quick analysis of very small specimens.

(ii) Some years ago advances in experimental techniques permitted the development of the diffraction of slow electrons (10 to 1000 eV). Penetration is negligible and diffraction occurs only at the first atomic layer of the crystal which behaves as a plane diffraction grating. Owing to the possibility of achieving very high vacua (10^{-10} torr) allowing surface layer contamination to be avoided, reproducible results have been obtained for the diffraction of slow electrons; this has come to be an essential tool for the study of surfaces.

(*b*) *Neutron diffraction.* Neutrons, like X-rays, penetrate matter very easily, because they are insensitive to electrostatic forces. Therefore they give information about volume structure. However, there are very important differences between neutron diffraction and X-ray diffraction:

(i) The scattering of a neutron by an atom is essentially a nuclear phenomenon and often varies considerably from one element to another whereas X-ray scattering is related to the number of electrons in the atom. When a crystal is formed from different atoms of nearly equal atomic number, it is impossible to distinguish them by X-ray studies, but these different atoms can scatter neutrons very distinctively, and therefore in general neutron diffraction will give more information.

(ii) X-Ray scattering by light atoms having few electrons is very weak. On the other hand, the scattering cross-section for neutrons by certain light nuclei like hydrogen is large, and in these cases neutron diffraction proves to be more advantageous; it is a very important technique in the study of the structures of organic compounds.

(iii) To the process of nuclear scattering may be added a magnetic phenomenon resulting from the interaction between the magnetic moment of the neutron and the magnetic moment of the atom if this is not zero. The scattering pattern of neutrons by an atom will have a symmetry related to the direction of the magnetic moment of the atom. If the specimen studied has a *magnetic order*, the symmetry observed in individual scattering patterns will be retained in the diffraction pattern. Study of the latter reflects the magnetic order and provides the solid-state physicist with a valuable method of investigating magnetic materials.

Part 2

Planetary Model and Principal Quantum Number

5

The Classical Planetary Model of the Atom

The first part of this book was devoted to the study of energy exchanges between electromagnetic radiation and the atom considered as a whole. In this chapter we shall start to consider problems relating to the internal structure of atoms.

This chapter will be devoted to the description of atomic structure in terms of newtonian classical mechanics. We shall see how Rutherford's experiment helps to prove the existence of a very small positively charged nucleus. Then from this knowledge we shall draw conclusions concerning the motion of the internal electrons. However, we shall first review some general laws of newtonian mechanics which will be used in these two discussions. The reader familiar with these ideas could omit the next section.

5.1 The General Two-body Problem

We shall now consider the motion of two particles exerting reciprocal interaction forces on one another. This is what applied mathematicians call the two-body problem. More exactly, let us postulate that the two forces f_1 and f_2 exerted by the two particles on one another have the following characteristics:

(i) they are equal and opposite (principle of action and reaction)

$$f_1 + f_2 = 0$$

(ii) they are parallel to the radius vector joining the two particles;
(iii) their magnitude depends only on the separation r of the two particles.

5.1.1 Reduction to the Centre of Mass

As is usual for completely isolated systems the solution of the problem can be simplified by separating it into two distinct problems: (a) the motion of the centre of mass in an absolute or laboratory frame of reference; and (b) the motion of the system about the centre of mass.

(a) The centre of mass C of the two particles M_1 and M_2 is defined by

$$(m_1 + m_2)\,\mathbf{OC} = m_1\,\mathbf{OM}_1 + m_2\,\mathbf{OM}_2$$

By differentiating with respect to time, the relation between the velocity $\boldsymbol{V}$ of the centre of mass C and the velocities $\boldsymbol{V}_1$ and $\boldsymbol{V}_2$ of the points M_1 and M_2 can be obtained in the laboratory frame

$$(m_1 + m_2)\,\boldsymbol{V} = m_1\,\boldsymbol{V}_1 + m_2\,\boldsymbol{V}_2 = \boldsymbol{p}$$

where $\boldsymbol{p}$ is the total momentum of the system; by virtue of the equation for the rate of change of momentum, it is constant since the system is isolated (by differentiating again with respect to time, one obtains $\mathrm{d}\boldsymbol{p}/\mathrm{d}t = \boldsymbol{f}_1 + \boldsymbol{f}_2 = 0$). Hence C is in uniform motion, and can be taken as the origin of a new galilean frame of reference, in motion with respect to the laboratory frame.

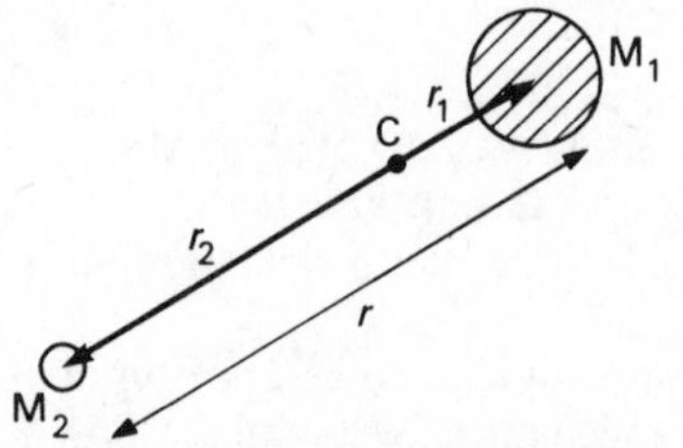

Figure 5.1 Centre of mass

(b) With the centre of mass C as the origin (figure 5.1), let $\boldsymbol{r}_1$, $\boldsymbol{r}_2$ and $\boldsymbol{v}_1$, $\boldsymbol{v}_2$ be the radius vectors and the velocities of the two particles. From the definition of C

$$m_1\boldsymbol{r}_1 + m_2\boldsymbol{r}_2 = 0$$

and by putting

$$\mathbf{M}_1\mathbf{M}_2 = \boldsymbol{r} = \boldsymbol{r}_2 - \boldsymbol{r}_1$$

it is found that

$$\begin{cases} \boldsymbol{r}_1 = -\dfrac{m_2}{m_1 + m_2}\boldsymbol{r} \\[2ex] \boldsymbol{r}_2 = +\dfrac{m_1}{m_1 + m_2}\boldsymbol{r} \end{cases}$$

Thus the problem in six variables (the three components of $\boldsymbol{r}_1$ and the three components of $\boldsymbol{r}_2$) has been reduced to a problem in three variables (the three components of $\boldsymbol{r}$).

The fundamental equation of dynamics gives

$$f_2 = m_2 \frac{d^2 \boldsymbol{r}_2}{dt^2} = \frac{m_1 m_2}{m_1 + m_2} \frac{d^2 \boldsymbol{r}}{dt^2} = \mu \frac{d^2 \boldsymbol{r}}{dt^2}$$

(the same result is obtained for the particle M_1 by writing $f_1 = -f_2$).

In other words, *our two-body problem is reduced to the simpler problem of the motion of a single particle of mass* μ under the influence of a central force f_2 depending on the distance r between the centre of force and the particle. μ is called the *reduced mass* such that

$$\mu = \frac{m_1 \, m_2}{m_1 + m_2} \quad \text{or} \quad \frac{1}{\mu} = \frac{1}{m_1} + \frac{1}{m_2}$$

From the second equation, it may be seen that μ is always less than the smaller of the two masses m_1 and m_2. The variation of μ as a function of m_2, with m_1 constant, is shown below

m_2	0	m_1	∞
μ	0	$m_1/2$	m_1

If one of the particles is much lighter than the other, for example $m_2 \ll m_1$, the reduced mass μ is nearly equal to the smaller of the two masses, and the centre of mass C nearly coincides with the heavier of the two particles. The reduced motion is practically identical to the actual motion of the light particle around the nearly stationary heavy particle.

Comment I The velocity $\boldsymbol{v}$ of the reduced motion is equal to the relative velocity of the two particles

$$\boldsymbol{v} = d\boldsymbol{r}/dt = \boldsymbol{v}_2 - \boldsymbol{v}_1 = \boldsymbol{V}_2 - \boldsymbol{V}_1$$

(the velocities $\boldsymbol{v}_2$ and $\boldsymbol{v}_1$ are in the centre of mass frame, $\boldsymbol{V}_1$ and $\boldsymbol{V}_2$ in the laboratory frame).

Comment II In the following sections we shall use the angular momentum vector $\boldsymbol{\sigma}$ of a material system. This vector is calculated at a point, and depends on the choice of this point. If it is calculated at the centre of mass C, it has the same value in both the centre of mass and laboratory frames

$$\boldsymbol{\sigma}(C) = m_1 \boldsymbol{r}_1 \times \boldsymbol{v}_1 + m_2 \boldsymbol{r}_2 \times \boldsymbol{v}_2 = m_1 \boldsymbol{r}_1 \times \boldsymbol{V}_1 + m_2 \boldsymbol{r}_2 \times \boldsymbol{V}_2$$

(since $\boldsymbol{V}_1 - \boldsymbol{v}_1 = \boldsymbol{V}_2 - \boldsymbol{v}_2 = \boldsymbol{V}$ and $m_1 \boldsymbol{r}_1 + m_2 \boldsymbol{r}_2 = 0$).

This total angular momentum $\boldsymbol{\sigma}(C)$ calculated at the centre of mass is equal to the angular momentum of the reduced motion. The relations between the velocities $\boldsymbol{v}_1$ and $\boldsymbol{v}_2$ and the velocity $\boldsymbol{v}$ of the reduced motion are identical to the relations already written between the radius vectors $\boldsymbol{r}_1$, $\boldsymbol{r}_2$ and the radius vector $\boldsymbol{r}$ of the reduced motion (one need only derive the latter in order to obtain the former)

$$\boldsymbol{\sigma}(C) = m_1 \boldsymbol{r}_1 \times \boldsymbol{v}_1 + m_2 \boldsymbol{r}_2 \times \boldsymbol{v}_2 = m_1 \left(\frac{m_2}{m_1 + m_2} \right)^2 \boldsymbol{r} \times \boldsymbol{v} + m_2 \left(\frac{m_1}{m_1 + m_2} \right)^2 \boldsymbol{r} \times \boldsymbol{v}$$

or

$$\sigma(\mathrm{C}) = \frac{m_1 m_2}{m_1 + m_2} \boldsymbol{r} \times \boldsymbol{v} = \mu \boldsymbol{r} \times \boldsymbol{v}$$

5.1.2 First Integrals of the Motion Under a Central Force

(*a*) *Application of the equation for the rate of change of angular momentum.* Let $\boldsymbol{\sigma}$ be the angular momentum vector, following the notation of a number of mechanics texts. In the whole of this chapter and in the following chapter σ will designate the magnitude of the vector $\boldsymbol{\sigma}$ and care must be taken not to confuse it with the cross-sections defined in chapter 3. With this notation the angular momentum equation for our problem may be written

$$\frac{\mathrm{d}\boldsymbol{\sigma}}{\mathrm{d}t} = \boldsymbol{r} \times \boldsymbol{f} = 0$$

since $\boldsymbol{f}$ is parallel to $\boldsymbol{r}$. Hence the angular momentum vector $\boldsymbol{\sigma} = \mu \boldsymbol{r} \times \boldsymbol{v}$ is a constant of the motion and so (i) the direction of $\boldsymbol{\sigma}$ is fixed and therefore the motion is in a plane; (ii) the magnitude σ of the vector $\boldsymbol{\sigma}$ is a constant.

Having defined the position of the particle in the plane by its polar angle θ one can write $\sigma = \mu r^2 \mathrm{d}\theta/\mathrm{d}t$ and thereby deduce *the areal law*

$$\boxed{r^2 \frac{\mathrm{d}\theta}{\mathrm{d}t} = 2 \frac{\mathrm{d}S}{\mathrm{d}t} = \frac{\sigma}{\mu}}$$

where S is the area swept out by the radius vector. It will be noted that $\mathrm{d}\theta/\mathrm{d}t$ has a constant sign; the rotational motion always has the same direction about the centre of force C.

(*b*) *Introduction of the law for the dynamic force.* Since the force $\boldsymbol{f}_2$ depends only on the distance r, it is derived from a potential energy $W(r)$; in other words a function $W(r)$ can always be found such that $\boldsymbol{f}_2 = -\mathrm{grad}\, W(r)$.

In polar co-ordinates and by applying the areal law, the kinetic energy may be written

$$\frac{1}{2} \mu v^2 = \frac{\mu}{2} \left[\left(\frac{\mathrm{d}r}{\mathrm{d}t} \right)^2 + r^2 \left(\frac{\mathrm{d}\theta}{\mathrm{d}t} \right)^2 \right] = \frac{\mu}{2} \left(\frac{\mathrm{d}r}{\mathrm{d}t} \right)^2 + \frac{\sigma^2}{2\mu r^2}$$

The total energy E is a constant of the motion

$$\frac{1}{2} \mu v^2 + W(r) = \frac{\mu}{2} \left(\frac{\mathrm{d}r}{\mathrm{d}t} \right)^2 + \frac{\sigma^2}{2\mu r^2} + W(r) = E$$

hence

$$\frac{\mathrm{d}r}{\mathrm{d}t} = \pm \sqrt{\left[\frac{2}{\mu} (E - W) - \frac{\sigma^2}{\mu^2 r^2} \right]}$$

Generally, before the temporal variation is found, the trajectory of the particle is determined, in other words a direct relationship between the two co-ordinates θ and r, by writing

$$\frac{\mathrm{d}\theta}{\mathrm{d}r}=\frac{\mathrm{d}\theta}{\mathrm{d}t}\cdot\frac{\mathrm{d}t}{\mathrm{d}r}$$

giving

$$\boxed{\frac{\mathrm{d}\theta}{\mathrm{d}r}=\pm\frac{\sigma}{\mu r^2\sqrt{\left\{\dfrac{2}{\mu}[E-W(r)]-\dfrac{\sigma^2}{\mu^2 r^2}\right\}}}}$$

The trajectory is found by a simple integration as a function of r once $W(r)$ is known. The values of the two constants σ and E are imposed by the initial conditions. The $\pm$ sign also depends on the initial conditions, but it can change during the motion.

The range of possible values of the distance r is encompassed by the values which make the quantity under the square root positive

$$\boxed{W(r)+\frac{\sigma^2}{2\mu r^2}\leqslant E}$$

When the square root is zero, $\mathrm{d}\theta/\mathrm{d}r$ becomes infinite, so that the inverse $\mathrm{d}r/\mathrm{d}\theta=0$; r goes through an extremum and therefore its derivative $\mathrm{d}r/\mathrm{d}\theta$ changes sign. By choosing the radius vector corresponding to this extremum as the polar axis, the change from θ to $-\theta$ does not change the value of r (since the $\pm$ sign in the equation for $\mathrm{d}\theta/\mathrm{d}r$ changes as well): the radius vector corresponding to an extremum of r is an axis of symmetry for the trajectory.

Depending on the circumstances the inequality framed above limits the values of r to a restricted interval or alternatively allows r to become infinite. In the case where r becomes infinite, $\mathrm{d}\theta/\mathrm{d}r=0$ so that r tends towards infinity at a particular value of θ: the trajectory has an asymptote parallel to this particular direction.

By choosing the arbitrary additive constant in the potential energy $W(r)$ so that the latter is zero where r is infinite, in accordance with current practice, then the first term of the above inequality is zero for infinite r. The trajectory exists at infinity only if the constant E, the total energy, is positive; it does not exist if E is negative.

5.2 The Rutherford Scattering Experiment

The experiment carried out in 1911 by the New Zealand physicist Rutherford was one of the most important events in the history of atomic physics because it provided proof of the existence of atomic nuclei. The experiment consisted of bombarding a thin metallic foil with α-particles emitted by a radioactive source: the α-particles passed through the metal foil but were deflected in all directions; they are described as having been scattered by passing through the metal foil.

Many of the ideas introduced to describe this experiment are valid for other scattering experiments carried out by nuclear or high-energy physicists, with projectiles other than α-particles and even with targets other than nuclei. The experiment therefore has more than a historic importance, and justifies a detailed description.

Accordingly, we shall start by presenting general concepts applicable to all scattering experiments whatever the law of interaction force concerned. We shall show how the theory of a scattering experiment may be derived whatever the function $W(r)$ representing the interaction energy of two particles as a function of their separation r. Only then shall we apply this to the particular case of the Rutherford experiment.

The interpretation of a scattering experiment involves two successive steps:

(i) calculation of the trajectory of a 'projectile' particle, taking into account its interaction with a single 'target' particle in a particular position (we shall assume throughout that the 'target' particle is stationary before interacting with the projectile);

(ii) statistical analysis of all possible particular positions, in order to obtain an averaged result which may be compared with experiment.

5.2.1 Deflection of a Projectile Passing near a Single Target

This is a two-body problem and everything said in the preceding section applies. All that is necessary is to define the initial conditions.

Let $\boldsymbol{v}_\infty$ be the initial velocity of the projectile in the laboratory frame. Since the target is initially stationary in this same frame, $\boldsymbol{v}_\infty$ is also the relative velocity $\mathrm{d}\boldsymbol{r}/\mathrm{d}t$ of the two particles before their interaction. It is also, therefore, the initial velocity of the reduced motion (see section 5.1). The motion occurs in the plane determined by the centre of mass C and the velocity $\boldsymbol{v}_\infty$.

It is also necessary to define the relative positions of the two particles before their interaction. If no force were exerted on the projectile, it would have a straight line trajectory D, which would pass the target at a minimum distance b (see figure 5.2); b is called the *impact parameter* and from it the two constants of the motion, the angular momentum σ and energy E may be found

$$\sigma = \mu b v_\infty \quad \text{and} \quad E = \tfrac{1}{2}\mu v_\infty{}^2$$

(imposing the convention that the potential energy $W(r)$ is zero at infinity). The trajectory starts from infinity and has the straight line D as an asymptote.

The derivative $\mathrm{d}r/\mathrm{d}\theta$ can change sign only when the square root in its numerator is zero; the distance r therefore decreases to the minimum value

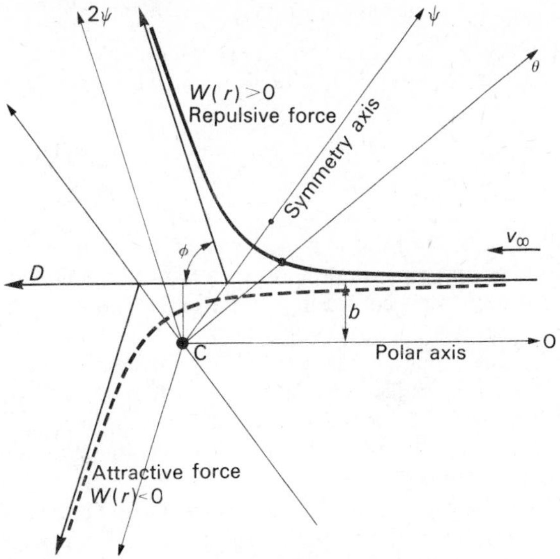

Figure 5.2 Rutherford's scattering experiment—case of a single projective

ρ, which makes the square root zero; it is given by

$$W(\rho) = E - \frac{\sigma^2}{2\mu\rho^2} = \frac{\mu v_\infty^2}{2}\left(1 - \frac{b^2}{\rho^2}\right)$$

Comment If $W(r)$ is positive (repulsive force) $\rho > b$, while if $W(r)$ is negative (attractive force), $\rho < b$.

During this time, the polar angle will have changed by an amount

$$\psi = \int_\infty^\rho \frac{d\theta}{dr}\, dr = \int_\rho^\infty -\frac{d\theta}{dr}\, dr$$

so

$$\boxed{\psi = \int_\rho^\infty \frac{b\, dr}{r^2\sqrt{\left[1 - \frac{b^2}{r^2} - \frac{2W(r)}{\mu v_\infty^2}\right]}}}$$

This integral allows ψ to be calculated as a function of b and v_∞, once the function $W(r)$ is known. (With the conventions adopted in figure 5.2, $d\theta/dr$ is negative and ψ is positive.)

At the end of the preceding section, we saw quite generally that the trajectory is symmetric with respect to the radius vector corresponding to an extremum of r. After passing through the minimum distance ρ from the centre of force, the trajectory sets off towards a new asymptote at infinity, symmetric to the line D, that is to say in a direction having a polar angle 2ψ.

Hence, in relation to its initial direction D, *the projectile will have been deflected through an angle*

$$\boxed{\phi = \pi - 2\psi}$$

(see figure 5.2), and it will have regained a velocity equal in magnitude to its initial velocity. The integral framed above allows us finally to calculate the angle of deflection ϕ in terms of the impact parameter b, the velocity v_∞ and the function $W(r)$.

The trajectory thus calculated is that of the reduced motion in the centre of mass frame. The actual trajectory of the projectile in the laboratory frame is little different from this if the projectile is much lighter than the target, because the target and the centre of mass are nearly coincident and stationary. This is precisely the situation in Rutherford's experiment.

This central force problem is in fact an elastic collision problem, in which the law of interaction between the projectile and the target is known. The general results obtained by writing the conservation equations can be applied to it: if the target is much heavier than the projectile, there is hardly any transfer of kinetic energy between the two particles (see section 2.5 and appendix 4).

5.2.2 Statistics of an Assembly of Particles—Differential Cross-Section

The statistics of all the possible particular positions have still to be discussed. This statistical analysis must be carried out for both the assembly of projectiles and the assembly of targets.

(*a*) *Statistics of the assembly of projectiles assuming that there is only one target particle.* In an actual experiment it is possible to obtain nearly mono-energetic projectiles all having the same initial velocity v_∞ (the same magnitude and direction). However, these projectiles are distributed at random within a beam of a certain width. In one second, the number of projectiles crossing an elementary area $\mathrm{d}S$ of the cross-section of the beam, is proportional to $\mathrm{d}S$; it is equal to $\mathscr{D}\,\mathrm{d}S$. The coefficient of proportionality $\mathscr{D}$ is the number of projectiles crossing unit area per unit time; it is a measure of the intensity of the beam of projectiles (see section 3.1.1).

Let us consider all the projectiles whose incident trajectories (before scattering) pass through an annulus centred on the target, contained between circles of radii b and $b + \mathrm{d}b$, and of area $\mathrm{d}\sigma = 2\pi b\,\mathrm{d}b$ (see figure 5.3). On account of the relation between the impact parameter b and the angle of deflection ϕ, all the projectiles are deflected into directions having an angle between ϕ and $\phi + \mathrm{d}\phi$, within the solid angle $\mathrm{d}\Omega = 2\pi \sin\phi\,\mathrm{d}\phi$ contained between two cones of semi-vertical angles ϕ and $\phi + \mathrm{d}\phi$.

By generalising the idea of a cross-section (see section 3.1.1) $\mathrm{d}\sigma$ may be described as the cross-section for the partial scattering of projectiles within an angle between ϕ and $\phi + \mathrm{d}\phi$.

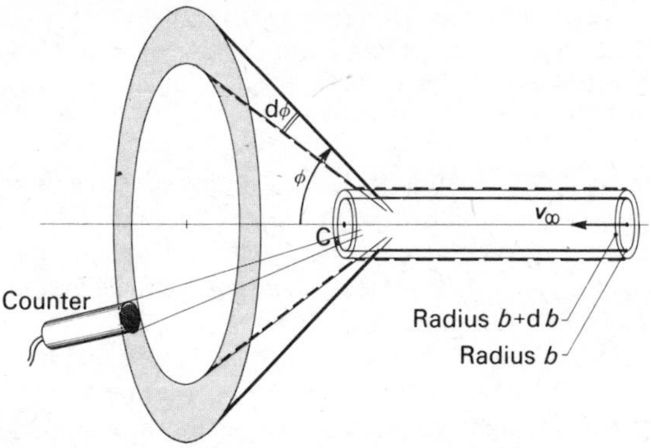

Figure 5.3 Scattering of a beam of projectiles by a target C (only the asymptotes of the trajectories are shown)

This cross-section $d\sigma = 2\pi b\, db$ is proportional to the angular aperture $d\phi$ of the partial process, or alternatively to the solid angle $d\Omega$, which is itself proportional to $d\phi$. A parameter independent of the angular aperture $d\phi$ is obtained by dividing $d\sigma$ by $d\Omega$, and it characterises the scattering process in directions of angle ϕ

$$\boxed{\frac{d\sigma}{d\Omega} = \frac{b}{\sin\phi} \times \frac{db}{d\phi}}$$

This parameter is called the *differential cross-section* of the scattering process for trajectories deflected through an angle ϕ. If the function $W(r)$ representing the interaction energy of the particles is known, the relation between b and ϕ can be calculated (see section 5.2.1); the formula framed above then permits this differential cross-section to be calculated.

Comment In this chapter it is impossible to confuse the magnitude of the angular momentum with the cross-section, since the latter is always in differential form.

Once the intensity of projectiles $\mathscr{D}$ crossing a unit area is known, the number of projectiles passing through the cross-section $d\sigma$ in one second is given by

$$dN = \mathscr{D}\, d\sigma = \mathscr{D}(d\sigma/d\Omega)\, d\Omega$$

The introduction of the differential cross-section clearly shows the proportionality between the solid angle $d\Omega$ and the number of projectiles scattered into it. Initially all these projectiles are distributed uniformly within an annulus of radius b; finally, therefore, they are also distributed uniformly within the solid angle included between the two cones of semi-vertical angles ϕ and $\phi + d\phi$. The same formula may therefore be used when only a fraction of a solid angle is considered; it gives the number of particles received in one second by a counter, provided that $d\Omega$ is identified as the solid angle subtended by the target C at the entrance window of the counter (see figure 5.3).

Comment Strictly speaking, the asymptotes of the final trajectories (after scattering) are not concurrent, but our concern is with those that pass within a very small distance (less than 100 pm) of the target C, which is altogether negligible compared with the distance of the counter.

(*b*) *Statistics of the assembly of targets.* Until now we have assumed that there is only one target particle but a large number of them exist in an actual experiment. If a beam of projectiles of cross-sectional area S passes through a scattering foil of thickness l, containing n target particles per unit volume, the total number of target particles is nlS.

However, owing to the relatively small distance over which the interaction forces act, and to the rather high velocity v_∞ of the projectiles, the latter must pass very close to a target particle before they are appreciably deflected; in other words, their impact parameter b must be very small. It is then possible to reduce the total number of target particles sufficiently so that (i) a projectile deflected by one target has only a very small probability of being deflected by a second target; and (ii) different targets do not mask one another. (These conditions are achieved by using extremely thin foils, whose thickness l is of the order of several tens of micrometres, that is about 1000 interatomic distances.)

Under these conditions, each target acts by itself without interfering with other targets; the total number of scattered particles is obtained by simply adding the identical contributions of each of the targets. It is sufficient to multiply the results obtained with one target by the number of targets nlS. That is to say, in an actual experiment, the number of particles received by the counter in one second is

$$\mathrm{d}N = nlS\mathscr{D} \times \frac{\mathrm{d}\sigma}{\mathrm{d}\Omega}\,\mathrm{d}\Omega$$

In other words, knowing that the entrance window of the counter subtends a solid angle $\mathrm{d}\Omega$ at the scattering foil, and knowing the number $\mathrm{d}N$ of particles received by the counter in one second, an experimental measurement of the differential cross-section may be found

$$\boxed{\frac{\mathrm{d}\sigma}{\mathrm{d}\Omega} = \frac{1}{nlS\mathscr{D}} \times \frac{\mathrm{d}N}{\mathrm{d}\Omega}}$$

To summarise, we have obtained two different expressions enabling the differential cross-section to be calculated: the first allows a theoretical calculation from an assumed law for the potential energy $W(r)$, and the other allows a practical determination from an experimental measurement of the number of scattered particles. The validity of the hypothesis concerning the interaction energy $W(r)$ may be tested by comparison of the theoretical and experimental values thus obtained. *The relation between theory and experiment is thus established through the differential cross-section.*

5.2.3 The Particular Example of a $1/r$ Potential Energy—the Rutherford Experiment

The experiment carried out by Rutherford was designed to investigate the electric charge distributed within atoms. At the beginning of this century something was already known of the atomic number Z (= the number of electrons within each atom). To ensure overall electric neutrality of an atom, it was also necessary to assume that it contained a positive charge Ze (e is the elementary charge $= 1{\cdot}60 \times 10^{-19}$ C). However, different 'models' could be envisaged for the distribution of this positive charge within the atom. Rutherford proposed the α-particle scattering experiment in order to differentiate between these different models.

α-Particles are helium nuclei of atomic mass 4, carrying two elementary positive charges. In the interaction between an α-particle and an electron, the centre of mass is practically coincident with the α-particle, which is about 7000 times heavier than the electron; by reason of the argument given in section 5.1, the centre of mass is in uniform motion, so that the motion of the α-particles is virtually unaffected by the interaction with electrons.

However, suppose that the mass of the atom resides predominantly in its positive charge. In the interaction between the positive charge of a heavy atom (mass number ≈ 200) and an α-particle, the centre of mass is almost coincident with the centre of gravity of the atom, and the motion of the α-particle is similar to the reduced motion around a fixed centre of force. In conclusion, by virtue of its intermediate mass, the α-particle's motion is sensitive only to the positive charge of the atom.

The electrostatic potential energy $W(r)$ differs according to the distribution law of this positive charge, giving rise to different calculated cross-sections. Comparison between measured and calculated cross-sections allows the actual function $W(r)$, and thus the distribution law of the positive charge, to be determined.

Here we shall confine ourselves to the calculation corresponding to the Rutherford model, by assuming the existence of a positive point charge. In this model, the electrostatic potential energy between the point charge Ze of the nucleus and the point charge $2e$ of the α-particle, both of them positive, is inversely proportional to their separation r

$$W(r) = \frac{1}{4\pi\varepsilon_0} \times \frac{2Ze^2}{r} = \frac{C}{r}$$

We can then complete the general calculation started in section 5.2.1. The ratio

$$\frac{W(r)}{\mu v_\infty^2} = \frac{C}{\mu v_\infty^2} \times \frac{1}{r}$$

comes into the calculation of the angle between the asymptote and the symmetry axis and the rest of the calculation is simplified by introducing the length $a = C/\mu v_\infty^2 = C/2E$

$$\psi = \int_\rho^\infty \frac{b\,\mathrm{d}r}{r^2 \sqrt{\left(1 - \dfrac{2a}{r} - \dfrac{b^2}{r^2}\right)}} = \int_\rho^\infty \frac{-\mathrm{d}(b/r)}{\sqrt{\left[1 + \dfrac{a^2}{b^2} - \left(\dfrac{b}{r} + \dfrac{a}{b}\right)^2\right]}}$$

This standard integral may be evaluated by using the new variable

$$u = \frac{b/r + a/b}{\sqrt{(a^2/b^2 + 1)}}$$

and noting that $u = 1$ when $r = \rho$ (values which reduce the square root to zero), which gives

$$\psi = \int_1^{\frac{a}{b}\sqrt{(1+a^2/b^2)}} \frac{-\mathrm{d}u}{\sqrt{(1-u^2)}} = \cos^{-1}\left[\frac{a/b}{\sqrt{(1+a^2/b^2)}}\right]$$

hence $\cos\psi = a/b/\sqrt{(1 + a^2/b^2)}$ and $\tan\psi = b/a$.

By writing $\psi = \pi/2 - \phi/2$, the relation between the impact parameter b and the angle of deflection ϕ is obtained

$$\boxed{b = a \cot \phi/2}$$

Comment I The calculation is exactly the same for an interaction between charges of opposite signs; the force constant C is then negative (attractive force), but the length a may be defined algebraically. For the same modulus C a supplementary angle ψ and an opposite angle ϕ are then obtained. This is shown in figure 5.2.

Comment II We need not determine the trajectory of the particle completely in this problem. However, if we want to calculate it, the definite integral for ψ may be replaced by an indefinite integral: $\theta = \cos^{-1} u + \theta_0$.

A suitable rotation of the polar axis allows the constant of integration θ_0 to be eliminated from the equation, and so we may write $u = \cos\theta$. Hence

$$\frac{1}{r} = \frac{a}{b^2}\left[\sqrt{\left(1 + \frac{b^2}{a^2}\right)} \cos\theta - 1\right] = \frac{a}{b^2}(\varepsilon \cos\theta - 1)$$

This is the equation of a hyperbola in polar co-ordinates, with the focus at the centre of force, eccentricity $\varepsilon > 1$, and whose major axis has length $2a$. In order to render meaningful the sign of the force constant C only that part of the hyperbola for positive r is considered: if the force is repulsive ($C > 0$ and $a > 0$) this is the arc of the hyperbola farthest from the centre of the mass, but on the other hand if the force is attractive ($C < 0$, and $a < 0$) it is the arc nearest the centre of mass (see figure 5.2).

Knowing the relation between b and ϕ, we may complete the calculation of the differential cross-section, started for a general case in section 5.2.2

$$\frac{\mathrm{d}\sigma}{\mathrm{d}\Omega} = \frac{b}{\sin\phi}\frac{\mathrm{d}b}{\mathrm{d}\phi} = \frac{a^2}{4\sin^4\phi/2}$$

so

$$\frac{d\sigma}{d\Omega} = \left(\frac{Ze^2}{4\pi\varepsilon_0 \mu v_\infty^2}\right)^2 \frac{1}{\sin^4 \phi/2}$$

The careful measurements made by Geiger and Marsden showed that the number of scattered particles were in good accord with the $1/\sin^4 \frac{\phi}{2}$ law as a function of the angle of deviation ϕ, and with the $1/v_\infty{}^4$ law as a function of velocity v_∞ of the α-particles (α-particles produced by radioactive emission are monoenergetic and they can be slowed down more or less uniformly by passing them through an obstacle such as a layer of air). *They demonstrated the accuracy of the formula for the potential energy $W = C/r$ used by Rutherford and also the hypothesis of the quasi point nucleus.*

However, if measurements are performed at large angles ($\phi \approx \pi$) the $1/\sin^4 \frac{\phi}{2}$ law no longer appears to be true when ϕ becomes larger than a certain value ϕ_0, that is to say when the impact parameter b becomes smaller than the value

$$b_0 = a \cot \frac{\phi_0}{2}$$

Hence the law of force changes when the distance r becomes equal to or less than b_0; this length b_0 is interpreted as an indication of the size of the nucleus. *Thus sizes of nuclei are measured to be between* 10^{-14} *and* 10^{-15} m (10^{-4} to 10^{-5} of the size of the atom).

Comment I At the beginning of the century Thomson's model was commonly accepted. He assumed that the positive charge was distributed with uniform density inside a sphere of the size of the whole atom. The advantage of this model was that it gave rise to elastically bound electrons, consistent with the classical theory of radiation (see appendix 2). Let us assume that the atomic electrons move within this sphere; according to a theorem of electrostatics, an electron situated at a distance r from the centre C of the sphere is attracted towards the centre C with the same force as a point charge of charge Q contained within a sphere of centre C and radius r, passing through the electron (see figure 5.4).

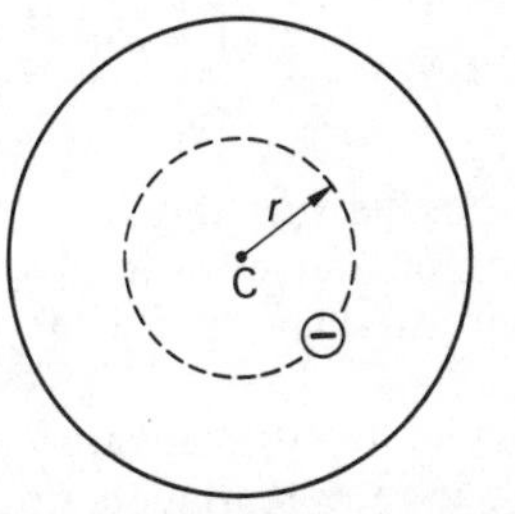

Figure 5.4 Thomson model of the atom

This attractive charge Q depends on the volume of the sphere, in other words, on an r^3 law; the central attractive force proportional to Q/r^2 is therefore proportional to the distance r. This is what is called an elastic force. The Thomson model gives rise to an electrostatic potential energy $W(r) = Cr^2$. *The Rutherford experiment demolished the Thomson model.*

Comment II The measurements made by Geiger and Marsden in 1911 gave only relative values accurately, thus allowing observation of variations of cross-sections. The experiment was repeated and improved by Chadwick in 1920 in order to allow absolute measurements of cross-sections; he was able to find the accurate value of the atomic number Z of the target atom. This was the first time that a *direct* measurement of the positive charge of the nucleus was carried out. The number Z of elementary charges measured in this way agreed with (i) the position of the element in the Periodic Table, (ii) the order of magnitude of the number of atomic electrons obtained by Barkla from X-ray scattering experiments (see appendix 2), and (iii) the value of Z deduced by means of the Moseley law from the interpretation of X-ray spectra (see chapter 7).

Comment III If light elements are used as targets when the variation of cross-section with atomic number Z is studied, it is then necessary to take account of the recoil of the nucleus in the interpretation of the experiment. The importance of this recoil was clearly demonstrated in an experiment carried out by Rutherford himself: taking a thin layer of paraffin (that is a saturated hydrocarbon with a general formula C_nH_{2n+2}) as a target for the α-rays he showed that the former emits a 'radiation' more penetrating than the α-rays. These are nuclei of hydrogen, or protons, which are targets four times lighter than the α-projectiles, and in the interaction the targets can acquire velocities much higher than those of the incident projectiles.

5.3 Planetary Motion of the Internal Electrons of the Atom

Having demonstrated the existence of a point nucleus, the Rutherford description of the atom can be completed. The internal atomic electrons are attracted by the positive nuclear charge according to the $1/r^2$ Coulomb law of electrostatics, just as the planets of the solar system are attracted by the sun according to Newton's law of attraction between masses, which also depends on $1/r^2$. Below, we review briefly the entirely classical properties of this planetary motion.

Let us first complete the calculation of the trajectory started in section 5.1 by using the electrostatic potential energy between a nucleus of charge Ze and an electron of charge $-e$

$$W(r) = -\frac{1}{4\pi\varepsilon_0}\frac{Ze^2}{r} = \frac{C}{r}$$

(We disregard the interactions between the electrons.) In dealing with electrons internal to the atom, we confine ourselves to trajectories that do not go out to infinity, so that the constant total energy E is negative (see the last paragraph of section 5.1.2).

A negative constant E which obeys the framed inequality in section 5.1.2 is compatible only with a negative potential energy $W(r)$. This is the case here: the constant C is negative corresponding to an attractive force.

The calculation is carried out in the same way as in section 5.2.3, the only difference being the sign of the constant total energy

$$\boxed{E<0}$$

By analogy, the two constants a and b are introduced having dimensions of length: $b^2=-\sigma^2/2\mu E$ and $a=C/2E$ (positive since C and E have the same sign) allowing one to write

$$\frac{\mathrm{d}\theta}{\mathrm{d}r}=\pm\frac{\sigma}{\mu r^2\sqrt{\left[-\dfrac{2E}{\mu}\left(-1+\dfrac{2a}{r}-\dfrac{b^2}{r^2}\right)\right]}}=\pm\frac{b}{r^2\sqrt{\left(-1+\dfrac{2a}{r}-\dfrac{b^2}{r^2}\right)}}$$

by changing to the variable $u=(b/r-a/b)/\sqrt{(a^2/b^2-1)}$ one obtains

$$\mathrm{d}\theta=\pm\frac{\mathrm{d}u}{\sqrt{(1-u^2)}}$$

giving $\theta-\theta_0=\cos^{-1}u$ or $u=\cos(\theta-\theta_0)$. Rotation of the polar axis through an angle θ_0 allows the constant in the equation to be suppressed; one can then write $u=\cos\theta$ and hence

$$\boxed{\frac{1}{r}=\frac{a}{b^2}\left[\sqrt{\left(1-\frac{b^2}{a^2}\right)}\cos\theta+1\right]=\frac{a}{b^2}(\varepsilon\cos\theta+1)}$$

This is the equation of an ellipse in polar co-ordinates whose focus is at the centre of force and eccentricity $\varepsilon<1$.

The length of the major axis of the ellipse may be calculated very simply. The major axis is a symmetry axis passing through the centre of force and the two extremities of the major axis correspond to extreme values of r, the minimum value ρ_1 and maximum value ρ_2 (see figure 5.5).

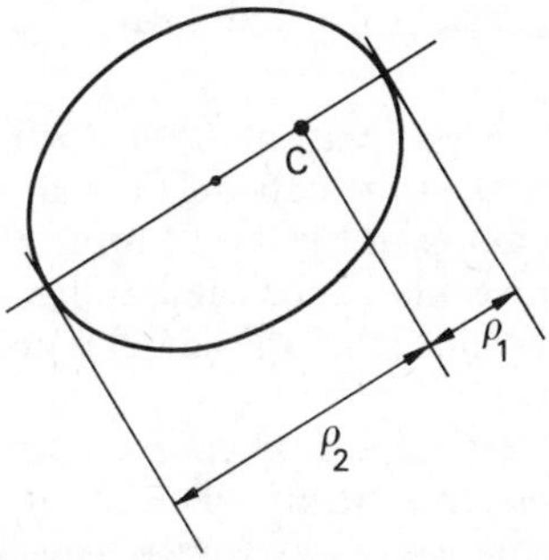

Figure 5.5 The major axis of an elliptic orbit

Now ρ_1 and ρ_2 are values of r that make the square root zero in the formula for $\mathrm{d}\theta/\mathrm{d}r$; they are therefore the roots of the quadratic equation: $-r^2 + 2ar - b^2 = 0$. The sum of the roots may be found immediately

$$\boxed{\rho_1 + \rho_2 = 2a = C/E}$$

The major axis of the ellipse depends only on the constant C in the law of force and on the total energy E; it is independent of the angular momentum σ.

Conversely, the total energy E of the motion depends only on C and on the major axis of the ellipse; it is independent of the eccentricity ε of the ellipse.

From the relation between a, b and the eccentricity ε, the length of the minor axis of the ellipse may be shown to be $2b$.

Comment All this may be checked easily by changing to cartesian co-ordinates

$$x = r\cos\theta \quad \text{and} \quad y = r\sin\theta$$

The above equation in polar co-ordinates can be rewritten in the form

$$b^2 = \varepsilon ar\cos\theta + ar \quad \text{or} \quad ar = b^2 - \varepsilon ax$$

By squaring the second equation

$$a^2 r^2 = a^2(x^2 + y^2) = (b^2 - \varepsilon ax)^2$$

Rearranging, using the relation between a, b and ε, one obtains

$$a^2 y^2 + b^2(x + \varepsilon a)^2 - a^2 b^2 = 0 \quad \text{or} \quad \frac{y^2}{b^2} + \frac{(x + \varepsilon a)^2}{a^2} = 1$$

Having determined the trajectory of an electron, one turns to the temporal properties of its motion. We can calculate the period T of the motion using the areal law (see section 5.1.2), knowing that the total area within the ellipse $S = \pi ab$

$$\frac{\sigma}{\mu} = 2\frac{\mathrm{d}S}{\mathrm{d}t} = 2\frac{S}{T} = 2\frac{\pi ab}{T}$$

from which

$$T^2 = 4\pi^2\frac{a^2 b^2 \mu^2}{\sigma^2} = \frac{2\pi^2 a^2 \mu}{|E|} = \frac{\pi^2}{2}\frac{\mu C^2}{|E|^3} = \frac{4\pi^2 \mu a^3}{|C|}$$

(this is Kepler's third law of planetary motion: for C/μ constant, the square of the period is proportional to the cube of the major axis). So the period T depends only on the constant C of the law of force and on the total energy E of the motion; or what amounts to the same thing, on the major axis $2a$ of the ellipse; but it is independent of the angular momentum σ and of the eccentricity ε of the ellipse.

These formulae have a very important consequence. Applying the classical theory of radiation to an electron within an atom, since its motion is accelerated, it produces an electromagnetic wave of the same frequency as its motion, and it radiates electromagnetic energy continuously (see appendix 2). To assure conservation of energy, its total energy E must decrease and since the energy E is negative, its modulus $|E|$ must increase. *When the energy of the electron decreases, the period T of its motion should become less and the frequency of the radiated electromagnetic wave should increase. This is in complete contradiction to the precise spectral frequencies measured experimentally and proves that the laws of classical physics are not applicable on an atomic scale.* This led Bohr to formulate new laws for spectral transitions (see chapter 1).

Comment The Thomson model which we have discussed led to elastically bound electrons, and as the period of the latter was independent of the energy and amplitude of their motion, the radiation problem could be explained. Historically, it was Rutherford's experiment which prepared the way for Bohr, by destroying the Thomson model.

6

The Bohr Model of the Hydrogen Atom

6.1 Explanation of the Spectrum of Hydrogen in Terms of Circular Orbits

In chapter 1 we saw very generally how to interpret the existence of spectral lines in terms of the energy levels of atoms, by means of the Bohr condition $h\nu_{np} = E_p - E_n$. However, we have not yet attempted an interpretation of the exact numerical values of the spectral frequencies ν, that is to say, the exact values of energy that the atom can take. We shall now start to do so in respect of the particularly simple case of the hydrogen atom.

We saw that the wavelengths λ_{np} of the hydrogen spectrum are interrelated by the experimental Balmer–Rydberg law

$$\frac{1}{\lambda_{np}} = R\left(\frac{1}{n^2} - \frac{1}{p^2}\right)$$

where n and p are integers and R is the so-called Rydberg constant.

This experimental law is easily interpreted due to the fact that the hydrogen atom has only one electron. This interpretation was given by Bohr in 1913 at the same time as the general law for spectral transitions.

The picture of the atom proposed by Bohr at that time can no longer be considered as very satisfactory. However, it merits description because it is particularly simple; it is a good example of what is known in physics as a 'model' and of the extremely useful role played by these models.

6.1.1 The Quantum Condition

Amongst the possible motions of the electron around the nucleus, Bohr confined himself *a priori* to those of circular trajectory. Imposing a circular trajectory equally imposes a relation between the two constants of motion, the total energy E and the angular momentum σ (in the calculations of the preceding chapter, both axes of the ellipse are written as equal, $a = b$). This relation is easily derived by using the general theory of the preceding chapter.

The electrostatic potential energy for a separation r between the charge $-e$ of the electron and the charge Ze of the point nucleus is

$$W(r) = -\frac{1}{4\pi\varepsilon_0}\frac{Ze^2}{r} = \frac{C}{r}$$

(the constant C is negative). Applying Newton's law, relating force and acceleration, along the radius vector

$$f = -\frac{\mathrm{d}W}{\mathrm{d}r} = \frac{C}{r^2} = -\frac{mv^2}{r}$$

(where m and v are the mass and velocity of the electron). Thus we find the following two relations:

(i) by introducing the angular momentum $\sigma = mvr$, then $v = -C/\sigma$;
(ii) the kinetic energy $\frac{1}{2}mv^2 = -C/2r = \frac{1}{2}W(r)$,

and hence the total energy of the motion

$$E = \tfrac{1}{2}mv^2 + W(r) = C/2r = -\tfrac{1}{2}mv^2$$

and so

$$\boxed{E = -\frac{mC^2}{2\sigma^2}}$$

Denoting two particular atomic states by n and p and using the Balmer–Rydberg law, then

$$E_p - E_n = \frac{mC^2}{2}\left(\frac{1}{\sigma_n^2} - \frac{1}{\sigma_p^2}\right) = h\nu_{np} = \frac{hc}{\lambda_{np}} = Rhc\left(\frac{1}{n^2} - \frac{1}{p^2}\right)$$

For this equality to be satisfied whatever the integers n and p, it is simplest to make the following hypothesis: the values of the angular momentum σ_n corresponding to each atomic state are proportional to the integer n; that is to say, the possible values of angular momentum are multiples of a certain constant.

By using experimental numerical values, this constant is found to be $\hbar = h/2\pi$, in other words

$$\boxed{\sigma = n\hbar}$$

If this value of the angular momentum is substituted into the expression for the energy, one obtains an expression for the Rydberg constant for $Z = 1$

$$R = \frac{mC^2}{4\pi c\hbar^3} = \frac{me^4}{(4\pi\varepsilon_0)^2\, 4\pi c\hbar^3}$$

Since the number of elementary charges carried by the hydrogen nucleus is $Z = 1$, then from this formula one may calculate theoretically $R = 109\,737{\cdot}3$ cm^{-1}, whereas one measures experimentally $R = 109\,677{\cdot}7$ cm^{-1}. The two values are in agreement within an accuracy better than 1/1000.

6.1.2 Interpretation of the Series Spectrum

By adopting this model to interpret the spectral terms $T_n = R/n^2$ of hydrogen, we have defined the sign and arbitrary constant of the energy. We have justified the assignment of a negative energy to the different states of the atom

$$E = -Rhc/n^2$$

and have defined zero energy: it corresponds to the situation where the electron is at infinity with zero velocity, in other words, where the atom is ionised.

In chapter 1 we set out a diagram representing the energy levels as well as the spectral transitions by means of which the atom passes from one to the other (see figure 1.3). Spectral lines of a particular series are emitted when the atom carries out all the transitions which result in it falling to the same lower energy level.

The lower level $n = 1$ gives rise to the ultraviolet series

$$\frac{1}{\lambda} = R\left(1 - \frac{1}{p^2}\right) \qquad \text{(Lyman, 1906)}$$

The lower level $n = 2$ gives rise to the visible series

$$\frac{1}{\lambda} = R\left(\frac{1}{4} - \frac{1}{p^2}\right) \qquad \text{(Balmer, 1885)}$$

The lower level $n = 3$ gives rise to the infrared series

$$\frac{1}{\lambda} = R\left(\frac{1}{9} - \frac{1}{p^2}\right) \qquad \text{(Paschen, 1908)}$$

The lower level $n = 4$ gives rise to the infrared series

$$\frac{1}{\lambda} = R\left(\frac{1}{16} - \frac{1}{p^2}\right) \qquad \text{(Brackett, 1922)}$$

The lower level $n = 5$ gives rise to the infrared series

$$\frac{1}{\lambda} = R\left(\frac{1}{25} - \frac{1}{p^2}\right) \qquad \text{(Pfund, 1924)}$$

The wavelength limit of each series, for p infinite, corresponds to the transition in which an initially ionised atom falls to the corresponding level n.

The most stable state is the state of minimum energy, obtained for $n = 1$. This is the ground state of the atom. All the lines of the Lyman series are resonance lines. The wavelength limit of the Lyman series $\lambda_1 = 91{\cdot}2$ nm corresponds to the transition between the ionised state and the ground state where the atom is normally found; the energy involved in this transition is equal to the minimum energy that must be given to the atom to ionise it, namely the ionisation energy $W_i = eV_i$. Therefore

$$\boxed{-E_1 = Rhc = hc/\lambda_1 = eV_i}$$

From the given numerical values of R and λ_1, the ionisation potential of the hydrogen atom $V_i = 13{\cdot}6$ V may be easily verified. Furthermore this wavelength limit λ_1 is also the wavelength threshold for photoionisation of the same atom (which we have called λ_i in section 1.2.4).

6.1.3 Derivation of the Parameters of the Circular Motion

Using the quantum condition $\sigma = n\hbar$, we can calculate the various quantities characterising the circular motion of the electron around the nucleus. We retain the atomic number Z in these calculations to allow subsequent generalisation.

(*a*) *The speed of the electron*: $v = -C/\sigma = +Ze^2/4\pi\varepsilon_0\hbar n$. If the ratio v/c of the speed of the electron to the speed of light is calculated, we may introduce the dimensionless constant

$$\boxed{\alpha = \frac{e^2}{4\pi\varepsilon_0 \hbar c} = \frac{1}{137{\cdot}04}}$$

In the CGS system, $4\pi\varepsilon_0 = 1$; this is the reason why a different expression $\alpha = e^2/\hbar c$ is found in many books, but it has the same numerical value since the constant α is dimensionless. It is called the *fine-structure constant* for reasons that will appear later (see section 6.3), and its introduction allows many calculations to be simplified. It enables us to write the ratio

$$\boxed{v/c = \alpha(Z/n)}$$

For hydrogen, where $Z=1$, v is small compared with c, which justifies carrying out the calculation without taking relativity into account.
(*b*) *The energy of the motion*: $E=mv^2/2$, which may be simplified by using the fine structure constant

$$\boxed{E=-mc^2\frac{\alpha^2}{2}\frac{Z^2}{n^2}=-Rhc\frac{Z^2}{n^2}}$$

($mc^2=511$ keV is the rest mass energy of the electron; by putting $n=Z=1$, the ground state of hydrogen $E_1=-13{\cdot}6$ eV is easily found).
The Rydberg constant can also be written

$$R=\frac{mc}{h}\times\frac{\alpha^2}{2}$$

(*c*) *The diameter of the orbit is* $2r$ *where*

$$2r=\frac{C}{E}=-\frac{2\sigma^2}{mC}=4\pi\varepsilon_0\frac{2\hbar^2}{me^2}\frac{n^2}{Z}$$

or alternatively

$$2r=\frac{h}{mc}\times\frac{1}{\pi\alpha}\times\frac{n^2}{Z}=\frac{\Lambda}{\pi\alpha}\times\frac{n^2}{Z}$$

(It will be recalled that $\Lambda=h/mc=2{\cdot}4$ pm is the Compton wavelength.) For the ground state of hydrogen ($n=Z=1$), it follows that $2r\approx 0{\cdot}1$ nm. So a length is obtained which is in agreement with the interatomic distances deduced from Avogadro's number or X-ray diffraction experiments.
(*d*) *The orbital frequency of the electron:* $f=v/2\pi r=-E/\pi\sigma=-2E/hn=2\nu_1/n$, where $\nu_1=E/h$ is the frequency limit of the series of lines which terminate on the energy level E. It may be seen that orbital frequencies are of the same order of magnitude as the frequencies of spectral transitions.
In particular, by considering the transition frequency between two neighbouring levels whose quantum numbers differ only by one unit

$$\nu_{n-1,n}=Rc\left[\frac{1}{(n-1)^2}-\frac{1}{n^2}\right]=Rc\frac{2n-1}{n^2(n-1)^2}=\frac{E_n}{h}\frac{2n-1}{(n-1)^2}$$

$$\frac{\nu_{n-1,n}}{f}=\frac{n(2n-1)}{2(n-1)^2}=\frac{2n^2-2n}{2n^2-4n+1}\to 1 \quad \text{as} \quad n\to\infty$$

When the quantum number n has large values, the frequency $\nu_{n-1,n}$ of electromagnetic radiation emitted by the atom becomes equal to the orbital frequency f of the electron in its orbit, conforming to the classical laws of radiation (see appendix 2). Here we find one of the many aspects of the *correspondence principle* stated by Bohr in 1923: when the quantum numbers are large the results of quantum physics become identical to those of classical physics.

6.2 Involvement of the Nucleus—Hydrogen-like Ions

In the calculations of the preceding section, we assumed that the electron of mass m moves around a stationary nucleus. This would be the case if the nucleus was infinitely heavy compared with the electron. It is a useful first approximation since the mass M_H of the proton is about 2000 times greater than the mass m of the electron.

However, as a result of the discussion at the beginning of the previous chapter, it is strictly the centre of mass of the proton–electron system that is fixed, so our calculations apply to the reduced motion around the centre of mass of a particle having the reduced mass $\mu = mM_H/(m + M_H)$. We have also seen that the total angular momentum of the system of two particles is equal to the angular momentum of the reduced motion (provided they are calculated about the centre of mass). If the Bohr quantisation condition $\sigma = n\hbar$ is applied to the total angular momentum of the atomic system, the calculations of the preceding section remain valid, except that the actual mass m of the electron is replaced by the reduced mass μ. Taking this into account we can calculate the small correction to be applied to the preceding results. Letting R_∞ be the Rydberg constant calculated previously using the actual mass of the electron, so that the nucleus is assumed to have infinite mass, and letting R_H be the Rydberg constant calculated with the reduced mass μ, involving the actual mass of the proton M_H

$$\frac{R_\infty}{m} = \frac{R_H}{\mu}$$

so

$$\boxed{R_H = R_\infty \frac{\mu}{m} = \frac{R_\infty}{1 + m/M_H}}$$

The new constant R_H calculated here is slightly less than the constant R_∞ of the preceding section, the relative difference being $m/M_H = 1/1836 = 0{\cdot}54 \times 10^{-3}$. If this correction is applied to the numerical value previously calculated the exact experimental value is obtained. This numerical agreement to within an accuracy of 10^{-5} between the calculated value of R_H and its experimental value contributed to the acceptance of Bohr's hypothesis.

The above calculations may be applied to other atomic systems. When an electron is detached from a helium atom of atomic number 2, or two electrons from a lithium atom of atomic number 3, the ions He^+ or Li^{2+} are formed having only one electron. They are atomic systems of two particles to which the theory derived for atomic hydrogen can equally well be applied by simply adjusting the charge and mass of the nucleus. In this way the spectra of the ions He^+, Li^{2+}, Be^{3+} and so on can be easily interpreted.

The energy values are proportional to the square of the constant C in the law of force, and so to the square of the electric charge on the nucleus. Thus the possible energy values for the helium ion are obtained

$$E = -Rhc(Z^2/n^2) = -4Rhc/n^2$$

(a) To a first approximation, the constant R is the same as that for hydrogen (it would be exactly the same if the masses of the nuclei of both helium and hydrogen could be considered as infinite) and *hence one may deduce coincidences between certain lines of the ion He^+ with lines of the hydrogen atom H.*

For example, the Pickering series of the ion He^+ corresponds to transitions to a lower level $n = 4$, and the wavelengths of the corresponding lines are given by

$$\frac{1}{\lambda} = 4R\left(\frac{1}{16} - \frac{1}{p^2}\right) = R\left(\frac{1}{4} - \frac{4}{p^2}\right)$$

The wavelengths corresponding to even values of the integer p are equal to the wavelengths of the Balmer series of the hydrogen atom. Historically, the observation of these coincidences was not understood immediately, and gave rise to some confusion.

(b) To a second approximation, we must take account of the fact that the constant R depends on the mass of the nucleus. Its value must be slightly different for the atom H and the ion He^+ and accurate measurements show that in reality *these coincidences we have mentioned are not absolutely exact.* For the ion He^+, the Rydberg constant is measured to be $R_{He} = 109\,722{\cdot}3\ \text{cm}^{-1}$, which is entirely explained by the formula $R_{He} = R_\infty/(1 + m/M_{He})$.

(The mass of helium is four times that of hydrogen; hence the difference between R_{He} and R_∞ must be four times smaller than in the case of hydrogen, the relative difference being $0{\cdot}14 \times 10^{-3}$.)

Comment Since Rydberg constants have been determined to accuracies of the order of 10^{-6}, the above formulae can give rise to a metrological application. Mass spectroscopic techniques allow an accurate measurement of the ratio M_H/M_{He} because these two masses are of the same order of magnitude, but they cannot measure the ratio m/M_H with the same precision because the two masses involved have very different orders of magnitude. This ratio m/M_H can be deduced from values of the Rydberg constants

$$\frac{R_H}{R_{He}} = \frac{1 + m/M_{He}}{1 + m/M_H} = \frac{1 + (m/M_H)(M_H/M_{He})}{1 + m/M_H}$$

hence

$$\frac{m}{M_H}\left(\frac{R_H}{R_{He}} - \frac{M_H}{M_{He}}\right) = 1 - \frac{R_H}{R_{He}}$$

Numerical calculation results in $m/M_H = 1/1836{\cdot}5$, compared with the accepted value 1836·1.

6.3 Generalisation to Elliptic Orbits—Fine Structure

After Bohr's discovery the German physicist, Sommerfeld, attempted to generalise his ideas. For symmetry reasons connected with the three spatial co-ordinates, he proposed three distinct quantum conditions. One of them describes the orientation in space of the plane of the trajectory and does not concern us here (see chapter 10, spatial quantisation). The other two may be expressed as two independent conditions imposed on the energy E and the angular momentum σ

$$E = -RhcZ^2/n^2 \quad \text{and} \quad \sigma = k\hbar$$

where n and k are two distinct integers; thus the relation between E and σ characteristic of circular orbits is abandoned, and this represents the point of departure from Bohr's theory.

Bohr's circular orbits reappear in the particular case where $k = n$; but when k is less than n, elliptical trajectories are obtained. The theory presented in section 5.3 shows that k cannot be greater than n (because the parameter b must be less than the parameter a); a proof was also given of the general relation $E = C/2a$ between the energy of the motion and the length $2a$ of the major axis of its trajectory. Several distinct trajectories whose major axes have the same

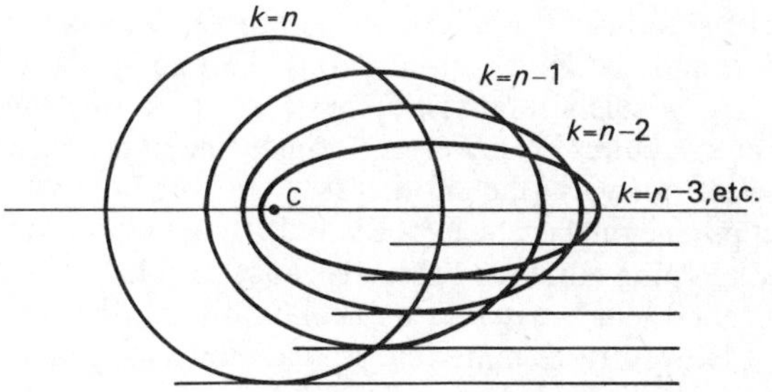

Figure 6.1 Sommerfeld's quantised ellipses corresponding to the same quantum number $n = 5$ (the minor axis is proportional to the angular momentum, that is to the quantum number k)

length but whose eccentricities are different now correspond to the same value of the principal quantum number n, and so to the same value of the energy E. They are shown in figure 6.1, taking account of the fact that they all have the same focus at the centre of mass C of the atomic system. This is an example of what is known as the degeneracy of an energy level: the existence of several atomic states having the same energy, but differing in other respects.

All this would have been pure speculation and of little practical interest were it not that Sommerfeld investigated this theory in greater depth by means of the laws of relativity which we have so far ignored. In section 6.1.3 we calculated the velocity v of an electron in its orbit and saw that it was small compared with the velocity c of light. However, it is not completely negligible, and Sommerfeld carried out the relativistic calculation by making a limited expansion as a function of the small quantity $v/c = (Z/n)$. For an orbit corresponding to the two quantum numbers n and k, he calculated the energy

$$E_{nk} = -Rhc\,\frac{Z^2}{n^2}\left[1 + \frac{\alpha^2 Z^2}{n^2}\left(\frac{n}{k} - \frac{3}{4}\right) + \text{terms in } \alpha^4 \text{ and higher order}\right]$$

Small corrections dependent upon the quantum number k, that is to say, on the ellipticity of the trajectory, must be made to the energy values calculated by Bohr. The relativistic calculation allows the degeneracy of the energy level E_n to be lifted, by separating it into several close but distinct states E_{nk}. This is described as the fine structure of the energy levels of hydrogen, and is the reason why the dimensionless constant $\alpha = e^2/4\pi\varepsilon_0 \hbar c$ introduced in section 6.1 is called the fine-structure constant. However, the energy still depends mainly on the quantum number n, called, for this reason, the principal quantum number.

Sommerfeld's relativistic theory has been checked experimentally. When the hydrogen spectrum is recorded with a high-resolution interferometer, many closely spaced lines are observed instead of the isolated lines observed with poor resolution. This fine structure in the spectral lines proves the existence of fine structure in the energy levels. (This structure in fact becomes very small when n becomes large; even the fine structure of the $n = 2$ level in lines of the Balmer series is hard to observe; see volume 2, chapter 4.)

For hydrogen and the hydrogen-like ions the fine structure is very small because its origin is solely relativistic; but for atoms with several electrons much larger fine structures are observed, originating in the energy of the interaction of the electrons with one another (see chapter 7).

We shall not pursue further the hypotheses formulated by Sommerfeld or his calculations concerning elliptical orbits, because they have not resulted in an entirely consistent theory, capable of explaining in all cases the observed energy levels. However, Sommerfeld's studies have had two important consequences.

(a) They showed that refinement and extension of classical mechanics did not provide a wholly satisfactory explanation of the internal structure of the atom, and thus the invention of quantum mechanics was necessitated.

(b) They formed an indispensable addition to Bohr's theory by treating separately and distinctly the energy on the one hand and angular momentum on the other: the relation between energy and angular momentum which is characteristic of circular orbits and was fundamental to Bohr's reasoning is useful only in special cases.

The principal quantum number n should be related only to energy and not to angular momentum.

6.4 General Conclusions

The description of the Bohr model of the atom, corrected by Sommerfeld, gives a clearer understanding of the precise part played by 'models' and of their usefulness.

(a) Even if it stems from partial or inaccurate hypotheses a model can often give some correct results and clarify important ideas, thus preparing the way for more elaborate theories. For example, the Bohr–Sommerfeld model gives us (i) the idea of quantisation of angular momentum; but this idea requires substantiation (see chapter 10, spatial quantisation); (ii) the interpretation of the negative energy $E_n = -hcT_n$ for each spectral term T_n as the binding energy of the electron, so that zero energy corresponds to an ionised atom; (iii) the law for quantisation of energy in a field of force whose potential energy is $W(r) = C/r$. It may be expressed as a function of the constant C of the law of force and of the principal quantum number n by the formula

$$E_n = -\frac{\mu}{2}\frac{C^2}{\hbar^2}\frac{1}{n^2}$$

A sound theory of atomic structure must accommodate this law and this is done in the treatment of the hydrogen atom in books on quantum mechanics (see volume 2, chapter 1).

In the next chapter we shall see how these two latter ideas are generalised to atoms having many electrons.

(b) Even if the limitations and inaccuracies of a model are fully recognised, a sufficiently simple one often provides a useful semblance of reality, facilitating rapid qualitative argument and calculations of orders of magnitude. For example the electronic orbits of the Bohr–Sommerfeld model are not an exact description of reality; quantum theory results in a far less precise definition of the positions of the electrons (see volume 2, chapter 1). Nevertheless, consideration of these orbits is convenient in order (i) to calculate the order of magnitude of the dimensions of the atom (see section 6.1.3); (ii) to indicate the origin of atomic magnetism (see chapter 8 and volume 2, chapter 5); (iii) to explain the variation of electronic interactions with the eccentricity of the orbits, in other words, with their angular momentum (see chapter 7).

7

X-Ray Spectra

The interpretation of optical spectra for atoms having many electrons is rather complicated because in addition to the electrostatic electron–nucleus interaction, it is also necessary to take into account the electrostatic interaction between the electrons, as well as the magnetic interactions. Furthermore the energies corresponding to these various interactions may be of similar orders of magnitude.

X-rays, with wavelengths of the order of 100 pm, have much higher frequencies than visible light rays; X-ray spectra involve the deeper energy levels of heavy atoms (having large atomic number Z). In these circumstances, the energy of the electron–nucleus interaction is clearly dominant compared with the energies corresponding to other interactions, and to a first approximation, it is possible to give a relatively simple account of X-ray spectra (Kossel, 1920).

In the first three sections of this chapter, we shall see how three different experimental techniques using X-rays allow measurement of the deep energy levels of atoms. Then in the final section we shall give an interpretation of these measured values.

7.1 The Absorption Spectra of X-Rays

It is well known that X-rays can pass through matter but that their intensity is reduced while doing so. This attenuation of X-rays depends upon the medium through which they pass, and is the basis of medical and industrial radiography. It also depends on the frequency of the X-rays: high-frequency X-rays (energy $h\nu \gtrsim 100$ keV or short wavelength $\lambda \lesssim 10$ pm) are called hard X-rays because they are attenuated far less than low frequency X-rays (energy $h\nu \lesssim 1$ keV or long wavelength $\lambda \gtrsim 1$nm), called soft X-rays. However, this attenuation is not a monotonic function of frequency; it is this function of frequency that we wish to study.

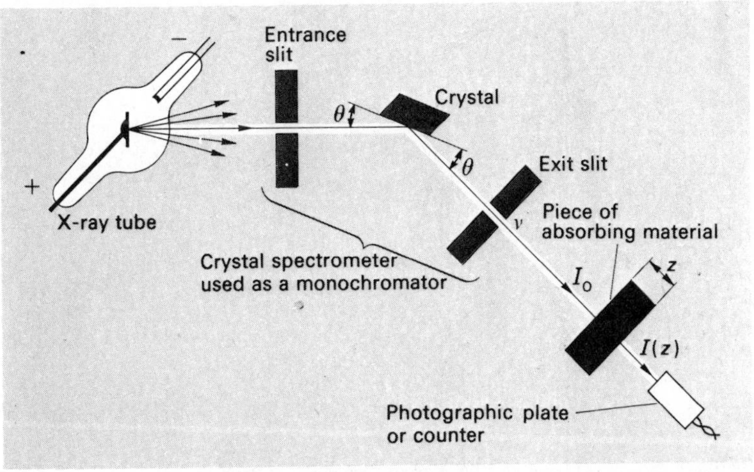

Figure 7.1 Apparatus for studying the absorption of X-rays as a function of their frequency

Figure 7.1 shows the experimental arrangement for carrying out this study. An X-ray tube emits radiation having a continuous distribution over a very wide frequency range (the variation of intensity with frequency is not important here, and we shall postpone the discussion of this to section 7.3). A crystal spectrometer serves as a monochromator; it receives all the radiation, but only lets through monochromatic radiation of a particular frequency ν (which can be calculated, knowing the lattice spacing d, from Bragg's diffraction law $2d\sin\theta = k\lambda = kc/\nu$). By means of a photographic plate or a counter the intensity of the monochromatic X-ray beam may be measured before and after passing through the piece of material being studied.

By comparing the intensity I_0 of the incident X-ray beam with the intensity $I(z)$ of the same beam after passing through a thickness z of material, it may be shown that the absorption of the X-ray beam by the material obeys an exponential law characteristic of random phenomena

$$\frac{dI}{I} = -K\,dz \quad \text{giving} \quad I(z) = I_0\,e^{-Kz}$$

Thus the absorption coefficient K, characteristic of the absorbing material used and independent of its thickness, may be measured.

The crystal spectrometer enables the frequency of the incident beam to be varied, so that the curve of the absorption coefficient K as a function of the frequency ν can be drawn. This is known as the absorption spectrum of the material used. Experimentalists nearly always express their results as a function of the wavelength λ. We have converted them so as to express them as a function

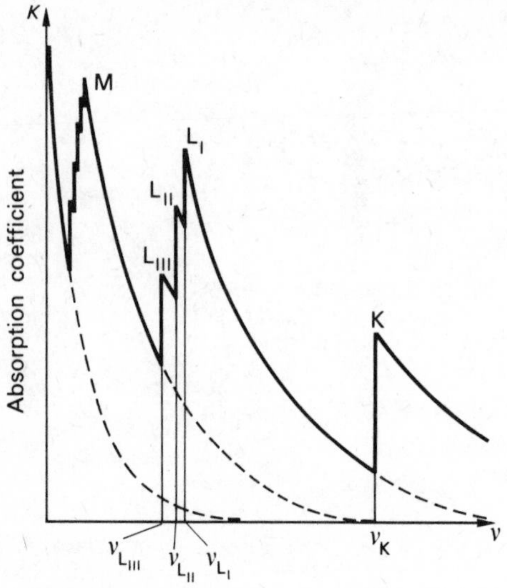

Figure 7.2 An X-ray absorption spectrum

of frequency ν, which is a more convenient parameter for interpreting the phenomena. Figure 7.2 is a typical example of the shape of the curves obtained: smooth parts of the curve, varying nearly as C/ν^3, are separated by sharp discontinuities where the coefficient of proportionality C suddenly increases as the frequency goes through certain values.

More precisely, figure 7.2 illustrates the type of curve obtained when the absorbing screen is made up entirely of a single chemical element. A discontinuity, called the K discontinuity, is observed at a frequency much higher than all the others; as the frequency decreases three closely separated discontinuities called L_I, L_{II} and L_{III} may be observed; at still lower frequencies, five discontinuities M_I, M_{II}, M_{III}, M_{IV} and M_V; then seven other discontinuities designated by the letter N, and so on, which are observable only with sufficiently heavy elements (of sufficiently large atomic number). When the absorbing screen is composed of several elements, its spectrum contains all the frequency discontinuities seen in each of the separate elements; this is so regardless of the chemical combinations of the elements. *The frequencies at which discontinuities occur are characteristics of the chemical elements, that is to say of different types of atoms, independent of the molecules of which they form part.*

For example for the atom of gold, atomic number 47, one observes

$\lambda_K \simeq 50$ pm, corresponding to photons of energy $h\nu_K \simeq 25\,000$ eV
$\lambda_L \simeq 350$ pm, corresponding to photons of energy $h\nu_L \simeq 3600$ eV

Comment I These statements are true only to a first approximation.

(a) Each of the discontinuities described above are in fact composed of several very closely spaced discontinuities. Refer to comment II in section 7.3.4.

(b) If the spectra are examined under high resolution and if sufficiently low frequencies are studied (sufficiently long wavelengths, or soft X-rays) small frequency shifts are observed according to the molecule in which the element occurs. In recent years, a technique has even been developed based entirely on the observation of these small frequency shifts (changes of energy of the order of an electron-volt or a fraction of an electron-volt).

Comment II An idea of the order of magnitude of absorption coefficients is obtained from Bragg and Pierce's experimental law which gives the mass absorption coefficient (the absorption coefficient K divided by the density ρ of the medium) for an element of atomic number Z

$$K/\rho \simeq CZ^3\lambda^3$$

If K/ρ is expressed in SI units and λ in nanometres

$C \approx 0{\cdot}1$ for $\lambda > \lambda_K, \nu < \nu_K$,
$C \approx 1$ for $\lambda < \lambda_K, \nu > \nu_K$.

It is in fact necessary to add to this term $CZ^3\lambda^3$ another term independent of λ and representing the attenuation of the incident X-ray beam due to scattering.

The interpretation of these characteristic frequencies follows from the form of the observed curves. X-Rays may be absorbed by many simultaneous processes acting in competition and the measured absorption coefficient K can then be considered as the sum of partial coefficients that would correspond to each of the processes acting in isolation. We have symbolised this in figure 7.2 by extending the $1/\nu^3$ portions of the experimentally determined curves with dotted lines. Each of the frequency discontinuities behaves as a *threshold* above which a new process contributes to the absorption; this particular process cannot occur when the frequency is below this threshold.

For example, let us consider the K discontinuity of a particular atom A; a mechanism exists whereby the atoms A can destroy incident photons if their energy $h\nu$ is greater than $h\nu_K$, that is $h\nu \geqslant h\nu_K$.

This threshold in frequency and in energy is one of the main characteristics of the photoelectric effect in metals (see chapter 1); it arises because the electron must be given a greater energy than the escape energy (or extraction energy) of the metal; that is $h\nu_K \geqslant W_S = h\nu_S$.

If we assume that there is an electron in atom A whose binding energy is $E_K = -h\nu_K$ (we keep to the convention that a stationary electron at infinity has zero energy), then in order to detach this electron from the atom, it has to be given the extraction energy $W_K = -E_K = h\nu_K$. Assuming that each electron interacts only with a single photon, the condition $h\nu \geqslant h\nu_K$ results.

Thus each of the absorption discontinuities may be interpreted as a photoelectric threshold; each of the discontinuities corresponds to a different escape energy, in other words to the detachment of an electron of different binding energy.†

In the following two sections we shall see how this interpretation is confirmed.

7.2 Velocity Spectra of X-Ray Photoelectrons

The interpretation of the mechanism for the absorption of X-rays as a photoelectric effect was confirmed by the experiments of the French physicist Maurice de Broglie (1922):

(a) first he observed the trajectories of charged particles produced in a cloud chamber by a beam of X-rays (photoelectrons are distinguished from Compton electrons because they go off in the direction of the electric field of the wave, perpendicular to the direction of propagation of the beam of X-rays);

(b) then he verified that these charged particles were indeed electrons by measuring their q/m ratio;

(c) finally he measured the velocity of these photoelectrons by developing the velocity spectrum analyser, which is still used in laboratories today, and which we shall now describe.

It has been established that the kinetic energy of the photoelectrons enters into the energy conservation equation for the photoelectric effect, and this is why measurement of their velocity is of great importance. The method of measurement used for the photoelectric effect in metals, with ordinary light rays, cannot be applied to the photoelectrons produced by X-rays because their kinetic energies are very much larger, in accordance with the energies of X-ray photons. The method commonly used to measure the velocities of high-energy charged particles consists of measuring the radius of curvature of their trajectory in a magnetic field; this method, which is used in mass spectrum analysis or in nuclear physics in cloud chambers, may be adapted to X-ray photoelectrons.

The apparatus used is shown in figure 7.3. The absorption of X-rays is produced in a very thin metal wire M (perpendicular to the plane of the diagram) so that (i) the electrons detached from an atom experience minimal retardation before escaping from the metal by the shortest possible path, and (ii) the origin of the electron trajectories is well defined. The metal wire is placed in an evacuated chamber where the electrons can move freely.

† *Translator's note:* Strictly, an absorption discontinuity arises whenever a vacancy is created within an atom by excitation of an electron. The electron is not necessarily detached from the atom, though to simplify the discussion in this chapter, the authors generally assume that it is (see comment II in section 7.3.4).

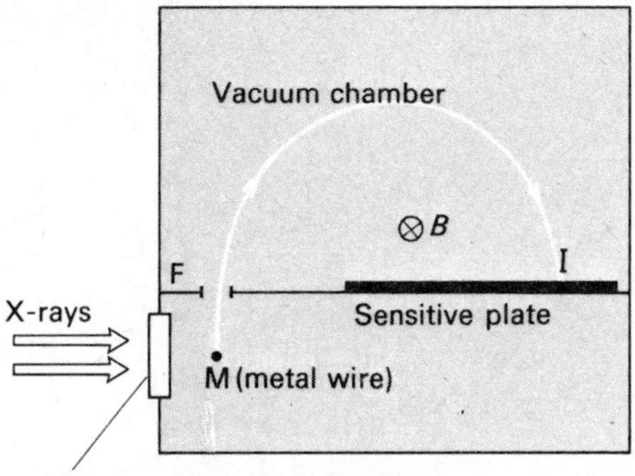

Figure 7.3 An electron velocity spectrum analyser

The whole chamber is placed in the air gap of an electromagnet which creates a magnetic induction $\boldsymbol{B}$ (perpendicular to the plane of figure 7.3). A small slit F, situated very near to the metal wire M, selects photoelectrons whose velocity vector is in a particular direction MF, chosen as perpendicular to the induction vector $\boldsymbol{B}$. Under these conditions, the photoelectrons of charge $-e$ describe a circle in a plane perpendicular to the field (in the plane of the diagram) whose radius R is proportional to their momentum

$$\boxed{R = \kappa mv/eB}$$

(This formula may be derived by writing down the equation which relates the Laplace force to the normal component of the acceleration: $evB/\kappa = mv^2/R$; we recall that the coefficient κ is unity if SI units are used, see the preface.)

To measure the radius of this circle, the electrons are stopped by a sensitive plate on which the track at their point of impact I can be observed. If the plate is placed in the plane perpendicular to MF passing through F, the electrons describe a semicircle before striking it, and the diameter of the circle FI = 2 R may be measured.

Comment This arrangement has the advantage of allowing some focusing of those electrons that leave the metal wire M with the same speed v but slightly different directions, and which nevertheless are selected by the slit F because of its finite width. This is shown in figure 7.4 where three trajectories are illustrated having the same diameter (the same speed), starting from the point M, with tangents making small angles to one another. These circles intersect again in the region of a point diametrically opposed to M, and since they intersect at small angles, their separation remains small in this region opposite M. In other words, by considering a family of circles passing through M, and concentrating on their point of intersection I on the axis Mx, the distance MI passes through a maximum when the circle is tangent to the axis My, perpendicular to Mx, and the distance MI therefore changes very little when the tangent at M is slightly different from My.

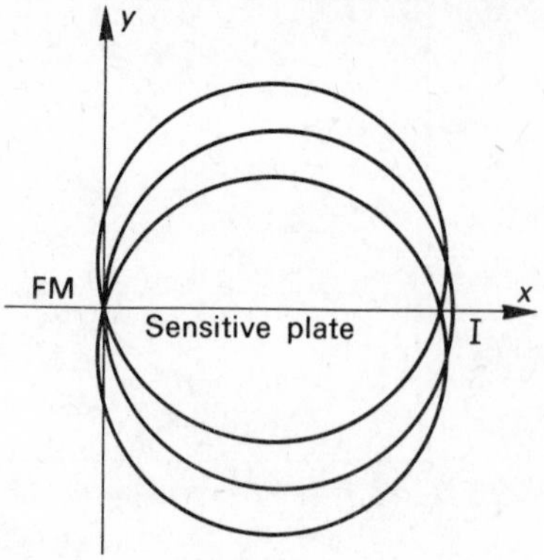

Figure 7.4 Focusing of electrons in a velocity spectrum analyser

Since the slit F is elongated parallel to the direction of the vector $\boldsymbol{B}$, impacts of electrons having the same velocity v form a small straight segment parallel to this direction. *When the sensitive plate is developed after an experiment, a series of short parallel lines is observed, entirely analogous to an optical spectrum; this is called a velocity spectrum.* The distance of each line from the slit F is proportional to the velocity v of the electrons, which can thus be calculated.

In this way, it may be ascertained that the velocities of photoelectrons detached from a metal by X-rays can have only certain particular values. So each of these velocities v can be associated with one of the X-ray absorption discontinuities of the metal used (studied in the preceding section), and knowing the frequency ν of the incident X-rays, it may be confirmed that conservation of energy is satisfied in the photoelectric effect (Einstein's photoelectric equation)

$$\boxed{h\nu = W_K + \tfrac{1}{2}mv^2 = h\nu_K + \tfrac{1}{2}mv^2}$$

(and similarly for the other discontinuities, denoted by L_I, L_{II}, L_{III}, M, N and so on).

This demonstrates the validity of the interpretation of each X-ray absorption discontinuity as a photoelectric effect threshold, and confirms the existence within the atom of electrons whose binding energy is

$$E_K = -h\nu_K,\ E_L = -h\nu_L,\ E_M = -h\nu_M,\ \text{or}\ E_N = -h\nu_N$$

and so on. Conversely, if the absorption discontinuities are unknown, this technique provides a much easier method of measuring the extraction energy $W_K = h\nu - \tfrac{1}{2}mv^2$, that is, the binding energy $E_K = -W_K$ of the internal electrons of the atom.

Comment I If the frequency ν of the incident X-radiation is changed, the kinetic energy of the photoelectrons is altered and the lines of the velocity spectrum are shifted. However, it can sometimes happen that the positions of some lines of the velocity spectrum are independent of the frequency of the incident radiation, so that certain electrons leave the metal with a kinetic energy independent of ν. This is the Auger effect which will be explained later.

Comment II The velocity spectrum analyser is also used in nuclear physics to measure the frequencies of γ-photons. If the extraction energy of the electrons is known, measurement of their kinetic energy allows the energy $h\nu$ of the incident γ-photons to be found.

7.3 X-Ray Emission Spectra

The mode of operation of an X-ray tube is as follows: a piece of metal called an anticathode is bombarded with highly accelerated electrons, and the piece of metal then becomes a source of X-radiation. This radiation can be analysed

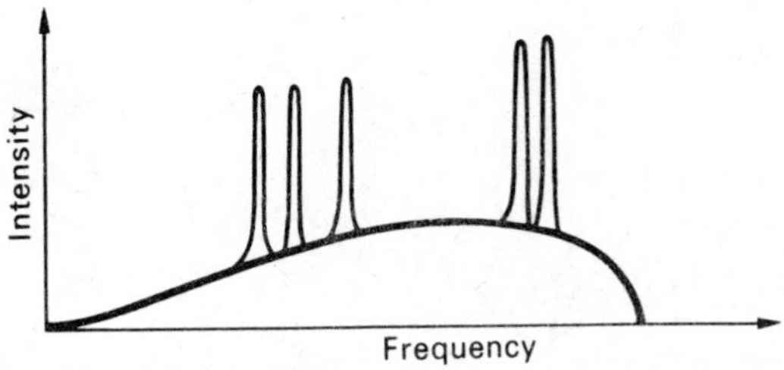

Figure 7.5 An X-ray emission spectrum showing a line spectrum superimposed on a continuous spectrum

with a crystal spectrometer. The spectrum obtained appears as a superposition of a continuous spectrum and a line spectrum; in the continuous spectrum, part of the X-ray intensity is distributed continuously over a band of frequencies, whereas in the line spectrum a relatively high intensity is emitted at certain special frequencies (see figure 7.5).

The lines and the background continuum of the spectrum have completely different explanations as we shall see in the two following sections.

The difference in origin between the continuous spectrum and the line spectrum is well-illustrated by the phenomenon of *X-ray fluorescence producing X-radiation having only a line spectrum, without a background continuous spectrum.* When a substance is irradiated with a beam of X-rays of a given frequency, it absorbs part of this beam (see section 7.1) and in turn becomes a source of X-rays. Spectral analysis of this new radiation allows a distinction to be made between:

(i) the radiation scattered elastically { without change of frequency (Thomson scattering), and with change of frequency (Compton scattering)

(ii) the proper fluorescence radiation, forming a line spectrum characteristic of the substance. This line spectrum is identical to that produced in an X-ray tube by bombarding the same substance with electrons.

Comment The term fluorescence is used whenever atoms themselves become emitters of radiation after absorbing an incident electromagnetic wave. The internal structure of the atom is involved in the phenomenon of fluorescence; this is what distinguishes it from elastic scattering (see section 1.3).

7.3.1 The Continuous Spectrum

The main characteristics of the continuous spectrum are shown in the figures 7.6(a) and (b) where curves representing the intensity I of the X-radiation

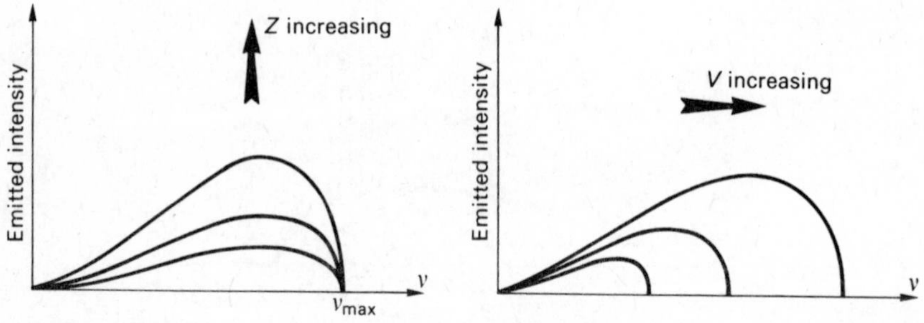

(a) (b)

Figure 7.6 (a) Spectra obtained by bombarding various anticathodes with electrons of the same energy; (b) continuous spectra emitted by the same anticathode for various values of the voltage V used to accelerate the electrons

emitted by the anticathode are drawn as a function of the frequency ν (omitting the line spectra).

As shown in figure 7.6(a), the continuous spectrum depends little on the metal used for the anticathode; the height of the curve increases with the atomic number Z of this metal, but the shape of the curve is independent of Z. The shape of this curve is characterised by the existence of a maximum, and a very rapid descent to zero at a certain maximum frequency ν_{max}, above which no radiation is emitted. This frequency ν_{max} is completely independent of the metal used for the anticathode.

On the other hand the curve is strongly dependent on the voltage V used for accelerating the electrons, as shown in figure 7.6(b); the maximum frequency ν_{max} increases proportionally with the voltage V.

If the constant of proportionality is measured, it is found to be numerically equal to the constant h/e which comes into the photoelectric effect (see chapter 1): this implies that $eV = h\nu_{max}$.

Since the continuous spectrum depends mainly on the velocity of the incident electrons, it may be surmised that the corresponding X-radiation is emitted by these electrons. When electrons moving at high velocities reach the anticathode, they are subjected to strong electrostatic forces arising mainly from the nuclei of the constituent atoms. The electrons are therefore strongly accelerated (in the vectorial sense of the term); according to the classical theory of radiation the accelerated charges emit electromagnetic waves and the greater the acceleration, the higher their frequency (see appendix 2). It is the sudden slowing down of the electrons when they penetrate the anticathode which is responsible for the continuous spectrum; this may be described as *deceleration radiation* but often the German term *Bremsstrahlung* is used.

Although the classical theory of radiation explains the principle of Bremsstrahlung, it is impossible to explain its mechanism in detail without reference to quantum theory. The problem can be treated as an elastic collision between an incident electron and a stationary target nucleus, in which part of the kinetic energy of the electron is transferred into kinetic energy of a photon. (We consider all the energy of the photons as kinetic energy, see sections 2.2.2 and 2.3.2.) The energy of the projectile is much smaller than the rest energy of the target nucleus and, as always in such collision problems, the target nucleus takes up the momentum but receives hardly any energy (see section 2.3.4).

To summarise, each photon is emitted by a single electron, and the maximum energy that an electron may provide in the form of a photon is its entire kinetic energy when it is brought to a complete stop. It may therefore be understood why the photons have a maximum energy

$$\boxed{h\nu_{max} = \tfrac{1}{2}mv^2 = eV}$$

In fact, measurement of the maximum frequency ν_{max} of the continuous X-ray spectrum has been used in metrology, and for a long time has provided perhaps the most exact measurement of the ratio h/e. (The first measurement of h by this method was carried out by Duane and Hunt in 1915.)

7.3.2 The Line Spectrum—Comparison with the Absorption Spectrum

In contrast to the continuous spectrum, the line spectrum depends mainly on the material from which the X-rays originate, this being either the anticathode of the X-ray tube or the absorbing material used in a fluorescence experiment. In the first case, the frequencies of the spectral lines are independent of the voltage which accelerates the electrons and in the second case independent of the frequency of the incident radiation. They depend only on the chemical elements of which the material is composed. The frequencies are characteristic of the atoms of the chemical elements and the latter may be either in a pure state or part of an alloy or chemical compound. This is equally true of the frequencies of the absorption discontinuities (but see comment I(b) of section 7.1).

The line spectrum characteristic of an element is similar in many ways to the line spectra commonly observed in the optical region:

(a) the lines are grouped in series, appearing in distinct regions of frequency;

(b) the lines obey the Ritz combination principle; the frequencies or wavenumbers can be expressed as differences of spectral terms

$$\frac{1}{\lambda_{np}} = \frac{1}{c}\nu_{np} = T_p - T_n$$

in other words they can also be interpreted as transitions between energy levels of that element.

These very general laws, which apply to all line spectra, can be supplemented in the case of X-rays by comparison with the absorption spectra.

(i) The various series of the spectrum occur at frequencies close to each absorption discontinuity, just below such a discontinuity. This is shown in figure 7.7 where the absorption spectrum (figure 7.7(a)) and the emission spectrum (figure 7.7(b)) are drawn on the same frequency scale. Accordingly each series is designated by the same letter as its related discontinuity, this being its frequency limit.

(ii) The spectral terms T_n that define the spectral lines are identical to the wavenumbers $1/\lambda_K$, $1/\lambda_L$, $1/\lambda_M$ and so on, corresponding to the absorption discontinuities. In other words, *the frequency of each emission line is equal to the difference of frequencies between two absorption discontinuities.*

This confirms the interpretation of the absorption discontinuities in terms of binding energy, as we have seen in previous sections

$$E_K = -h\nu_K, \ E_L = -h\nu_L, \ E_M = -h\nu_M, \ E_N = -h\nu_N$$

and so on. For the K series, one observes

the K_α line: $\nu_{K_\alpha} = \nu_K - \nu_L$, so that $h\nu_{K_\alpha} = E_L - E_K$
the K_β line: $\nu_{K_\beta} = \nu_K - \nu_M$, so that $h\nu_{K_\beta} = E_M - E_K$
the K_γ line: $\nu_{K_\gamma} = \nu_K - \nu_N$, so that $h\nu_{K_\gamma} = E_N - E_K$

The K series is formed from all transitions in which an electron falls back to the same lower energy level E_K.

The same is observed in the L series:

the L_β line: $\nu_{L_\beta} = \nu_L - \nu_M$ so that $h\nu_{L_\beta} = E_M - E_L$
the L_γ line: $\nu_{L_\gamma} = \nu_L - \nu_N$ so that $h\nu_{L_\gamma} = E_N - E_L$

The L series is formed from all transitions in which an electron falls back to the same lower energy level E_L, and similarly for each of the series.

To simplify the examples given above, we have not distinguished between the energy levels L_I, L_{II}, L_{III} or M_I, M_{II} and so on. A more detailed study would do this, and enable consideration of the L series, for example, as three distinct series L_I, L_{II} and L_{III} whose lines are unresolved because in the corresponding transitions the electrons fall back to energy levels E_{L_I}, $E_{L_{II}}$ or $E_{L_{III}}$ which are very close to one another.

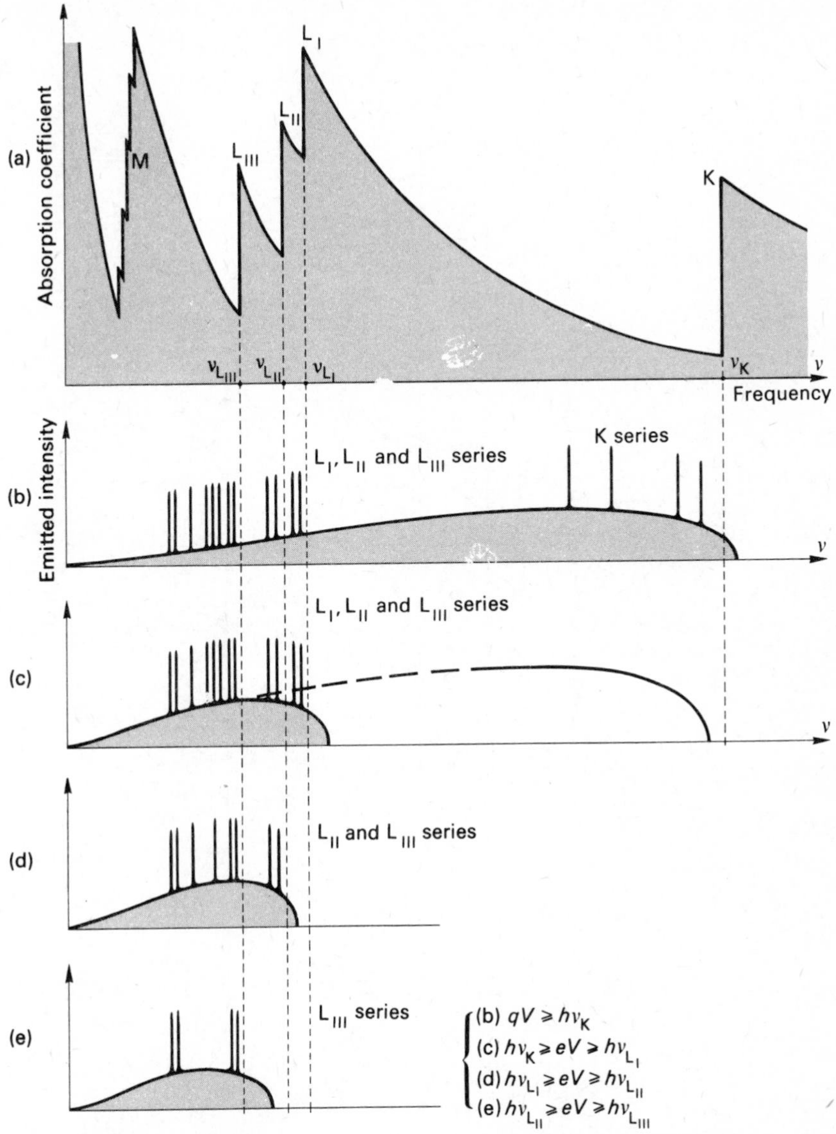

Figure 7.7 Comparison of (a) absorption and (b)–(e) emission spectra

Comment Table 7.1 gives examples of transitions corresponding to observed lines in the series K, L_I, L_{II}, L_{III}. The upper energy levels of these transitions are shown horizontally, and the lower levels vertically.

The great number of blank spaces in this table will be noted. Only transitions between certain pairs of levels are observed, and not between others. For the hydrogen atom, transitions between all pairs of levels are observed, but this is an exceptional

Table 7.1 Conventional symbols for various X-ray transitions

Lower levels of the transitions	↓	Upper energy levels of the transitions														
		L_I	L_{II}	L_{III}	M_I	M_{II}	M_{III}	M_{IV}	M_V	N_I	N_{II}	N_{III}	N_{IV}	N_V	N_{VI}	N_{VII}
K series	K		$K_{\alpha 2}$	$K_{\alpha 1}$		$K_{\beta 2}$	$K_{\beta 1}$				$K_{\gamma 2}$	$K_{\gamma 1}$				
L_I series	L_I					$L_{\beta 4}$	$L_{\beta 3}$				$L_{\gamma 2}$	$L_{\gamma 3}$				
L_{II} series	L_{II}				L_η			$L_{\beta 1}$		$L_{\gamma 5}$			$L_{\gamma 1}$			
L_{III} series	L_{III}				L_l			$L_{\alpha 2}$	$L_{\alpha 1}$	$L_{\beta 6}$			$L_{\beta 2'}$	$L_{\beta 2}$		

case. In general, 'selection rules' exist which determine the allowed transitions from amongst all those that would seem possible *a priori* (see volume 2, chapter 4).

To summarise, physicists employ three different experimental techniques for studying the deep energy levels of atoms, E_K, E_{L_I}, $E_{L_{II}}$, $E_{L_{III}}$, E_{M_I} *and so on.*

In describing these energy levels, slightly different terms are often used: one speaks of the K shell, of the L shell or of the M shell and so on, and for the L shell for example, three subshells are distinguished, L_I, L_{II} and L_{III}.

7.3.3 Conditions for Observation of X-rays—the Exclusion Principle

We have emphasised that the spectral frequencies of X-ray lines depend only on the atom studied and not on the experimental conditions. On the other hand, the possibility of observing any particular spectral line does depend on the experimental conditions.

When the radiation from an X-ray tube is analysed, the observed spectral lines are always superimposed on a continuous spectrum (see figure 7.5). Frequencies greater than the maximum limit ν_{max} of the continuous spectrum are never observed, and if this is too low, because the accelerating voltage V for the electrons is small, certain lines are missing from the observed spectrum (see figure 7.7(c), (d) and (e)).

> More precisely, the lines of a given series are either all missing simultaneously, or all present together in the observed spectrum. The complete series appears in the spectrum *on condition that the maximum frequency* ν_{max} *of the continuous spectrum is greater than the frequency of the corresponding absorption discontinuity.*

For example, the complete K series appears on condition that the accelerating voltage V is high enough so that $\nu_{max} > \nu_K$, in other words

$$\tfrac{1}{2}mv^2 = eV = h\nu_{max} > h\nu_K = W_K$$

This indicates that the kinetic energy of the accelerated electrons that bombard the anticathode must be greater than the extraction energy $W_K = -E_K$ of the electrons bound to the atom with an energy E_K. On giving up their kinetic energy to an atom, the accelerated electrons are able to detach an electron in the energy level E_K.

The same considerations arise in fluorescence emission. We have seen that atoms irradiated by a primary beam of X-rays emit a fluorescence radiation whose spectrum is the line spectrum typical of those atoms. However, if the frequency ν of the primary radiation is too small, the whole spectrum is not observed; certain series are absent. Exactly as before, the lines of a particular series are all simultaneously absent or all present together in the observed spectrum. *A complete series appears in the spectrum on condition that the frequency* ν *of the primary X-radiation is greater than the frequency of the corresponding absorption discontinuity.*

For example, for the K series

$$\boxed{\nu \geqslant \nu_K}$$

Thus, from what we have seen in sections 7.1 and 7.2, the K series appears only when the primary radiation is able to ionise the atom by detaching from it an electron of binding energy E_K. To summarise, whatever the technique used, K *series transitions, corresponding to the falling of an electron to the* K *level, are possible only if an electron in the* K *level has previously been detached from the atom.*

The same phenomenon is observed for each of the series corresponding to the other energy levels. This is a way of clearly distinguishing, for example, between the three L series, (see figure 7.7). By varying the energy eV of the incident electrons (for emission from an X-ray tube) or the energy $h\nu$ of the primary X-radiation (for emission by fluorescence), then depending on its magnitude, it may be sufficient to detach an electron from one of the L levels or from two of these levels, or indeed from all three:

if eV (or $h\nu$) $\geqslant h\nu_{L_I}$, the three series L_I, L_{II}, L_{III} are observed (figure 7.7(b) and (c));

if $h\nu_{L_I} \geqslant eV$ (or $h\nu$) $\geqslant h\nu_{L_{II}}$, the L_I series disappears; only the two series L_{II} and L_{III} are observed (figure 7.7(d));

if $h\nu_{L_{II}} \geqslant eV$ (or $h\nu$) $\geqslant h\nu_{L_{III}}$, only the L_{III} series is observed (figure 7.7(c));

if $h\nu_{L_{III}} \geqslant eV$ (or $h\nu$), none of the L series is observed.

Comment We shall not try to interpret in detail the position of the lines drawn in figure 7.7. We merely state the following: a line of frequency between ν_{L_I} and $\nu_{L_{II}}$ certainly appears in the L_I series; a line of frequency between $\nu_{L_{II}}$ and $\nu_{L_{III}}$ certainly does not appear in the L_{III} series.

All this may be explained by making the hypothesis that the number of places available for the electrons in each energy level is limited.

Since the most stable states are those of minimum energy, electrons in an atom normally accumulate in the deepest levels where they occupy all the available places. Thus for nearly all atoms (with the exception of the lightest) the K and L levels are usually completely occupied; this is described by saying that the K and L shells are full. Therefore, an electron in higher levels can fall to the K level (with the emission of a K series line) or to an L level (with the emission of an L series line) only if a vacant place has been created there beforehand.

The emission of X-ray photons by an atom occurs by spontaneous re-arrangement of its electrons, *following the creation of an empty space in one of its deeper energy levels by ionisation* (see footnote on page 172).

Comment The spontaneous rearrangement of an ionised atom sometimes proceeds in a very different way, which was observed and explained in 1925 by the French physicist Auger. When an electron from an L level, for example, loses energy and takes up an available place in the K level, it liberates a quantity of energy

$$E_L - E_K = h(\nu_K - \nu_L)$$

This energy may appear in the form of a K_α line photon, but it can also be utilised within the atom by giving another electron, in the L level for example, the extraction energy $W_L = -E_L = +h\nu_L$ necessary to detach it from the atom (that is, to place it at infinity with zero velocity). To conserve energy, this electron must carry away any excess energy in the form of kinetic energy

$$\tfrac{1}{2}mv^2 = (E_L - E_K) - W_L = h(\nu_K - 2\nu_L)$$

Thus the appearance in the velocity spectrum of photoelectrons whose velocities are independent of the frequency of the incident radiation is explained (see comment I at the end of section 7.2). To summarise, the Auger effect is an internal photoelectric effect produced by the characteristic emission frequencies of the atom. The atom that undergoes the Auger effect becomes doubly ionised; the effect can occur many times in cascade and thus gives rise to highly ionised atoms.

In conclusion, the study of X-ray spectra has led us to formulate one of the most fundamental hypotheses for the understanding of atomic structure. It remains to determine the number of places available in each level (or alternatively in each shell). It was this research that resulted in the enunciation in 1925 of the *Pauli exclusion principle*, the cornerstone of the theory of the structure of the periodic system of the elements (see volume 2, chapter 2). However, to appreciate the significance of this exclusion principle, we must study the magnetic properties of atoms, a topic on which the third part of this book will concentrate, starting in the next chapter.

7.3.4 Comparison with Optical Spectra

We can describe the energy levels of a heavy atom and the transitions between them in the same way as we described in the previous chapter the levels of the hydrogen atom and the corresponding transitions. This is shown in a simplified way in part (a) of figure 7.8 (the energy scale has been distorted; in comparison with the energy of level K, all the higher levels ought to be much closer together).

For the hydrogen atom, the only level normally occupied by the single electron is the lowest level. In the case of a heavy atom, having a much larger number of electrons Z, all the deepest levels are usually completely occupied. The highest of the occupied levels forms the ground level for the optical spectra: all levels higher than this are excited levels normally unoccupied.

The lines of the optical spectrum correspond to transitions between normally free excited levels; the energies involved are of the order of a few electron-volts at the most. The optical resonance lines in particular are transitions between an excited level and the ground level. X-ray lines on the other hand, correspond to transitions between deep energy levels that are normally occupied, and the energies involved are from a few hundreds of electron-volts to more than 100 000 eV.

It can now be seen why it is almost impossible to carry out an experiment

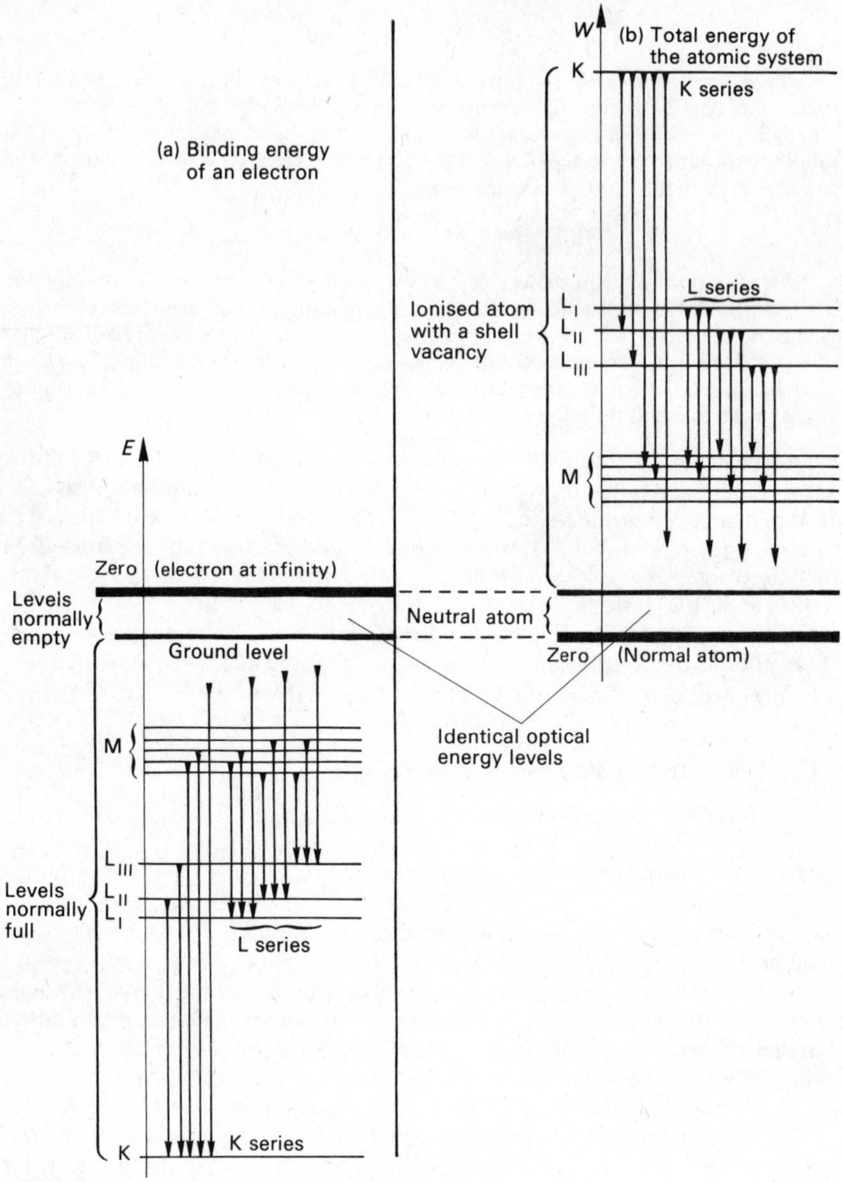

Figure 7.8 Energy levels and X-ray transitions. (To clarify the figure, only the K, L and M levels and the ground level are shown)

with X-rays analogous to that of optical resonance. For example, suppose that atoms are irradiated with the K_α line from their own spectrum such that $h\nu_{K_\alpha} = E_L - E_K$; it is impossible for the electrons in the K level to absorb this energy and be raised to the higher L level because all the available places in the L level are normally occupied.

For an optical resonance experiment to be possible with a given spectral line, it is insufficient that the lower energy level of the corresponding transition is normally occupied by electrons; there must also be places available in the higher energy level of the transition.

In the optical region, it is the first condition, with respect to the lower level, that is important and that distinguishes resonance lines from other spectral lines, because the higher energy levels are normally unoccupied excited levels. On the other hand, the lines usually observed in X-ray spectra correspond to transitions between deep energy levels that are normally completely occupied by electrons, and the second condition, with respect to the higher level, is not fulfilled. A resonance experiment is possible only for lines corresponding to a transition between a deep energy level and an excited level; but the frequency of such a line is very close to the frequency of an absorption discontinuity, near to a series limit, so its intensity is usually negligible.

However, if the absorption spectra are examined with high resolution, it is observed that the absorption discontinuities are not as well defined as indicated on the distorted frequency scale; they show a genuine 'structure' over a small frequency interval, with peaks and troughs in the absorption coefficient K. The actual limit, which represents the ionisation of the atom, is in fact preceded on the low-frequency side by several *resonance absorption lines* representing transitions between the deep level corresponding to this limit and certain excited states.

Comment I Many books use a different notation to represent X-ray transitions, leading to energy diagrams differing from those we have used. This is shown in figure 7.8(b).

In our discussion we have used the individual binding energy of each electron E_K, E_L and so on. This energy is defined by adopting the convention that the energy is zero when the electron is at infinity (outside the atom) with zero velocity. In these circumstances, binding energies are all negative. However, the discussion could also be based on the total energy of the atom, and then the atom in its normal state is assumed to have zero energy. This does not alter the description of the optical transitions—the energy scale is simply translated by shifting the zero—but the X-rays on the other hand are emitted by an ionised atom, and this completely inverts the order of the levels. When an electron in the K level is detached from an atom, it is given an extraction energy $W_K = -E_K + h\nu_K$. An atom with a vacancy in the K level is then represented by a very large positive energy equal to $h\nu_K$. An atom having a vacancy in the L level is represented by a much smaller positive energy

$$W_L = h\nu_L$$

and so on. To represent a spectral transition, instead of saying (as we have done above) that an electron falls from a higher level L to a lower level K, it should now be said that an ion of high energy having a vacancy in the K level is transformed into an ion of lesser energy having a vacancy in the L level. This is how the diagram shown in figure 7.8(b) should be read.

Comment II Apart from the final paragraph of this section, we have disregarded the structure in the absorption coefficient K at each absorption discontinuity. For simplicity, we have assumed that an inner electron of an atom must be detached (the atom must be ionised), thereby creating a vacancy in a deep energy level, before X-ray lines can be observed. However, all that is necessary is to excite an inner electron to any unoccupied level, of which there are many within a few electron-volts of the ionisation limit.

7.4 The Moseley Law

In the first three sections of this chapter we have shown how it is possible to identify and measure deep energy levels of atoms. However, we have not concerned ourselves with the actual values that can be measured; we shall now do this.

7.4.1 Experimental Results

Let us compare the energies E_K corresponding to the deepest level of each atom. The modulus of the energy $E_K = h\nu_K$ increases with atomic number Z of the atom, and very nearly follows a parabolic law in Z^2.

Comment Historically, this parabolic law in Z^2 was first suggested in relation to the frequencies of emission lines (Moseley, 1913); but the law is more readily confirmed from the frequencies of the absorption discontinuities.

The first of these energy levels for $Z = 1$, is the ground state of the hydrogen atom, and we have calculated this in the preceding chapter from the Rydberg constant R: $|E| = Rhc$. This allows the constant in the parabolic law to be determined as follows

$$\frac{|E_K|}{Rhc} = \frac{\nu_K}{Rc} \simeq Z^2$$

Graphically, it is easier to check a linear law than a parabolic law. For this reason, it is convenient to draw, as a function of the atomic number Z, the curve representing the dimensionless quantity

$$y = \sqrt{\left(\frac{|E_K|}{Rhc}\right)} = \sqrt{\left(\frac{\nu_K}{Rc}\right)} = \sqrt{\left(\frac{1}{R\lambda_K}\right)}$$

which should be very nearly a straight line. On the same graph analogous curves may be drawn for the other energy levels E_{L_I}, $E_{L_{II}}$ and so on. In this way the Bohr–Coster diagram, reproduced in figure 7.9, is obtained.

From this diagram, it may be seen that the curves thus drawn are to a *first approximation* nearly straight lines, but that these lines do not pass through the origin:

for the K level one obtains: a line of slope 1: $y = Z - s_K$, with $s_K \approx 1$ to 2;

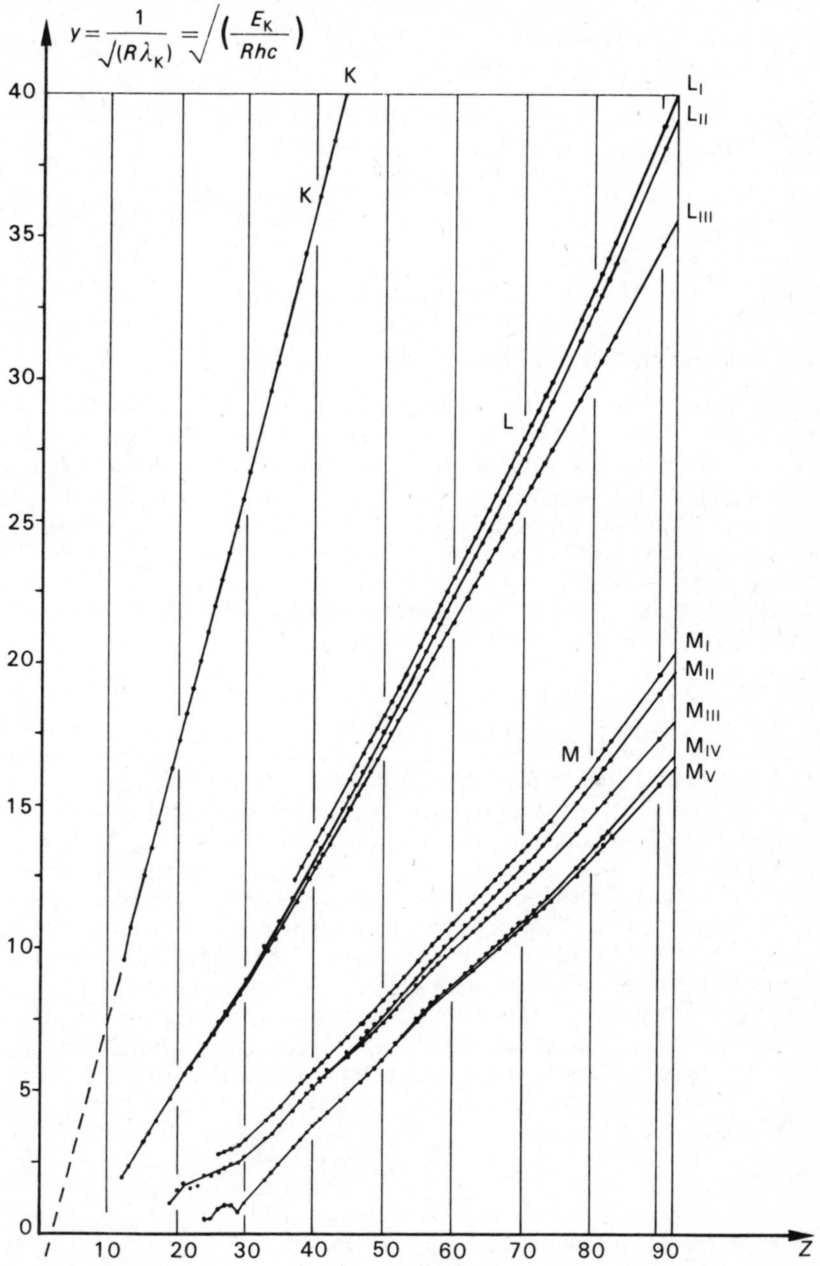

Figure 7.9 The Bohr–Coster diagram (revised in 1930)

for the L level: three lines of slope $\frac{1}{2}$: $y = \frac{1}{2}(Z - s_L)$, with $s_L \approx 10$;
for the M level: five lines of slope $\frac{1}{3}$: $y = \frac{1}{3}(Z - s_M)$, with $s_M \approx 20$.

Thus to a *first approximation* the values of the deep energy levels of the different atoms as a function of their atomic number Z are given by the same formula

$$\boxed{E_n = -Rhc \frac{(Z - s_n)^2}{n^2}} \qquad \text{where for} \quad \begin{cases} \text{level K, } n = 1 \text{ and } s_1 \approx 1 \text{ to } 2 \\ \text{level L, } n = 2 \text{ and } s_2 \approx 10 \\ \text{level M, } n = 3 \text{ and } s_3 \approx 20 \end{cases}$$

At the interior of each atom the energy levels are determined mainly by the principal quantum number n, as in the case of hydrogen.

7.4.2 Interpretation by the Bohr Theory

This formula differs only in the coefficient s from the formula $E = -RhcZ^2/n^2$ derived from Bohr's theory (see chapter 6). This may be interpreted quite easily by taking into account the screening of the nucleus from the electron being considered by the other electrons that are much closer to the nucleus; this is why s is called the screening coefficient. In reality, Bohr's theory is useful only for a single electron system (the hydrogen atom and hydrogen-like ions). For atoms having several electrons, the interaction energy between electrons must be added to the electron–nucleus electrostatic interaction energy which retains the same form

$$W(r) = \frac{C}{r} = -\frac{1}{4\pi\varepsilon_0} \frac{Ze^2}{r}$$

However, the action of the other electrons on one particular electron can be roughly accounted for in terms of average values. Since the electrons orbit about the nucleus at very high frequencies, their average effect may be estimated to a first approximation as if their charge were uniformly distributed over a sphere centred on the nucleus, of radius equal to their mean separation. Application of Coulomb's law of electrostatics (which we discussed in section 5.2.3 in relation to the Thomson model) shows that electrons farther from the nucleus than the electron being considered do not have any effect on it, whereas the electrons nearer to the nucleus have the same average effect as if they were at the centre of the sphere, coincident with the nucleus.

By means of this very rough model, the electron interactions can be accounted for by subtracting from the charge on the nucleus that of the s electrons that are nearer to it than the electron being considered, the latter having a mean total potential energy

$$W(r) = -\frac{1}{4\pi\varepsilon_0} \frac{(Z - s)e^2}{r} = \frac{C}{r}$$

If Bohr's formula for quantised energies is applied

$$E = \frac{\mu C^2}{2\hbar^2 n^2} = -Rhc \frac{(Z - s)^2}{n^2}$$

the experimental formula framed above is obtained. Such agreement validates the generalisation of this quantisation formula to many atoms having electrons.

Without attaching too much importance to this crude model, it may, however, be admitted that the screening coefficient s_n represents an order of magnitude for the number of electrons nearer to the nucleus than those whose binding energy is being calculated.

This screening coefficient can have several distinct values for the same principal quantum number n. In fact the energy depends mainly on n, but the levels L, M, N and so on show a fine structure which subdivides each of them into several closely spaced levels. This fine structure is relatively much larger than for hydrogen because it results from interactions between electrons; for large values of n or for small values of Z, the fine structure differences are of

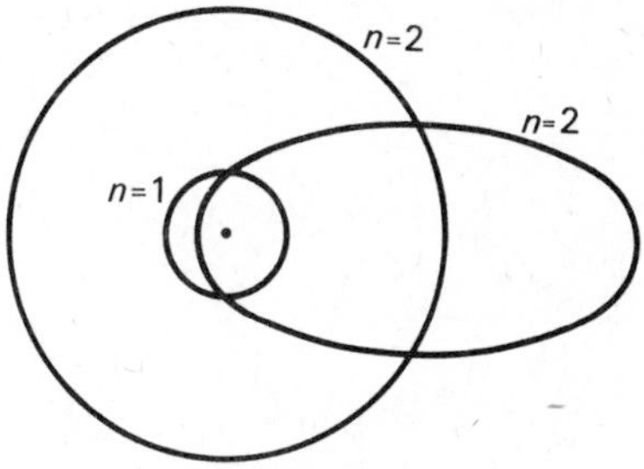

Figure 7.10 Penetration of the orbits

the same order of magnitude as the total energy. Sommerfeld's elliptic orbits enable some of these differences to be explained as an *orbital penetration effect* (figure 7.10): an electron describing a very oblate elliptic orbit passes very close to the nucleus in a region where other electrons do not screen it from the nucleus. The average screening coefficient of an elliptic orbit therefore must be smaller than that of a circular orbit of the same diameter (the same principal quantum number n (see section 6.3)). In other words the energy of an elliptic orbit is lower (greater in absolute value) than that of a circular orbit of the same diameter. This orbital penetration effect is not the only cause of fine structure; magnetic interactions must also be taken into account (see volume 2, chapter 4).

7.5 General Conclusions

To summarise, the study of X-ray spectra on which this chapter has concentrated provides three concurring methods for determining the deeper energy levels of atoms. Two main ideas arise from these measurements:

(a) the restriction on the number of electrons having the same binding energy—it follows that the deep shells of the heavier atoms are all normally occupied;

(b) the generalisation to atoms having many electrons of the idea of the principal quantum number n and of Bohr's formula for the quantised energy $E=-\mu C^2/2\hbar^2 n^2$ where C is the constant appearing in the equation for the potential energy $W(r)=C/r$.

These two ideas, firmly supported by experimental results, must be incorporated into the new quantum theory (see volume 2, chapters 1 and 2).

PART 3

Angular Momentum and Magnetic Moment

8

Classical Magnetism Arising from Orbital Motion

In chapters 5, 6 and 7, we have started to discuss the internal structure of atoms. Our interpretation of the observed energy levels was based, to a first approximation, on the law of electrostatic interaction between the various electric charges which make up the atom. However, when electric charges are in motion, magnetic interaction forces also exist between them. These forces are usually weak by comparison with electrostatic forces, and this is why we were able to disregard them in a first approximation. On the other hand, if we wish to achieve a deeper understanding of atomic structure, these magnetic interactions must be taken into account.

This chapter, and the three following, will familiarise the reader with the magnetic properties of atoms. In these four chapters we shall study mainly the overall magnetic properties of the atom that can be observed in experiments where the atom is subjected to an external magnetic field. Only in volume 2 do we come to study the magnetic interactions between the various internal constituents of the atom (volume 2, chapters 3, 4, 5 and 6).

*Henceforth we shall use the magnetic induction vector **B**, but no distinction will be made between induction and field.*

8.1 Microscopic Definition of Magnetic Moment

8.1.1 Review of the Classical Notion of Magnetic Moment

The magnetic moment vector is introduced into electromagnetism in order to characterise a small electric current (see figure 8.1): its direction is perpendicular to the plane of the circuit; its sense is such that the current flows around it

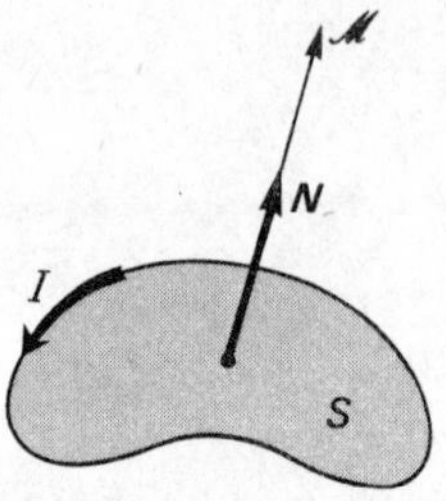

Figure 8.1 Illustration of the direction of the magnetic moment of a small electric circuit

in the trigonometric sense; its magnitude is equal to the product of the current I and the area S of the circuit. It is defined by the formula

$$\boxed{\mathcal{M} = \frac{1}{\kappa} ISN}$$

where N is a unit vector normal to the plane of circuit (with a direction chosen such that the current I flows around N in the trigonometric sense), and κ is a coefficient depending on the units. This coefficient is equal to the speed of light c in the gaussian system of units, often used in atomic physics, where the electric units of the c.g.s. electrostatic system are used in conjunction with the magnetic units of the c.g.s. electromagnetic system.

However, in the SI system of units, the coefficient κ is unity and may simply be omitted from all the formulae (see preface).

The importance of the moment vector $\mathcal{M}$ lies in the fact that it may be used to determine, to a first approximation, the interactions between the small circuit under consideration and other electric currents that are sufficiently far away.

(a) It is possible to calculate the magnetic induction field $\boldsymbol{B}$, produced by the small circuit C at a point P situated at a distance R far greater than the dimensions of the circuit (see figure 8.2) either from the vector potential

$$A(P) = \frac{\mu_0}{4\pi} \mathcal{M} \times \frac{\boldsymbol{u}}{R^2}$$

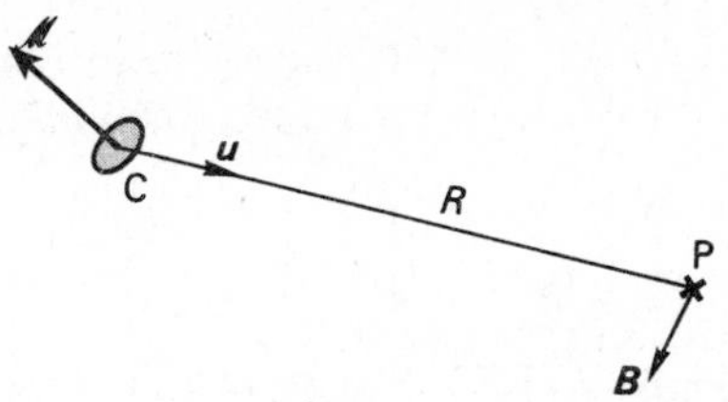

Figure 8.2 Magnetic field $\boldsymbol{B}$ at a distance R from a small circuit C

whereupon $\boldsymbol{B} = \text{curl}\,\boldsymbol{A}$ ($\boldsymbol{u}$ is a unit vector along CP directed from C to P); or from the pseudoscalar

$$U(P) = \frac{\mu_0}{4\pi}\mathcal{M} \cdot \frac{\boldsymbol{u}}{R^2}$$

whereupon $\boldsymbol{B} = -\,\text{grad}\,U$.

(b) The mechanical forces exerted on the small circuit when it is subjected to an external magnetic induction $\boldsymbol{B}^{\text{ext}}$ may be calculated from the interaction energy $W = -\mathcal{M} \cdot \boldsymbol{B}^{\text{ext}}$ (valid for variable or constant currents). Using the method of virtual work allows one to find (i) the resultant moment $\boldsymbol{\Gamma}$ of the forces applied to the small circuit $\boldsymbol{\Gamma} = \mathcal{M} \times \boldsymbol{B}^{\text{ext}}$; (ii) the resultant $\boldsymbol{F}$ of the forces applied to the small circuit, its components being

$$F_x = \mathcal{M} \cdot \frac{\partial \boldsymbol{B}^{\text{ext}}}{\partial x}, \qquad F_y = \mathcal{M} \cdot \frac{\partial \boldsymbol{B}^{\text{ext}}}{\partial y}, \qquad F_z = \mathcal{M} \cdot \frac{\partial \boldsymbol{B}^{\text{ext}}}{\partial z}$$

The force $\boldsymbol{F}$ exists only if the applied field $\boldsymbol{B}^{\text{ext}}$ is non-uniform.

Comment The vector $\boldsymbol{B}$ is locally (but not at all points in space) equal to a gradient: $B_z = -\partial U/\partial z$; hence

$$\frac{\partial}{\partial x}B_z = -\frac{\partial^2 U}{\partial x\,\partial z} = -\frac{\partial^2 U}{\partial z\,\partial x} = \frac{\partial}{\partial z}B_x$$

For example, two equivalent expressions may be obtained for the component F_z

$$F_z = \mathcal{M}_x\frac{\partial B_x}{\partial x} + \mathcal{M}_y\frac{\partial B_y}{\partial y} + \mathcal{M}_z\frac{\partial B_z}{\partial z} = \mathcal{M}_x\frac{\partial B_z}{\partial x} + \mathcal{M}_y\frac{\partial B_z}{\partial y} + \mathcal{M}_z\frac{\partial B_z}{\partial z}$$

This second expression can be summarised, for the three components, by writing

$$\boldsymbol{F} = (\mathcal{M} \cdot \text{grad})\,\boldsymbol{B}$$

For further details, textbooks on classical electromagnetism may be consulted.

8.1.2 Generalisation to a System of Point Charges in Motion

Electromagnetic theory, some results of which were summarised above, is based on the theory of magnetostatics, or in other words, on the existence of stationary currents. It will be recalled that conservation of electric charge leads to the condition div $\boldsymbol{j} = 0$ being imposed upon the current density vector

$$\boldsymbol{j} = \sum_n q_n \boldsymbol{v}_n$$

(q_n is the value of each point charge C_n, moving with velocity $\boldsymbol{v}_n$; the summation is over all charges contained within a unit volume.)

There is a problem in applying this theory to an atomic system because the number of charges in motion, according to the planetary model, is very small. However, the atom is a closed system where the charges are confined within a limited volume; for such a system it may be shown that in terms of mean values over time, the magnetostatic theory is on average correct (see volume 2, appendix 3). This theory, *in terms of mean values*, shows that the atom behaves as an element of zero current (in other words $\sum_n q_n \boldsymbol{v}_n = 0$). An expansion to first order gives, for the magnetic field produced at a large distance, an expression identical to that of a small electric circuit of magnetic moment $\mathcal{M}$, provided that

$$\boxed{\mathcal{M} = \frac{1}{\kappa} \sum_n \frac{1}{2} \boldsymbol{r}_n \times q_n \boldsymbol{v}_n}$$

where q_n is the value of the point charge C_n moving with velocity $\boldsymbol{v}_n$; $\boldsymbol{r}_n = OC_n$ is the radius vector between the origin and the charge C_n.

Taking into account that $\sum_n q_n \boldsymbol{v}_n = 0$, it may be shown that the vector $\mathcal{M}$ is independent of the origin O chosen for the calculation. $\mathcal{M}$ is called the magnetic dipole moment of the system of charges, abbreviated to magnetic moment. In an atomic system of point charges it plays the same role as the elementary magnetic moment vector defined for small electric circuits.

It should be remembered that this applies to mean values calculated over times of the order of several periods of the orbital motion. In terms of mean values over time, it is legitimate to reduce the magnetic interactions between the atom and the exterior to interactions with its magnetic moment.

Reference may be made to volume 2, appendix 3 for detailed proofs, but here we shall merely verify the equivalence of the above two definitions of magnetic moment in the particular case of an electric circuit confined to a plane.

We wish to find the vector

$$\kappa \mathcal{M} = \tfrac{1}{2} \sum_n \boldsymbol{r}_n \times q_n \boldsymbol{v}_n$$

by summing over all charges in motion within a circuit of copper wire. To do this, the circuit is divided up into small successive elements of length dl (see figure 8.3); it should be noted that $\boldsymbol{r}_n$ is practically the same for all charges within the same linear element. Let us first carry out the sum over all charges within the same element of length dl

$$\sum_{\substack{\text{within} \\ \mathrm{d}l}} (\tfrac{1}{2}\boldsymbol{r}_n \times q_n \boldsymbol{v}_n) = \tfrac{1}{2}\boldsymbol{r} \times \left[\sum_{\substack{\text{within} \\ \mathrm{d}l}} q_n \boldsymbol{v}_n \right] = \tfrac{1}{2}\boldsymbol{r} \times I \mathrm{d}\boldsymbol{l}$$

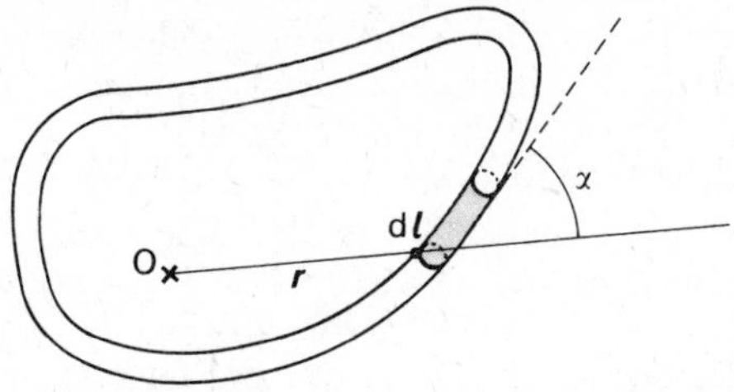

Figure 8.3 Calculation of the magnetic moment of an electric circuit in a plane

It remains to sum over the contributions of the different linear elements dl

$$\kappa \boldsymbol{\mathcal{M}} = \sum \begin{matrix}\text{contributions of} \\ \text{each linear element } \mathrm{d}\boldsymbol{l}\end{matrix} = \oint \tfrac{1}{2}\boldsymbol{r} \times I\,\mathrm{d}\boldsymbol{l} = \frac{I}{2}\oint \boldsymbol{r} \times \mathrm{d}\boldsymbol{l}$$

The vector product $\boldsymbol{r} \times \mathrm{d}\boldsymbol{l}$ is always perpendicular to the plane of the circuit and the vector sum reduces to an algebraic sum

$$|\boldsymbol{r} \times \mathrm{d}\boldsymbol{l}| = r\,\mathrm{d}l \sin\alpha = 2\,\mathrm{d}S$$

where dS is the area of the triangle constructed from the two vectors $\boldsymbol{r}$ and d$\boldsymbol{l}$. By integrating over the whole circuit, one obtains $\kappa\boldsymbol{\mathcal{M}} = IS\boldsymbol{N}$. We have thus established the equivalence that was postulated.

Similarly, as an example, the magnetic moment may be calculated for the particular case of a single charge q describing a circular orbit with a velocity of magnitude $v = 2\pi r/T$, where T is the period of the orbital motion. In this case $\boldsymbol{r}$ and $\boldsymbol{v}$ are perpendicular, and

$$|\boldsymbol{\mathcal{M}}| = \frac{1}{\kappa}\frac{q}{2} rv = \frac{1}{\kappa}\frac{q}{T}\pi r^2 = \frac{1}{\kappa}\frac{q}{T} S$$

The charge q always returns to the same point after a time T and is therefore equivalent to a current of magnitude $I = q/T$. In this particular case, we again establish the equivalence of the two expressions for the magnetic moment $\boldsymbol{\mathcal{M}}$.

It is the generalised definition of magnetic moment that is used throughout atomic physics. It will be used especially in section 8.3 for calculating diamagnetic susceptibilities. It also enables us to introduce at the beginning of the next chapter the fundamental concept of the gyromagnetic ratio.

8.2 Larmor's Theorem

Our aim is to evaluate the action of the magnetic field $\boldsymbol{B}$ on the electrons in orbital motion within an atom. A simple calculation for the particular case of a circular orbit whose plane is perpendicular to $\boldsymbol{B}$ shows that the effect of the field $\boldsymbol{B}$ is to change the angular velocity of the electron without changing the radius of its orbit. This result obtained for a simple case is a guide to the more general calculation that will be found further on.

We shall retain the usual symbol q to designate the algebraic value of electric charges, even for electrons. This algebraic notation is more convenient in magnetism problems and allows generalisation of certain results to charges other than electrons. We shall therefore distinguish between the algebraic charge q of the electron which is negative, and the elementary charge e which is a fundamental constant ($e = 1{\cdot}6 \times 10^{-19}$ C, and $q = -e$).

8.2.1 Introduction of a Rotating Frame of Reference (Larmor Frame)

When a magnetic field $\boldsymbol{B}$ is applied to an atom, the study of the motion of its electrons may be simplified by replacing the laboratory frame by a frame whose origin is at the centre of gravity of the atom and which rotates about an axis parallel to the magnetic field $\boldsymbol{B}$. To do this we have to solve the classical problem of transforming from one frame of reference to another; the laboratory frame is the inertial frame relative to which the moving frame rotates with an angular velocity $\boldsymbol{\omega}$ parallel to $\boldsymbol{B}$. The notation is given in table 8.1.

Table 8.1

	Absolute motion in the laboratory frame	Relative motion in the rotating frame	Motion of rotating frame
Velocity	$\boldsymbol{v}_{\text{lab}}$	$\boldsymbol{v}_r$	$\boldsymbol{v}_e = \boldsymbol{\omega} \times \boldsymbol{r} = \boldsymbol{\omega} \times \boldsymbol{r}_\perp$
Acceleration	$\boldsymbol{a}_{\text{lab}}$	$\boldsymbol{a}_r$	$\boldsymbol{a}_e = \boldsymbol{\omega} \times \boldsymbol{v}_e = \boldsymbol{\omega} \times (\boldsymbol{\omega} \times \boldsymbol{r}_\perp) = -\omega^2 \boldsymbol{r}_\perp$

$\boldsymbol{r}$ is the vector joining the centre of gravity of the atom to the electron being studied; $\boldsymbol{r}_\perp$ is the projection of this vector in a plane perpendicular to $\boldsymbol{\omega}$ (see figure 8.4).

We apply the laws for compounding velocities and accelerations

$$\boldsymbol{v}_{\text{lab}} = \boldsymbol{v}_r + \boldsymbol{v}_e$$

$$\boldsymbol{a}_{\text{lab}} = \boldsymbol{a}_r + \boldsymbol{a}_e + 2\boldsymbol{\omega} \times \boldsymbol{v}_r$$

The basic dynamic equation can be written only in the laboratory frame

$$m\boldsymbol{a}_{\text{lab}} = q\boldsymbol{E} + \frac{1}{\kappa} q\boldsymbol{v}_{\text{lab}} \times \boldsymbol{B} = q\boldsymbol{E} + \frac{q}{\kappa}(\boldsymbol{v}_r + \boldsymbol{v}_e) \times \boldsymbol{B}$$

$\boldsymbol{E}$ being the electrostatic field due to the nucleus and the other electrons. The force is $q\boldsymbol{E}$ only in absence of the magnetic field $\boldsymbol{B}$, but when a field $\boldsymbol{B}$ is applied,

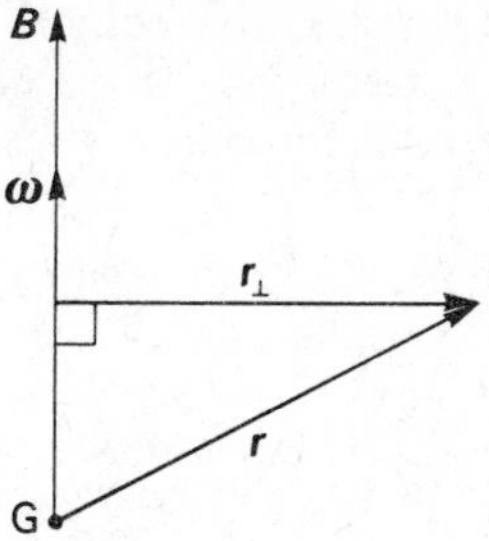

Figure 8.4 The vectors $\boldsymbol{B}$, $\boldsymbol{\omega}$, $\boldsymbol{r}_\perp$ and $\boldsymbol{r}$

the Laplace force must be added. We substitute for $\boldsymbol{a}_{\text{lab}}$ using the above formula for vector addition, and isolate $\boldsymbol{a}_r$ in the equation obtained

$$m\boldsymbol{a}_r = q\boldsymbol{E} - m\boldsymbol{a}_e - 2m\boldsymbol{\omega} \times \boldsymbol{v}_r + \frac{q}{\kappa}(\boldsymbol{v}_r + \boldsymbol{v}_e) \times \boldsymbol{B}$$

so

$$m\boldsymbol{a}_r = q\boldsymbol{E} - m\boldsymbol{a}_e + \frac{q}{\kappa}\boldsymbol{v}_e \times \boldsymbol{B} + \boldsymbol{v}_r \times \left(2m\boldsymbol{\omega} + \frac{q}{\kappa}\boldsymbol{B}\right)$$

$\boldsymbol{v}_r$ can be eliminated from the equation if $\boldsymbol{\omega}$ is given a particular value called the Larmor angular velocity

$$\boxed{\boldsymbol{\omega} = -\frac{1}{\kappa}\frac{q}{2m}\boldsymbol{B}}$$

(This angular velocity of the reference frame is chosen such that the Coriolis acceleration compensates for the Laplace force. *The formula is algebraic: if q is negative, $\boldsymbol{\omega}$ has the same direction as $\boldsymbol{B}$.*) By noting that

$$m\boldsymbol{a}_e = m\boldsymbol{\omega} \times \boldsymbol{v}_e = +\frac{1}{2}\frac{q}{\kappa}\boldsymbol{v}_e \times \boldsymbol{B}$$

then

$$m\boldsymbol{a}_r = q\boldsymbol{E} - m\boldsymbol{a}_e + \frac{q}{\kappa}\boldsymbol{v}_e \times \boldsymbol{B} = q\boldsymbol{E} + m\boldsymbol{a}_e$$

Thus, by transferring into a particular rotating frame, called the Larmor frame, we have, in the second term of the equation for the acceleration, replaced the Laplace force by the term $m\boldsymbol{a}_e$. The rest of the argument depends on a single approximation.

8.2.2 The Approximation leading to Larmor's Theorem

Bearing in mind the strengths of magnetic field that can be produced in the laboratory, it may be shown that the Laplace force $q\boldsymbol{v}_{\text{lab}} \times \boldsymbol{B}$ is always small compared with the electrostatic force $q\boldsymbol{E}$. This can be expressed in another way by introducing the mean angular velocity Ω of the electron in its normal orbital motion around the nucleus. For an estimate of orders of magnitude, we can express the electrostatic force as a function of this orbital angular velocity

$$\text{electrostatic force: } qE \approx ma_{\text{lab}} \approx m\Omega v_{\text{lab}}$$

Using the relation between B and ω, we can also express the Laplace force as a function of the Larmor angular velocity

$$\text{magnetic force: } \frac{q}{\kappa} v_{\text{lab}} B \approx m\omega v_{\text{lab}}$$

(In this order of magnitude estimate, the factor two is disregarded.)

$$\frac{\text{magnetic force}}{\text{electrostatic force}} \approx \frac{qv_{\text{lab}} B}{\kappa qE} \approx \frac{\omega}{\Omega} \approx \frac{v_e}{v_{\text{lab}}}$$

For a field $B = 1$ tesla (= 10 000 gauss), we calculate a Larmor frequency $\omega/2\pi$ of the order of 10^{10} Hz, whereas frequencies of orbital motion $\Omega/2\pi$ are comparable to or greater than optical frequencies, of the order of 10^{15} Hz. The Laplace force is thus very small compared with the electrostatic force.

The second term of the equation obtained above in the rotating frame involves moreover the term $m\boldsymbol{a}_e$; the same type of order of magnitude argument allows us to write

$$\frac{ma_e}{qE} \approx \frac{a_e}{a_{\text{lab}}} \approx \frac{\omega v_e}{\Omega v_{\text{lab}}} \approx \frac{\omega^2 r}{\Omega^2 r} = \left(\frac{\omega}{\Omega}\right)^2$$

If we make an approximate calculation by including terms of first order in relation to the small quantity ω/Ω, but disregarding terms of second order, we can neglect $m\boldsymbol{a}_e$ compared with $q\boldsymbol{E}$ and obtain the approximate equation in the Larmor frame $m\boldsymbol{a}_r = q\boldsymbol{E}$.

This equation is identical to that written in the laboratory frame in the absence of a magnetic field $m\boldsymbol{a}_{\text{lab}} = q\boldsymbol{E}$ (because the electric field $\boldsymbol{E}$ due to the nucleus possesses rotational symmetry about the axis of rotation of the Larmor frame). Hence Larmor's theorem may be stated

The motion in the Larmor frame after establishing a magnetic field $\boldsymbol{B}$ *is identical to the motion normally existing in the laboratory frame in the absence of a magnetic field.*

In other words, application of the magnetic field $\boldsymbol{B}$ has the effect of imposing an additional rotational motion on the atomic electrons, which causes all of them to rotate about an axis parallel to $\boldsymbol{B}$, passing through the centre of gravity of the atom, with the Larmor angular velocity

$$\omega = -\frac{1}{\kappa}\frac{q}{2m}B$$

that is to say half the cyclotron frequency (the angular velocity of a free electron in a magnetic field).

8.3 Application to the Calculation of the Diamagnetic Susceptibility

It is well known that substances may be classified by their magnetic properties into two groups (i) the paramagnetic substances whose induced magnetisation is in the same direction as the magnetic field (we shall discuss these in section 9.2); (ii) the diamagnetic substances whose magnetisation is in the opposite direction to the magnetic field. We shall see how the Larmor rotation explains this phenomenon of diamagnetism. We consider an individual atom or molecule, denoting its properties and those of its electrons by the subscript 0 in the absence of a magnetic field; their properties when subjected to a magnetic field $\boldsymbol{B}$ are denoted without a subscript.

In the absence of a field, the atom possesses a magnetic moment

$$\mathcal{M}_0 = \sum \frac{1}{2\kappa} \boldsymbol{r}_0 \times q\boldsymbol{v}_0$$

(this equation is written in the laboratory frame).

In the presence of a field, it has a magnetic moment

$$\mathcal{M} = \sum \frac{1}{2\kappa} \boldsymbol{r} \times q\boldsymbol{v}_{\text{lab}} = \sum \frac{1}{2\kappa} \boldsymbol{r} \times q(\boldsymbol{v}_r + \boldsymbol{v}_e)$$

using the same notation as in the preceding section.

From Larmor's theorem, the motion in the rotating frame in the presence of a field is identical to the motion in the laboratory frame in the absence of a field, so that $\boldsymbol{r} \times \boldsymbol{v}_r = \boldsymbol{r}_0 \times \boldsymbol{v}_0$. Hence

$$\mathcal{M} - \mathcal{M}_0 = \sum \frac{q}{2\kappa} \boldsymbol{r} \times \boldsymbol{v}_e$$

By introducing the vector $\boldsymbol{r}_\perp$ projecting the radius vector $\boldsymbol{r}$ in the plane perpendicular to the axis of rotation $\boldsymbol{\omega}$, and by expanding the double vector product, one may write

$$\boldsymbol{r} \times \boldsymbol{v}_e = \boldsymbol{r} \times (\boldsymbol{\omega} \times \boldsymbol{r}_\perp) = r_\perp^2\, \boldsymbol{\omega} - (\boldsymbol{\omega} \cdot \boldsymbol{r})\, \boldsymbol{r}_\perp$$

We remarked at the beginning of this chapter that the concept of magnetic moment can be applied to an atom as long as its mean value in time is used (calculated over time intervals of the order of periods of the orbital motion). Now the vector $\boldsymbol{r}_\perp$ takes all directions in the plane perpendicular to $\boldsymbol{\omega}$ during the orbital motion of an electron; it is therefore justifiable to assume that the contribution of the vectors $(\boldsymbol{\omega} \cdot \boldsymbol{r})\boldsymbol{r}_\perp$ to the mean value is zero. On the other hand, the vectors $r_\perp{}^2\boldsymbol{\omega}$ have a constant direction and their contribution is finite. Thus we obtain

$$\mathcal{M} - \mathcal{M}_0 = \sum \frac{q}{2\kappa} r_\perp{}^2 \omega = \frac{q}{2\kappa} \omega \sum r_\perp{}^2 = \frac{q}{2\kappa} \omega \cdot Z\overline{r_\perp{}^2}$$

where Z is the number of electrons in the atom (the atomic number) and $\overline{r_\perp{}^2}$ is the mean value of $r_\perp{}^2$ (calculated both over time and over all the atomic electrons). Substituting

$$\omega = -\frac{1}{\kappa}\frac{q}{2m} \boldsymbol{B}$$

one obtains

$$\mathcal{M} - \mathcal{M}_0 = -\frac{1}{\kappa^2}\frac{q^2}{4m} Z\overline{r_\perp{}^2}\,\boldsymbol{B}$$

The Larmor rotation always creates an additional magnetic moment opposed to the applied magnetic field.

However, for an atom having its own non-zero magnetic moment $\mathcal{M}_0$ the paramagnetic effect is much more important; the orientation of the moments $\mathcal{M}_0$ in the magnetic field $\boldsymbol{B}$ creates an overall magnetisation of the medium, in the same direction as $\boldsymbol{B}$, which masks the magnetisation in the opposite direction due to the diamagnetic effect.

When $\mathcal{M}_0 = 0$, only a diamagnetic effect is observed. Letting n be the number of atoms per unit volume, the medium has an intensity of magnetisation

$$\boldsymbol{M} = \sum_{\substack{\text{unit}\\ \text{volume}}} \mathcal{M} = -\frac{nq^2}{\kappa^2 4m} Z\overline{r_\perp{}^2}\,\boldsymbol{B}$$

If the mean distribution of atomic electrons, due to their motion, is spherically symmetric then by introducing the co-ordinates x, y, z of the electrons (the Oz axis being parallel to the field $\boldsymbol{B}$) we have

$$\overline{x^2} = \overline{y^2} = \overline{z^2}$$

and therefore

$$\overline{r_\perp{}^2} = \overline{x^2} + \overline{y^2} = (2/3)\,(\overline{x^2} + \overline{y^2} + \overline{z^2}) = (2/3)\,\overline{r^2}$$

If μ_r is the magnetic permeability of the medium (defined by $\boldsymbol{B} = \mu_r \mu_0 \boldsymbol{H}$) it will be recalled that $\boldsymbol{M} = (\mu_r - 1)\boldsymbol{B}/\mu_0$. Hence

$$\boxed{\mu_r - 1 = -\frac{\mu_0}{\kappa^2}\frac{q^2}{6m} nZ\overline{r^2} = -\frac{1}{\varepsilon_0 c^2}\frac{q^2}{6m} nZ\overline{r^2}}$$

Comment We have intentionally not used the magnetic susceptibility χ. Users of the various systems of units do not use the same convention to define it, and its value depends on the convention, although it is a dimensionless quantity. Users of the

SI. system use the convention $\chi = \mu_r - 1$, but numerical values given in tables of constants are more often calculated with the convention adopted by users of the CGS system $\chi = (\mu_r - 1)/4\pi$.

The number of atoms n per unit volume may be calculated from Avogadro's number $\mathcal{N}$, the atomic mass $\mathcal{A}$ and the density ρ of the medium, by noting that the mass of the atom is $\mathcal{A}/\mathcal{N} = \rho/n$. Hence

$$\mu_r - 1 = -\frac{\mu_0}{\kappa^2}\frac{\mathcal{N}q^2}{6m}\rho\frac{Z}{\mathcal{A}}\overline{r^2}$$

It is known that the value of the ratio $Z/\mathcal{A}$ hardly varies from one element to another: $Z/\mathcal{A} \simeq 500$. (Users of the CGS system should remember that the atomic mass $\mathcal{A}$ must be expressed in grammes and $Z/\mathcal{A} \simeq \frac{1}{2}$.)

Diamagnetic susceptibilities measured experimentally are all of the same order of magnitude: $|\mu_r - 1| \approx 10^{-5}$. By taking for example a material of density 2 on the CGS system and 2000 on the SI system one calculates $\sqrt{(\overline{r^2})} = 50$ pm. This agrees with the order of magnitude of atomic dimensions already estimated by other methods. Conversely, quantum mechanical calculations allow the quantity $\overline{r^2}$ to be evaluated, from which the susceptibility $\mu_r - 1$ may be calculated.

8.4 Application to the Classical Zeeman Effect

The classical theory of the Zeeman effect also makes use of Larmor's theorem. Although it does not give a correct explanation of all the phenomena observed experimentally, we present this classical theory here because it is a good introduction to the more detailed study of the Zeeman effect which will be carried out in a subsequent chapter.

8.4.1 Changes of Frequency due to Larmor Rotation

To simplify matters we shall limit ourselves to applying the Larmor theorem to particularly simple electron paths. Let us consider an electron moving in a circular orbit about an axis Oz with angular velocity Ω. When a magnetic field $\boldsymbol{B}$ is applied to it, parallel to Oz and in the same direction, a Larmor rotation around the same axis Oz with an angular velocity $\omega = -\frac{1}{\kappa}\frac{q}{2m}\boldsymbol{B}$ is superimposed on the normal motion of the electron. Since q is negative, ω is positive, so that the Larmor rotation has a trigonometric sense about Oz. The total angular velocity of the electron in the laboratory frame becomes $\Omega + \omega$.

(1) If the electron moves around $\boldsymbol{B}$ in the trigonometric sense ($\Omega > 0$), the Larmor rotation adds to the normal rotation, and the total rotational frequency of the electron

$$\nu = (\Omega + \omega)/2\pi$$

is increased (see figure 8.5(a)).

(2) If the electron moves around $\boldsymbol{B}$ in a negative sense ($\Omega < 0$), the Larmor rotation

is in the opposite sense and therefore detracts from the normal rotation. In other words, the total rotational frequency $\nu = |\Omega - \omega|/2\pi$ is diminished (figure 8.5(b)).

(3) On the other hand, let us consider a third type of electron having an oscillatory motion along the Oz axis, of frequency Ω. The Laplace force exerted on it by the magnetic field is zero, and its motion at frequency $\nu = \Omega/2\pi$ is not affected by application of the magnetic field.

According to the classical theory of radiation, a charge having a periodic motion of frequency ν emits an electromagnetic wave of the same frequency, and from this the following conclusion may be drawn: if a magnetic field is applied to a spectral lamp in which atoms normally emit a wave of frequency ν, a decomposition of this wave into three waves of different frequencies, called Zeeman components, should be observed: one wave of increased frequency $\nu + \Delta\nu$, one wave of diminished frequency $\nu - \Delta\nu$ and one other unchanged.

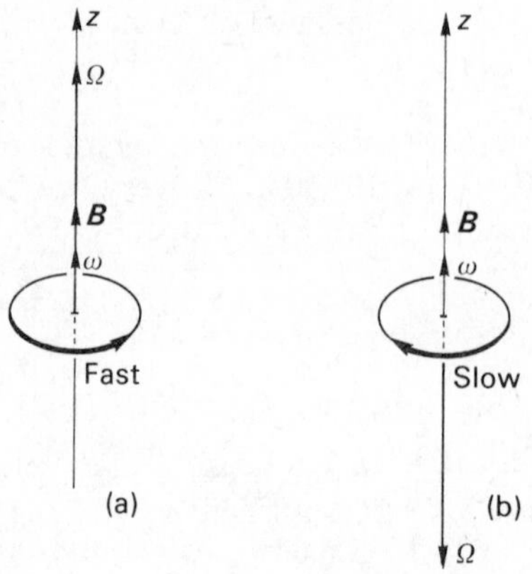

Figure 8.5 The effect of Larmor rotation on the rotational frequency of an electron

Comment A periodic motion in space can be obtained by adding vectorially three linear periodic motions corresponding to the x, y, z components. However, it can also be obtained by adding vectorially the three types of motion (1), (2) and (3) above. This allows generalisation of the above results.

8.4.2 Polarisation of the Emitted Waves

We shall now describe the phenomenon in more detail, in particular the dependence of the polarisation of the emitted waves on the direction of observation.

(*a*) *Transverse observation.* Light emitted in a direction Ox perpendicular to the magnetic field is focused on the slit of a spectrograph. The Ox components of electron motions of type (1) or (2) do not contribute to the emission of a wave in the Ox direction (an antenna does not radiate in a direction along its length); only Oy components emit a wave in the Ox direction. The two waves of increased or decreased frequency emitted by electrons of type (1) or (2) are therefore linearly polarised parallel to Oy, that is to say perpendicular to the magnetic field; this is called σ-polarisation. On the other hand the wave of unchanged frequency emitted by electrons of

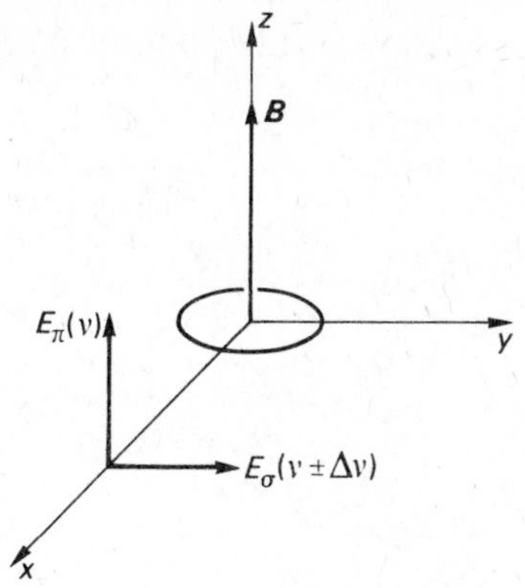

Figure 8.6 Transverse observation

type (3) is linearly polarised parallel to the magnetic field; this is called π-polarisation (see figure 8.6).

(*b*) *Longitudinal observation.* The light emitted in the Oz direction parallel to the magnetic field is focused on the slit of a spectrograph. Electrons of type (3) do not emit in the Oz direction, so the wave of unchanged frequency is not seen. Rotating electrons of types (1) and (2) emit circularly polarised waves in the Oz direction. The E_x and E_y components of the electric field of the wave are emitted by the x and y components respectively of the electron motion and are therefore out of phase with

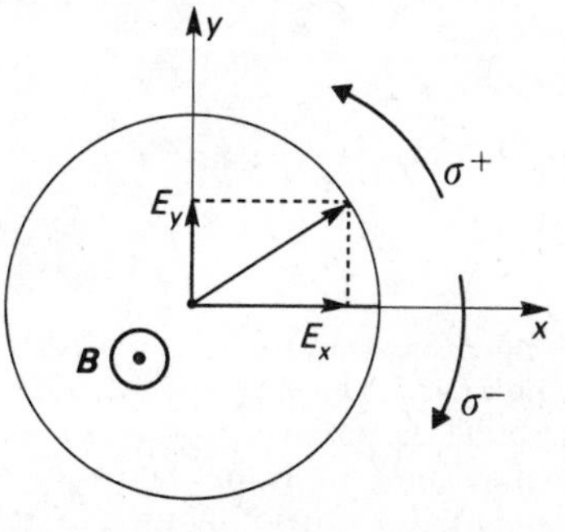

Figure 8.7 Longitudinal observation

one another by $\pi/2$; their resultant is a rotating field. The sense of the rotation is related to the + or − sign of the $\pi/2$ phase difference between the components. However, the sign of the phase difference is the same between x and y co-ordinates of the electron as between the E_x and E_y components of the electric field; therefore the resultant field $\boldsymbol{E}$ rotates in the same sense as the electron. The wave of increased frequency emitted by electrons of type (1) is polarised such that its electric field rotates in the trigonometric sense around the magnetic field $\boldsymbol{B}$, that is to say in the same sense as the magnetising current; this is called σ^+-polarisation. On the other hand the wave of diminished frequency has the opposite circular polarisation, in which the electric field rotates in the negative sense around $\boldsymbol{B}$, in a sense opposed to the magnetising current; this is called σ^--polarisation (see figure 8.7).

Figure 8.8 summarises these results, by showing in each case the lines observed in the spectrograph (intensity as a function of frequency ν) and their polarisation.

Comment I The sense of rotation of the electric field for circular polarisation is defined with respect to the direction of the magnetic field $\boldsymbol{B}$, and not with respect to the direction of propagation of the light. For a given frequency and polarisation of the light ($\nu + \Delta\nu$ and σ^+ for example) the observer sees the electric field rotating in opposite directions according as the longitudinal observation is in the direction of $\boldsymbol{B}$ or in the opposite direction to $\boldsymbol{B}$.

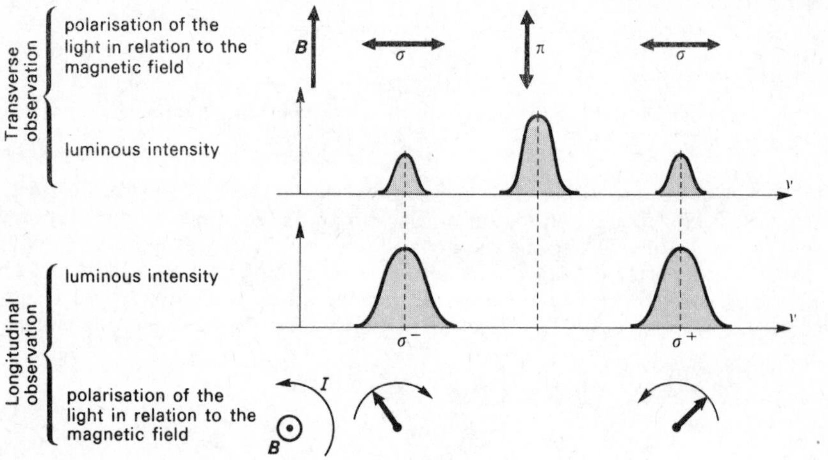

Figure 8.8 Frequencies, intensities and polarisations of lines emitted in a magnetic field (classical picture)

Comment II In any oblique direction the light emitted by circular orbits of types (1) and (2) is elliptically polarised. The shape of the ellipse and the sense of rotation may be obtained by projecting the circular orbit on to the plane of the wave.

Comment III As to the intensities of the lines, if the electron orbits of the emitting atoms are oriented at random, it is evident that the mean amplitudes of the oscillations parallel to the three axes Oxyz are equal. However, the power emitted by circular orbits of types (1) and (2) in a transverse direction Ox arises solely from the y component of the motion of the electrons. Therefore it can be expected that in a transverse direction orbits of types (1) and (2) together emit as much power as orbits of type (3). Thus for transverse observation each of the σ lines is half as intense as the π line.

In the direction Oz of the magnetic field, on the other hand, both x and y components contribute to the power emitted. Thus circular orbits of types (1) and (2) must radiate twice as much energy in the longitudinal direction Oz as in a transverse direction. This explains figure 8.8.

8.4.3 Experimental Observation of the Zeeman Effect and Conclusion

The magnitude of the frequency shift $\Delta\nu$ to be expected in a magnetic field $\boldsymbol{B}$ is

$$\frac{\Delta\nu}{B} = \frac{1}{\kappa}\frac{|q|}{4\pi m} = \frac{1}{\kappa}\frac{e}{4\pi m} = 14 \text{ GHz/tesla}$$

Optical frequencies are between 10^{14} and 10^{15} Hz. In a field $B \approx 1$ tesla, relative differences $\Delta\nu/\nu \sim 10^{-4}$ to 10^{-5} may be achieved. This is not large (the validity of Larmor's theorem rests on the small value of this ratio) but it is perfectly measurable, and in 1896 the Dutch physicist Zeeman and his students observed these frequency shifts under the action of a magnetic field.

In certain instances, the Zeeman effect may be observed as described above, that is to say the decomposition of the spectral line under study into three equidistant components of polarisation σ^-, π and σ^+; this is called the 'normal' Zeeman effect. However, more often the spectral lines split up differently under the action of a magnetic field: the number of components is not necessarily three; the frequency intervals between components may not be equal to one another, and the unshifted frequency component may be absent. At the beginning of the century this was called the 'anomalous' Zeeman effect. The existence of this 'anomalous' Zeeman effect is another example of the inability of classical physics to explain atomic phenomena correctly. This is not surprising since the explanation given above is contrary to the Bohr theory of spectral radiation and to the existence of stationary energy levels. In another chapter the correct explanation of the Zeeman effect is given with the aid of quantum theory.

In a similar way to the Bohr atom, the classical Zeeman effect is a good example of one of the 'models' used by physicists to describe real situations and which have played an important part in the discovery of atomic phenomena. Even if this model must now be considered imprecise, it is still a useful way of describing the polarisation of the light corresponding to the various Zeeman components (see section 11.3).

9

Gyromagnetic Effects

9.1 The Gyromagnetic Ratio and Larmor Precession

9.1.1 Comparison of Magnetic Moment and Angular Momentum

It is well known that the mechanics of systems of particles is based upon two important equations: the momentum equation and the angular momentum equation. This has led to great emphasis in mechanics on the concept of an angular momentum vector of a system of particles

$$\boldsymbol{\sigma} = \sum_n \boldsymbol{r}_n \times m_n \boldsymbol{v}_n$$

(m_n is the mass of a particle, having radius vector $\boldsymbol{r}_n$ and velocity $\boldsymbol{v}_n$).

It will be recalled that the angular momentum equation relates the resultant moment $\boldsymbol{\Gamma}$ of the forces applied to the system to the derivative of the angular momentum

$$\boldsymbol{\Gamma} = \mathrm{d}\boldsymbol{\sigma}/\mathrm{d}t$$

For the special case of a system that is isolated or subjected to a central force, $\boldsymbol{\Gamma} = 0$, so that $\boldsymbol{\sigma}$ is constant.

From our study of the planetary model of the atom we have seen that the motion of each electron takes place under the action of a central force directed towards the centre of gravity G of the nucleus (disregarding the interactions between the electrons themselves). By choosing the centre of force G as the origin, then once again $\boldsymbol{\Gamma} = 0$ and accordingly each electron in orbital motion has a constant angular momentum $\boldsymbol{\sigma}$.

Attention may be drawn to the similarity between the definitions of magnetic moment and of angular momentum: one passes from one to the other by replacing m_n by $\frac{1}{2}q_n$. Both definitions involve a weighted sum of the vectors $\boldsymbol{r}_n \times \boldsymbol{v}_n$, but in the one case the weighting coefficients are the masses m_n and in the other case they are the charges q_n.

If these ideas are applied to an atomic system where the only particles in motion are electrons all having the same mass m and the same charge q, the expressions for both moment vectors may be simplified by factorising m or q

$$\boldsymbol{\sigma} = m \sum \boldsymbol{r}_n \times \boldsymbol{v}_n \qquad \mathcal{M} = \frac{1}{\kappa}\frac{q}{2} \sum \boldsymbol{r}_n \times \boldsymbol{v}_n$$

Hence the magnetic moment and angular momentum vectors may be derived from one another simply by multiplying by a constant.

$$\boxed{\mathcal{M} = \gamma \boldsymbol{\sigma} \qquad \text{where} \qquad \gamma = \frac{1}{\kappa}\frac{q}{2m}}$$

γ is called the gyromagnetic ratio of the atomic system; classically it is the same for all atoms. The above formula is algebraic: since the charge q of the electron is negative ($q = -e$), the ratio γ is also negative, and the vectors $\mathcal{M}$ and $\boldsymbol{\sigma}$ are in opposite directions.

Comment I It would have been more natural to call γ the magnetogyric ratio, but the traditional name gyromagnetic has remained.

Comment II If the centre of gravity of the nucleus is chosen as the origin, then, as we have seen, the angular momentum of an electron is constant during its orbital motion. Hence the magnetic moment of each electron is also constant during its orbital motion.

The existence of this gyromagnetic ratio γ allows many phenomena to be explained, and it is of fundamental importance in atomic physics. In the following sections we shall see how the gyromagnetic ratio enables the Einstein–de Haas and Barnett experiments as well as magnetic resonance experiments to be explained; these experiments serve to justify its introduction. However, first we shall show how the use of the gyromagnetic ratio provides a simpler and more direct theory of Larmor rotation than that given in section 8.2.

9.1.2 The Influence of a Magnetic Field—the Gyroscopic Effect

In the preceding chapter, in order to calculate the effect of a magnetic field $\boldsymbol{B}$ on the charges in motion within the atom, we considered the instantaneous Laplace force $q\boldsymbol{v} \times \boldsymbol{B}$ applied to each electron. However, only average effects over time are measurable and these (as we saw at the start of the preceding chapter) may be calculated by using the magnetic moment $\mathcal{M}$ of the system of charges: the average of the forces due to a magnetic field $\boldsymbol{B}$ has a resultant moment $\boldsymbol{\Gamma} = \mathcal{M} \times \boldsymbol{B}$ (the resultant force is zero if the field $\boldsymbol{B}$ is uniform on the scale of atomic dimensions).

For the moment we will reduce the system of charges that we are considering to a single charge, a particular atomic electron. If its interaction with other electrons is disregarded and if the centre G of the nucleus is chosen as the origin, the electrostatic force has a moment of zero. The angular momentum equation may then be written

$$\mathrm{d}\boldsymbol{\sigma}/\mathrm{d}t = \boldsymbol{\Gamma} = \mathcal{M} \times \boldsymbol{B}$$

By multiplying both sides of the equation by the gyromagnetic ratio γ, and using the relation between the angular momentum and the magnetic moment $\mathcal{M}$ of the electron being considered, we obtain

$$\mathrm{d}\mathcal{M}/\mathrm{d}t = \gamma\mathcal{M} \times \boldsymbol{B}$$

The derivative of the vector $\mathcal{M}$ is perpendicular to the field $\boldsymbol{B}$ so changes in vector $\mathcal{M}$ are perpendicular to the force exerted on it (figure 9.1). This is the

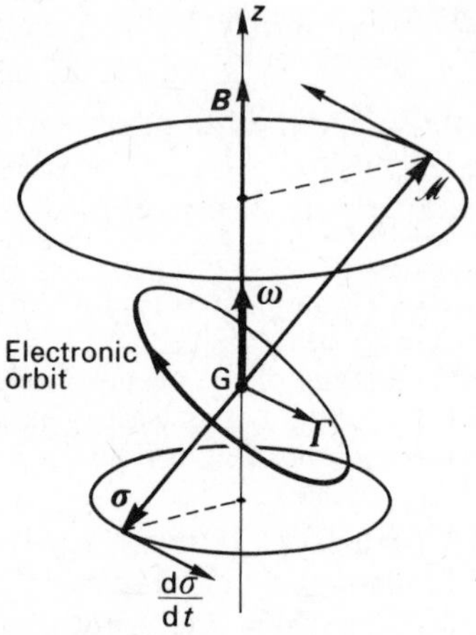

Figure 9.1 Larmor precession

classical property of gyroscopes; more precisely, the equation may be put in the form

$$\mathrm{d}\mathcal{M}/\mathrm{d}t = \boldsymbol{\omega} \times \mathcal{M}$$

where

$$\boxed{\boldsymbol{\omega} = -\gamma\boldsymbol{B}}$$

The equation therefore indicates that the vector $\mathcal{M}$ undergoes a rotational motion whose rotation vector $\boldsymbol{\omega}$ is parallel to the field $\boldsymbol{B}$. Since γ is negative, $\boldsymbol{\omega}$ is in the same direction as $\boldsymbol{B}$ and the rotation around $\boldsymbol{B}$ is in the trigonometric sense. This motion is identical with that of Larmor rotation in chapter 8, which allowed us to explain diamagnetism. However, in this section we have argued in terms of the colinear vectors $\mathcal{M}$ and $\boldsymbol{\sigma}$ which define the axis of rotation of the electron in circular motion. This investigation is therefore totally analogous to that of the precessional motion of the axis of rotation of a gyroscope, and is the reason why the expression *Larmor precession* is commonly used to designate this rotation of the vector $\mathcal{M}$, characterised by the rotation vector $\boldsymbol{\omega} = -\gamma \boldsymbol{B}$.

We may apply the above discussion for a single electron directly to an assembly of atomic electrons.

To summarise, the action of a magnetic field is not to orient, but to cause the magnetised gyroscope representing an atom to rotate around it.

Comment I Compared with the discussion of the preceding chapter (section 8.2, Larmor's theorem), the argument used in this section *has the advantage of postulating only the existence of the gyromagnetic ratio, whatever its exact origin*, and whatever its value, ignoring details of all the electric charges involved and their exact motion. We may apply it in particular to the nuclei of atoms whose values of γ are much smaller and may be either positive or negative.

Comment II As in the proof of Larmor's theorem, the argument based upon the gyromagnetic ratio relies also on an approximation. We have calculated only the average effect of the magnetic field $\boldsymbol{B}$, the average being determined over a time T of the order of several periods of orbital motion (see section 8.1 and volume 2, appendix 3). However, we may describe the action of the magnetic field in terms of its average value only if instantaneous effects occurring during times less than T are always very small.

This assumes that the action of the magnetic force qvB on the electron is much smaller than the electrostatic force qE which produces the orbital motion. In other words, the Larmor angular velocity ω due to the magnetic field is much smaller than the orbital angular velocity Ω of the electron due to the electrostatic force: $\omega \ll \Omega$. The validity of the arguments thus depends on the same conditions as the proof of the Larmor theorem. We have already mentioned that these conditions are satisfied in practice, but it may be useful to recall the orders of magnitude: using the fundamental constants, it may be calculated that for electrons in orbital motion $\gamma = -8{\cdot}8 \times 10^6$ in the CGS system and $\gamma = -8{\cdot}8 \times 10^{10}$ in the SI system. Hence the Larmor frequency in a magnetic field B may be found

$$\nu = \frac{|\omega|}{2\pi} = \frac{|\nu|}{2\pi} B$$

giving $\nu/B = |\gamma|/2\pi = 1{\cdot}4 \times 10^{10}$ Hz/tesla $= 1{\cdot}4 \times 10^6$ Hz/gauss.

9.2 Paramagnetism and Relaxation

It will be recalled that the interaction energy between a magnetic moment $\mathcal{M}$ and a magnetic field $\boldsymbol{B}$ is equal to the scalar product $W = -\mathcal{M} \cdot \boldsymbol{B} = -\mathcal{M}_z B$, where $\mathcal{M}_z$ is the component of the vector $\mathcal{M}$ in the direction Oz of the magnetic field, also called the longitudinal component. In Larmor precession, the longitudinal component $\mathcal{M}_z$ remains constant and therefore so does the energy of interaction between the field and the isolated atom. Because it is compatible with conservation of total energy on the one hand and of angular momentum (the Oz component) on the other hand, only the intervention of a third factor allows this state of affairs to change.

However, the microscopic interactions between neighbouring atoms in a material can cause changes in orientation of atomic magnets. In a vapour, these interactions occur in collisions between atoms of the vapour or with the walls of their container, the kinetic energy of the atoms of the vapour being conserved in these collisions. In a solid body, there are always interactions between neighbouring atoms and these are responsible for binding the atoms in a crystal lattice. In fact the atoms are in oscillatory motion around their equilibrium position, and these vibrations propagate in all directions, thus constituting a reserve of energy; this is the heat stored in the material.

These thermal agitation phenomena are governed by the laws of statistical thermodynamics. From Boltzmann's law, the number of atoms in thermal equilibrium, oriented in a particular direction, is proportional to the function

$$e^{-W/kT} = e^{+\mathcal{M}_z B/kT}$$

(k is Boltzmann's constant, T is the absolute temperature).

In thermal equilibrium, atoms whose longitudinal component has the same sign as B are therefore more numerous than others, and this results in an overall magnetisation of the medium, in the same direction as $\boldsymbol{B}$. In this way paramagnetism may be explained (induced magnetisation in the same direction as the applied field). With the aid of the exact equations describing this phenomenon, Langevin in 1905 was able to calculate the susceptibility of paramagnetic substances. (See books on classical thermodynamics; in the next chapter we shall show how quantum theory modifies Langevin's calculation.)

If the magnetic field is suddenly changed, another equilibrium will be established, in which the orientation of the atomic magnets is modified. However, the interaction processes that we have discussed above, which allow thermal equilibrium to be achieved, do not act instantaneously. The new thermal equilibrium is not obtained immediately, but evolves gradually. *This gradual evolution resulting in a new thermal equilibrium is called a relaxation phenomenon.*

This evolution is the result of a large number of individual microscopic phenomena and is governed by the laws of randomness. Its temporal variation ought therefore to follow the exponential law characteristic of random phenomena (for instance, radioactive lifetimes, excited state lifetimes and so on).

Let $\boldsymbol{M}_e(\boldsymbol{B},T)$ be the intensity of magnetisation at equilibrium, determined by Boltzmann's law as a function of magnetic field $\boldsymbol{B}$ and of the absolute temperature T. Also let $\boldsymbol{M}(t)$ be the actual intensity of magnetisation of the medium at time t. M and M_e represent the components of $\boldsymbol{M}(t)$ and $\boldsymbol{M}_e$ in any direction. The relative difference between the actual value M and the equilibrium value M_e decreases in proportion to the time dt in such a way that

$$\frac{d(M - M_e)}{M - M_e} = -\frac{1}{\tau}\,dt$$

so by integration: $M(t) - M_e = [M(0) - M_e]e^{-t/\tau}$ (for each component).

The time constant τ of this evolution is called the *relaxation time*. These relaxation times may be measured experimentally in various ways. Once they have been measured, they provide interesting information about the microscopic interactions existing within the medium. The simple exponential law that we have given above is sufficient to account for the observed phenomena in a majority of cases, but in certain cases the temporal variation is more complicated.

Depending on the type of magnetic moments studied (in particular, electronic or nuclear) and the material medium (solid, liquid or gas) in which they are embedded, relaxation times of widely different orders of magnitude may be measured; they can be much shorter than a microsecond or much longer than an hour.

Comment In most cases, it is necessary to distinguish clearly between the longitudinal component M_z of the magnetisation along the Oz axis parallel to the magnetic field $\boldsymbol{B}$, and the transverse components M_x and M_y along the Ox and Oy axes perpendicular to the field. A change in the longitudinal component M_z is accompanied by a change of energy, in contrast to the transverse components M_x and M_y; the processes that change M_x or M_y are not necessarily the same processes as those that changed M_z. Usually two different relaxation times are measured: a longitudinal relaxation time τ_1 corresponding to the longitudinal component M_z, and a transverse relaxation time τ_2 corresponding to the transverse components M_x and M_y. This difference between τ_1 and τ_2 will be disregarded in what follows.

> To summarise, the paramagnetic effect is an overall effect measured on a collection of atoms, and involving a preferential orientation of the atomic magnetic moments in the direction of the applied field $\boldsymbol{B}$. This overall effect is the result of the forces exerted on each individual atom by:
>
> (i) the applied magnetic field, which causes Larmor rotation (gyroscopic effect or diamagnetism);
>
> (ii) the random relaxation processes, capable of altering the longitudinal component of the magnetic moments.

These ideas are essential for detailed analysis and proper understanding of experiments involving magnetic fields, which we shall describe in the course of this chapter because they form the experimental basis of the concept of the gyromagnetic ratio.

9.3 The Einstein and de Haas Experiments (motion caused by a change of magnetisation)

9.3.1 Principle of the Experiment

This experiment was proposed in 1908 by Richardson and carried out in 1915. It is based on the following idea. The magnetism of paramagnetic or ferromagnetic substances is accounted for by assuming that their atoms (or molecules) have a magnetic moment. If these magnetic moments are oriented at random in all spatial directions, the total magnetisation of the substance is zero. However, when it is subjected to an external magnetic field, the atomic magnetic moments are oriented preferentially in a special direction and the vector sum of the individual magnetic moments is no longer zero; the substance then acquires a non-zero magnetic moment

$$\boldsymbol{\mathcal{M}}_{\text{total}} = \sum \boldsymbol{\mathcal{M}}_{\text{atomic}} \neq 0$$

In accordance with section 9.1 (gyromagnetic ratio), this preferential orientation of the atomic magnetic moments $\boldsymbol{\mathcal{M}}$ is accompanied by a preferential orientation of the atomic angular momenta $\boldsymbol{\sigma}$, in such a way that their vector sum also becomes non-zero. In other words, the overall microscopic electron motion within the atoms now has a non-zero angular momentum

$$\boldsymbol{\sigma}_{\text{el}} = \sum \boldsymbol{\sigma}_{\text{atomic}} = \frac{1}{\gamma} \sum \boldsymbol{\mathcal{M}}_{\text{atomic}} = \frac{1}{\gamma} \boldsymbol{\mathcal{M}}_{\text{total}}$$

In the study of the mechanics of our system of point particles, we must then distinguish between the following.

(*a*) *The microscopic motion of the electrons* within the atoms, relative to the centre of gravity G of each atom. The vector $\boldsymbol{\sigma}_{\text{el}}$ is the sum of the angular momenta $\boldsymbol{\sigma}_{\text{atomic}}$ calculated for each atom relative to its centre of gravity.

(*b*) *The motion of the centres of gravity of the atoms* in the laboratory frame of reference. For atoms bound to one another so as to form a solid body, the overall motion constitutes *the motion of the solid body*, that is to say, an experimentally observable macroscopic motion. We let $\boldsymbol{\sigma}_{\text{sol}}$ be the corresponding angular momentum.

From the laws for composition of angular momenta, the total angular momentum of a system of particles is equal to the sum of (i) the angular momentum calculated about the centre of gravity G in a frame of reference in which G is stationary (relative motion around G); (ii) the angular momentum calculated in an inertial frame, assuming that the total mass of the system is placed at the centre of gravity G (motion of centre of gravity). We apply this law to each isolated atom, then sum over all atoms in the solid; from this we find, with the notation defined above, that the total angular momentum of the whole system is equal to the sum

$$\boldsymbol{\sigma}_{\text{el}} + \boldsymbol{\sigma}_{\text{sol}}$$

On the other hand, as far as angular momentum is concerned the whole system behaves as if it were isolated. The only external force acting on the whole system is that of the magnetic field $\boldsymbol{B}$, and for the whole system the angular momentum equation may be written

$$\frac{\mathrm{d}}{\mathrm{d}t}(\boldsymbol{\sigma}_{\text{el}} + \boldsymbol{\sigma}_{\text{sol}}) = \boldsymbol{\Gamma} = \boldsymbol{\mathcal{M}}_{\text{total}} \times \boldsymbol{B} = 0$$

since the overall magnetic moment, $\boldsymbol{\mathcal{M}}_{\text{total}}$, is always parallel to the field $\boldsymbol{B}$.

Hence the total angular momentum remains constant, as in an isolated system

$$\boldsymbol{\sigma}_{\mathrm{el}} + \boldsymbol{\sigma}_{\mathrm{sol}} = \text{constant}$$

The appearance of a non-zero electronic angular momentum $\boldsymbol{\sigma}_{\mathrm{el}}$ at the instant of magnetisation must therefore be accompanied by an observable change in the angular momentum of the solid body $\boldsymbol{\sigma}_{\mathrm{sol}}$. Let us suppose that the solid body is initially stationary and demagnetised ($\boldsymbol{\sigma}_{\mathrm{sol}} = 0$ and $\boldsymbol{\sigma}_{\mathrm{el}} = 0$); magnetisation will give rise to an angular momentum $\boldsymbol{\sigma}_{\mathrm{sol}} = -\boldsymbol{\sigma}_{\mathrm{el}}$, so that it will be put into motion. This is confirmed by the experiment described below.

Comment I In relation to this phenomenon a purely mechanical analogy may be given. A person carrying a large spinning top stands on a disc which can turn freely around the vertical axis Oz. Initially (figure 9.2(a)) the disc is stationary and the axis

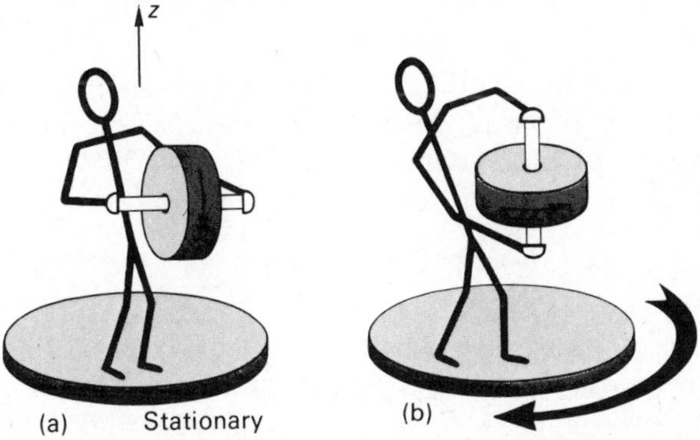

Figure 9.2 Mechanical analogy of the Einstein and de Haas experiment

of the top is horizontal. When the person alters the axis of the top so as to make it vertical (figure 9.2(b)) the disc starts to turn in the opposite sense to that of the top such that the vertical component σ_z of the angular momentum remains zero.

This same phenomenon occurs in the magnetised body; the large top is replaced by many small tops representing the atoms, and the magnetic field plays the part of the person, by changing the orientation of the tops with the help of the relaxation processes.

Comment II We have made use only of the law of conservation of angular momentum, and have not tried to use the law of conservation of energy. Because the equilibrium magnetisation is achieved by means of relaxation processes, the equation for the total energy must include the energy of thermal agitation (the heat stored in the material) in addition to the magnetic interaction energy.

9.3.2 The Ballistic Motion Experiment

The magnetised medium is a cylindrical rod with vertical axis Oz suspended by a wire whose torsional constant is C (figure 9.3). The moment of inertia of the rod about its axis is $I = mr^2/2$ (m is the mass and r the radius of the rod).

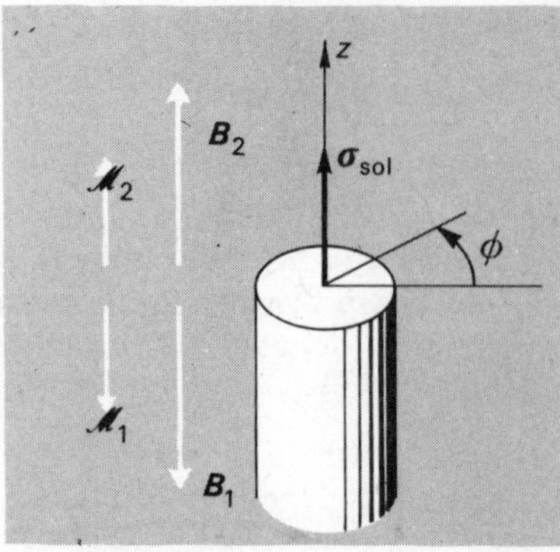

Figure 9.3 Magnetised cylindrical rod suspended in a magnetic field that is suddenly changed from $\boldsymbol{B}_1$ to $\boldsymbol{B}_2$. σ_{sol} is the angular momentum acquired by the rod

If the rod is turned away from its equilibrium position, it undergoes an alternating rotation of period $T = 2\,\pi\sqrt{(I/C)}$ in which the angle of rotation varies as a function of time t according to the equation

$$\phi = \phi_0 \sin \sqrt{\left(\frac{C}{I}\right)}\, t$$

(disregarding damping of the oscillations).

Initially the rod is subjected to a field $\boldsymbol{B}_1$ directed downwards, in which it has a total magnetic moment $\mathcal{M}_1$. Moreover, it is stationary, so that the total angular momentum of the system reduces to $\boldsymbol{\sigma}_{el} = (1/\gamma)\mathcal{M}_1$.

Suddenly, the sense of the magnetising current is reversed, so that the magnetic field and magnetic moment change direction and become

$$\boldsymbol{B}_2 = -\boldsymbol{B}_1, \qquad \mathcal{M}_2 = -\mathcal{M}_1$$

The electronic angular momentum becomes $\boldsymbol{\sigma}_{el} = (1/\gamma)\mathcal{M}_2 = -(1/\gamma)\mathcal{M}_1$. To ensure conservation of total angular momentum, the solid body must therefore acquire an angular momentum $\boldsymbol{\sigma}_{sol} = (2/\gamma)\mathcal{M}_1$, so choosing an initial state magnetised in the opposite direction doubles the effect.

It will be recalled that for a solid body in rotation, the angular momentum may be written $\boldsymbol{\sigma}_{sol} = I\boldsymbol{\omega}$, where I is the moment of inertia about the rotation axis, and $\boldsymbol{\omega}$ is the angular velocity vector about this axis, directed in such a way that the motion around it is in the trigonometric sense, and whose algebraic value along the axis is the angular velocity $\omega = \mathrm{d}\phi/\mathrm{d}t$. (Note that we have kept the symbol ω to designate the angular velocity of the rod; it should not be confused with its natural frequency $\sqrt{(C/I)}$.

Reversing the magnetic field therefore has given the rod an algebraic angular velocity $\omega_0 = (2/I)(\mathcal{M}_1/\gamma)$ (in figure 9.3, $\mathcal{M}_1$ is negative as is γ, therefore ω is positive). The rod is therefore caused to rotate by an instantaneous impulse (in the same way as a ballistic galvanometer). It oscillates with an amplitude ϕ_0 determined by the initial angular velocity

$$\left(\frac{d\phi}{dt}\right)_{t=0} = \phi_0 \sqrt{\frac{C}{I}} = \omega_0$$

therefore

$$\boxed{\phi_0 = \frac{2}{\sqrt{(IC)}} \times \frac{\mathscr{M}_1}{\gamma}}$$

The sense of the rotation allows the negative sign of γ to be confirmed. Measurement of the amplitude ϕ_0 then allows the value of γ to be determined.

9.3.3 The Sustained Oscillation Experiment

The amplitude of the deflection ϕ_0 may be increased considerably by the additive effect of several successive impulses. After half a period the rod passes through its equilibrium position in the opposite direction. If we choose this moment to reverse the magnetic field once again (by restoring it to its initial value B_1), we give the rod an impulse in the sense opposite to the first; however, since its angular velocity has changed sign, this new impulse will increase the magnitude of the angular velocity and therefore the amplitude of the oscillations. If the magnetic field is reversed on each pass through the equilibrium position, the effects of all the impulses received by the rod, alternately positive and negative, are added and the amplitude of the oscillations increases.

Until now we have disregarded the spontaneous damping of the oscillatory motion. The loss of velocity due to this damping increases as the amplitude of the motion becomes larger, and this limits the growth of the oscillations. They stabilise themselves at a constant amplitude when the increase of velocity due to an impulse is exactly balanced by the spontaneous decrease of velocity in half a period. A stationary state of sustained oscillations is then established, and its amplitude may be calculated.

By taking account of damping, the spontaneous motion of the torsional pendulum in the absence of external forces, that is to say, between two impulses, is given by

$$\phi = \phi_0 e^{-t/\tau} \sin \sqrt{\left(\frac{C}{I}\right)}\, t$$

where τ the damping time constant is longer than the period T.

In the following we consider only absolute values. Passing through equilibrium, the angular velocity is

$$|\omega_0| = \phi_0 \sqrt{\left(\frac{C}{I}\right)} e^{-t/\tau}$$

The loss of velocity between two successive passes through equilibrium is therefore

$$|\delta\omega_0| = \phi_0 \sqrt{\left(\frac{C}{I}\right)} (1 - e^{-T/2\tau}) \simeq \phi_0 \sqrt{\left(\frac{C}{I}\right)} \frac{T}{2\tau}$$

A stationary state is established when this loss of velocity is exactly balanced by the increase due to a reversal of the magnetic field

$$|\delta\omega| = \frac{1}{I}|\delta\sigma_{\text{sol}}| = \frac{2}{I}\left|\frac{\mathscr{M}_1}{\gamma}\right|$$

Equating these two values of $|\delta\omega|$

$$\phi_0 = \frac{2\tau}{T}\frac{2}{\sqrt{(IC)}}\left|\frac{\mathscr{M}_1}{\gamma}\right|$$

In relation to the ballistic method, the amplitude is larger by a factor $2\tau/T$, and the sensitivity of its measurement is accordingly improved.

Comment Substituting for the period $T = 2\pi\sqrt{(I/C)}$, a formula is obtained which includes neither the period nor the torsional constant C

$$\phi_0 = \frac{2\tau}{\pi I}\left|\frac{\mathscr{M}_1}{\gamma}\right|$$

The volume $\mathscr{V}$ of the rod comes into both the magnetic moment $\mathscr{M}_1$ and the moment of inertia I

$$\mathscr{M}_1 = \mathscr{V}M$$

where M is the intensity of magnetisation of the medium

$$I = \frac{r^2}{2}m = \frac{r^2}{2}\rho\mathscr{V}$$

where ρ is the density of the medium (disregarding the contribution to the moment of inertia of additions such as the supports or mirror). Hence

$$\phi_0 = \frac{4\tau}{\pi r^2 \rho}\left|\frac{M}{\gamma}\right|$$

The effect becomes larger as the damping of the torsional pendulum is reduced (τ large), and as the magnetic rod is made thinner (r small).

For soft iron, an induction $B \approx 1{\cdot}2$ tesla $= 12\,000$ gauss may be obtained. This would give an intensity of magnetisation $M \simeq B/\mu_0$. With a radius r of the order of 1 mm, and a damping constant of several minutes, an amplitude ϕ_0 of a few tens of degrees is obtained.

9.3.4 Results of Measurements and Conclusions

The mechanical effects corresponding to these reversals of magnetisation are weak compared with usual magnetic forces and care must be taken to eliminate all parasitic forces in order to draw any useful conclusion from an experiment of this type. Historically, some confusion has occurred in the interpretation of these early measurements. Despite all these uncertainties, the following conclusions can be drawn from these experiments:

(a) confirmation of the existence of a gyromagnetic effect, in other words the existence of a ratio γ between magnetic moments and atomic angular momenta;

(b) confirmation of the negative sign of this ratio γ;

(c) confirmation of the order of magnitude to be expected for this ratio γ;

(d) the exact numerical value measured does not, however, agree with the value calculated theoretcially $\gamma = \frac{1}{\kappa} \times \frac{q}{2m}$. Usually a value twice as large is obtained.

Comment It should be noted that for reasons of sensitivity these experiments are generally carried out only for those special ferromagnetic substances whose magnetisation is particularly strong.

For the moment we shall not try to interpret this numerical disagreement but the following inference may be drawn: within the atom there must exist more complicated phenomena than the simple orbital motion of the electrons considered in section 9.1 in order to calculate the gyromagnetic ratio. We shall discuss this subsequently in another chapter, but in the rest of this chapter and for the explanation of many other phenomena, it is sufficient to accept the properties (a), (b) and (c) stated above whatever their exact origin.

9.4 Barnett's Experiments (magnetisation caused by rotational motion)

The Einstein and de Haas experiments enable us to observe the rotation caused by changing the magnetisation of a solid body. In some respects Barnett's experiments represent the reciprocal phenomenon: one observes the magnetisation of the solid body caused by a forced rotation. However, the precise interpretation of these new experiments is more complex. It is based on the similarity between an atom and a gyroscope, described in section 9.1.

When a gyroscope is suspended from its centre of gravity (for example suspended by means of a universal joint, as in figure 9.4) such that no force is exerted on its axis of rotation, its support may be moved in any direction without changing the direction of this axis in relation to absolute (i.e. inertial) axes; this is one of the fundamental properties of gyroscopes. When a solid body is made to rotate with an angular velocity ω, the centres of gravity of the atoms are forced to rotate as well. However, the axes of the atomic gyroscopes must retain a fixed direction in relation to absolute axes which for this problem can be taken as laboratory axes.

Let us therefore place ourselves in a *rotating frame fixed in the solid body*. In this new frame, the solid body is stationary but the axis of the atomic gyroscope turns with an angular velocity $-\omega$. The derivative of the angular momentum $\boldsymbol{\sigma}$ of an atom relative to its centre of gravity is given by the usual

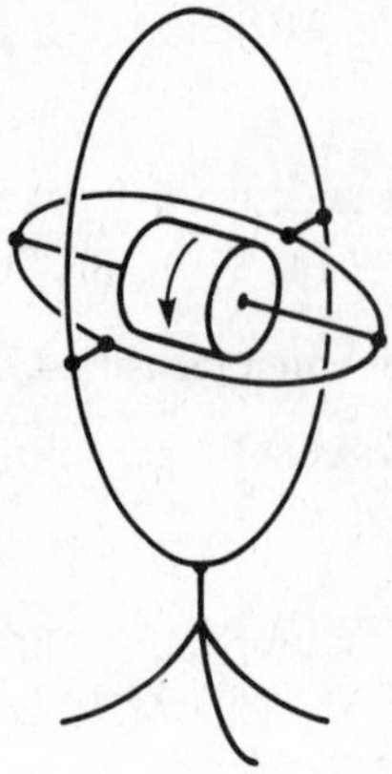

Figure 9.4 A gyroscope

formula for rotational motion

$$\frac{d\boldsymbol{\sigma}}{dt} = (-\boldsymbol{\omega}) \times \boldsymbol{\sigma} = \boldsymbol{\sigma} \times \boldsymbol{\omega} = \frac{\boldsymbol{\mathcal{M}}}{\gamma} \times \boldsymbol{\omega} = \boldsymbol{\mathcal{M}} \times \frac{\boldsymbol{\omega}}{\gamma}$$

Let us compare this with the formulae in section 9.1 for the effect of a magnetic field: the gyroscopic motion of the axis of the atom is identical to the Larmor precession that would be caused by a field $\boldsymbol{B} = \boldsymbol{\omega}/\gamma$.

In section 9.2 (paramagnetism and relaxation) we saw that the simultaneous action of Larmor precession at an angular velocity ω and relaxation processes results in a paramagnetic effect, that is to say a preferential orientation of the magnetic moments in the direction of the field.

In the present situation, and in terms of a frame of reference fixed in the solid body, the gyroscopic or magnetic axes of the atoms are subjected to:

(i) a rotation at an angular velocity $-\omega$, so as to create a magnetic field $\boldsymbol{B} = \omega/\gamma$,

(ii) the usual relaxation forces, assuming that the relaxation phenomena, provided that they are observed in a frame fixed in the solid body, do not depend on the motion of this frame.

The simultaneous action of (i) and (ii) must result in a paramagnetic effect, so that the medium must acquire a magnetisation $\boldsymbol{M}$ identical to that acquired under the action of a field $\boldsymbol{B} = \omega/\gamma$ capable of causing the same rotational motion.

The experiment was carried out by Barnett in 1914 (before the Einstein and de Haas experiment). He used a cylindrical magnetic rod and made it turn around its axis of revolution Oz. A long thin rod was chosen in order to overcome the problems of non-uniform magnetisation phenomena at its extremities. The magnetic field produced by the rod at a short distance from it allowed its magnetisation M to be determined. He measured M for a particular velocity of rotation ω; then he sought the magnetic field B that had to be applied to the rod when stationary for it to have the same intensity of magnetisation M. He deduced that $\gamma = \omega/B$.

The measurements are extremely difficult because the effect is very small: even for a fairly high velocity of rotation of 6000 revolutions per minute, so that $\omega/2\pi = 100$ Hz, a value of the apparent magnetic field $B = \omega/\gamma = 0{\cdot}7 \times 10^{-8}$ tesla is calculated, assuming the normal value of the gyromagnetic ratio $\gamma/2\pi = 14$ GHz/tesla. In order to obtain useful measurements with this type of experiment, it is necessary to compensate for the earth's field and to correct carefully for parasitic effects due to rotation of the apparatus.

Nevertheless, Barnett achieved this, and the conclusions he was able to draw from these experiments are in every way similar to those drawn in the preceding section from the Einstein and de Haas experiment:

(a) the existence of the gyromagnetic effect,
(b) the negative sign of the gyromagnetic ratio,
(c) the expected order of magnitude of γ,
(d) the actual numerical value for a ferromagnet is twice the value expected normally.

(see S. J. Barnett, *Review of Modern Physics*, 7 (1935), 129.)

Addendum In 1861 Maxwell investigated the existence of a gyromagnetic effect when a force is exerted on the axis of a coil of an electromagnet, around which electric charges are moving. When nothing could be observed, he concluded that the electric charge carriers had an extremely small mass.

Measurement of the mass of the electron at the end of the nineteenth century and the success of Langevin's theory of magnetism in 1905 promoted the idea of microscopic motion of the electrons and gave rise to the idea of a close relationship between magnetic moment and angular momentum. This is why Barnett repeated this investigation into the existence of a gyromagnetic effect, by applying it this time to a magnetic medium, and in 1914 he succeeded in carrying out the experiment described above. Barnett used a different argument that did not involve a change of reference frame and that is rather more complicated. However, we shall present it for interested readers or those who might be uneasy about the change of reference frame.

This argument is based on the fact that the relaxation processes tend to interrupt the precessional motion of an atomic magnet and on average render it stationary in relation to the solid body, in other words force it into the rotational motion of the solid. Until now we have thought in terms of the instantaneous motion of an atomic gyroscope; Barnett on the other hand used mean values, the averages being calculated over times long compared with the relaxation times. Two parts of the experiment have to be distinguished: the establishment of rotation and then the equilibrium state of steady rotation.

(*a*) *Establishment of rotation.* Since the relaxation forces tend to make the atomic magnets rotate, the gyroscopic axis of each atom is subjected *on average*, to a force $\boldsymbol{F}$ perpendicular to the axis of rotation Oz. From the theory of the gyroscope (reviewed in section 9.4) the derivative $\mathrm{d}\boldsymbol{\sigma}/\mathrm{d}t$ of the angular momentum is perpendicular to this force ($\mathrm{d}\boldsymbol{\sigma}/\mathrm{d}t = \boldsymbol{\Gamma} \approx \boldsymbol{\sigma} \times \boldsymbol{F}$; see figure 9.5(a)) that is, the displacement of the gyroscopic axis is perpendicular to the force $\boldsymbol{F}$ and it is therefore drawn nearer to the axis of rotation Oz.

The orientation of the atomic magnetic moments with respect to Oz is thus altered, and their vector sum is no longer zero: an experimentally observable overall magnetisation appears in the medium, parallel to the rotation axis Oz. If the rotation has a trigonometric sense around the axis Oz, then, remembering that the gyromagnetic ratio γ is negative, the induced magnetisation $\boldsymbol{M}$ is in the opposite direction to that of the Oz axis (see figure 9.5(a)).

When the atomic magnetic moments are drawn towards the Oz axis, this destroys their normal distribution of thermal equilibrium, and thus repulsion forces appear that result in the re-establishment of this equilibrium. These new forces, perpendicular to the preceding ones, are capable of instigating the rotation of the axes of the gyroscopes. A new equilibrium is established when the departure from normal equilibrium is great enough for the new repulsion forces to drive the axes of the gyroscopes at the same angular velocity ω as the whole solid body.

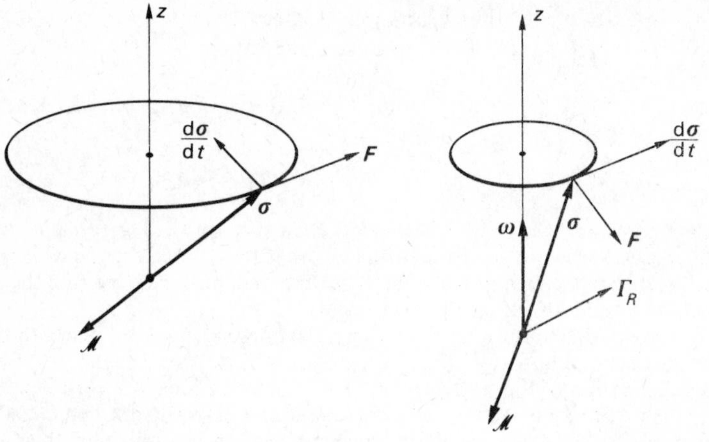

Figure 9.5 (a) Establishment of rotation; (b) steady-state rotation

(*b*) *Steady rotation.* We leave aside the transient phenomenon which occurs at the start of the rotation; we are now concerned only with steady-state regime in which the atomic axes remain on average fixed in relation to the solid body. In what follows we consider the average motion of the atoms in order to exclude fluctuations corresponding to thermal agitation; the argument is valid if one sums over a large assembly of atoms.

The angular momentum σ of an atom with respect to its centre of gravity undergoes an average variation with time, given by the usual formula for rotational motion

$$\mathrm{d}\boldsymbol{\sigma}/\mathrm{d}t \doteq \boldsymbol{\omega} \times \boldsymbol{\sigma}$$

By using the angular momentum equation, we may find the average resultant moment of the relaxation forces that drive the axis of an atomic gyroscope

$$\boldsymbol{\Gamma}_R = \frac{\mathrm{d}\boldsymbol{\sigma}}{\mathrm{d}t} = \boldsymbol{\omega} \times \boldsymbol{\sigma}$$

(see figure 9.5(b)). We now recall the phenomenon that occurs when a magnetic field $\boldsymbol{B}$ is applied to a stationary body:

(i) the field exerts a couple of moment $\boldsymbol{\Gamma}_B = \mathcal{M} \times \boldsymbol{B}$;

(ii) the relaxation processes give rise to an equilibrium distribution of the magnetic moments, in other words a state in which the magnetic moments are *on average* stationary. For this to occur, all the microscopic forces that bring about relaxation must have an average resultant moment $\boldsymbol{\Gamma}_R$ opposite to $\boldsymbol{\Gamma}_B$, so that the total average resultant moment $\boldsymbol{\Gamma}_R + \boldsymbol{\Gamma}_B = 0$. Hence

$$\boldsymbol{\Gamma}_R = -\boldsymbol{\Gamma}_B = -\mathcal{M} \times \boldsymbol{B} = \boldsymbol{B} \times \mathcal{M} = \gamma \boldsymbol{B} \times \boldsymbol{\sigma}$$

A comparison of the two formulae which give the average resultant moment $\boldsymbol{\Gamma}_R$ of the relaxation forces shows that the relaxation forces acting in the rotation experiment are identical to those that would exist in a stationary body subjected to a magnetic field $\boldsymbol{B} = \omega/\gamma$. Since these are the relaxation forces that cause orientation of the magnetic moments, they are a function of the magnetisation of the body; hence the magnetisation $\boldsymbol{M}$ acquired by a solid body in rotation is identical to that which it acquires when stationary under the influence of a magnetic field $\boldsymbol{B} = \omega/\gamma$. This is in agreement with the earlier result.

9.5 Magnetic Resonance Experiments (evidence for Larmor rotation; measurement of gyromagnetic ratios)

Because of the importance of the phenomenon of magnetic resonance in the laboratory, it will be given a full explanation here, much of which could be omitted on a first reading.

9.5.1 Principles of the Experiment (calculation in the absence of relaxation)

If we disregard relaxation phenomena, the motion of an atomic magnetic moment $\mathcal{M}$ subjected to a fixed magnetic field $\boldsymbol{B}_0$, reduces to a Larmor rotation whose rotation vector is $\boldsymbol{\omega}_0 = -\gamma \boldsymbol{B}_0$ (see section 9.1). Let us also apply a small magnetic field $\boldsymbol{B}_1$ to the medium under study, perpendicular to $\boldsymbol{B}_0$ and rotating around it with an angular velocity ω, which, depending on the situation, could be different from ω_0 or equal to ω_0.

(a) When the two angular velocities ω and ω_0 differ sufficiently the relative position of the magnetic moment $\mathcal{M}$ and of the rotating field $\boldsymbol{B}_1$ changes frequently and rapidly and its average may be expected to be zero.

(b) If, on the other hand, the two angular velocities ω and ω_0 are equal, the relative position of the magnetic moment $\mathcal{M}$ and of the rotating field $\boldsymbol{B}_1$ remains fixed in time. It is then possible for the force exerted by the small field $\boldsymbol{B}_1$ on the magnetic moment $\mathcal{M}$ to have a considerable effect. If such an effect is experimentally observable, it must provide evidence for the Larmor rotation at an angular velocity ω_0.

The calculation is similar to that in section 9.1 for finding the Larmor rotation. We write the angular momentum equation, but now take into account both the fields $\boldsymbol{B}_0$ and $\boldsymbol{B}_1$

$$d\boldsymbol{\sigma}/dt = \boldsymbol{\Gamma} = \boldsymbol{\mathcal{M}} \times (\boldsymbol{B}_0 + \boldsymbol{B}_1)$$

and because $\boldsymbol{\mathcal{M}} = \gamma\boldsymbol{\sigma}$, we obtain

$$\frac{d\boldsymbol{\mathcal{M}}}{dt} = \boldsymbol{\mathcal{M}} \times (\gamma\boldsymbol{B}_0 + \gamma\boldsymbol{B}_1) = (\boldsymbol{\omega}_0 + \boldsymbol{\omega}_1) \times \boldsymbol{\mathcal{M}}$$

by using the Larmor rotation vector

$$\boldsymbol{\omega}_0 = -\gamma\boldsymbol{B}_0$$

and introducing by analogy the vector

$$\boldsymbol{\omega}_1 = -\gamma\boldsymbol{B}_1$$

The solution of the problem may be simplified by using, instead of the laboratory frame Oxyz, *a new frame of reference in which the field* $\boldsymbol{B}_1$ *appears fixed.* This frame has the same Oz axis (parallel to the fixed field $\boldsymbol{B}_0$) as the laboratory frame, but its OX axis coincides with the direction and the sense of the rotating field $\boldsymbol{B}_1$ (see figure 9.6). In relation to the laboratory frame,

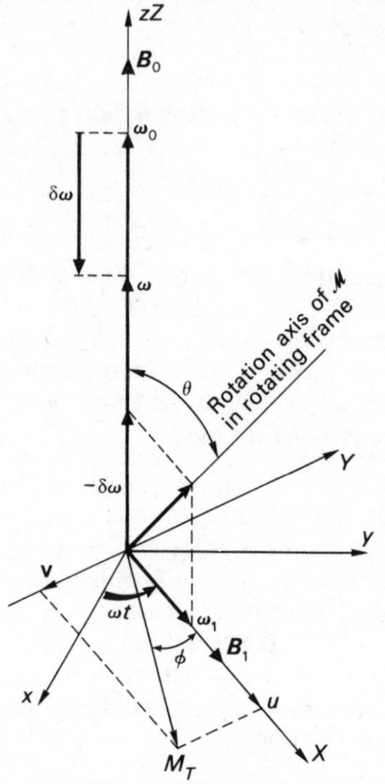

Figure 9.6 Magnetic resonance: transformation to a rotating frame

this frame rotates with the angular velocity ω of the field $\boldsymbol{B}_1$. (This should not be confused with the Larmor frame, defined in section 8.2, which rotates with a different angular velocity $\omega_0 = -\gamma B_0$.)

The angular momentum equation must be written in an absolute (i.e. inertial) frame, and in this problem the laboratory frame serves as an absolute frame. However, we can apply the formula for compounding velocities to the derivative of the vector $\mathcal{M}$ by distinguishing its absolute variation in the laboratory frame and its relative variation in the rotating frame

$$\left(\frac{\mathrm{d}\mathcal{M}}{\mathrm{d}t}\right)_{\text{lab}} = \left(\frac{\mathrm{d}\mathcal{M}}{\mathrm{d}t}\right)_{\substack{\text{rel to}\\ \text{rotating}\\ \text{frame}}} + \left(\frac{\mathrm{d}\mathcal{M}}{\mathrm{d}t}\right)_{\substack{\text{from frame}\\ \text{motion}}}$$

which gives

$$(\boldsymbol{\omega}_0 + \boldsymbol{\omega}_1) \times \mathcal{M} = \left(\frac{\mathrm{d}\mathcal{M}}{\mathrm{d}t}\right)_{\text{lab}} = \left(\frac{\mathrm{d}\mathcal{M}}{\mathrm{d}t}\right)_{\substack{\text{rel to}\\ \text{rotating}\\ \text{frame}}} + \boldsymbol{\omega} \times \mathcal{M}$$

In the rotating frame, one may deduce the equation

$$\boxed{\left(\frac{\mathrm{d}\mathcal{M}}{\mathrm{d}t}\right)_{\substack{\text{rel to}\\ \text{rotating}\\ \text{frame}}} = (\boldsymbol{\omega}_1 + \boldsymbol{\omega}_0 - \boldsymbol{\omega}) \times \mathcal{M} = (\boldsymbol{\omega}_1 - \delta\boldsymbol{\omega}) \times \mathcal{M}}$$

by introducing the vector $\delta\boldsymbol{\omega} = \boldsymbol{\omega} - \boldsymbol{\omega}_0$ which represents the difference between the rotation vector of the field $\boldsymbol{B}_1$ and the Larmor rotation vector. Since the vector $\boldsymbol{\omega}_1$ is fixed in the rotating frame, the interpretation is simple: the moment $\mathcal{M}$ carries out a rotation whose rotation vector is $\boldsymbol{\omega}_1 - \delta\boldsymbol{\omega}$. This rotation vector has a magnitude $\sqrt{(\omega_1{}^2 + \delta\omega^2)}$ and its angle θ to the Oz axis is given by $\tan\theta = -\omega_1/\delta\omega$ (see figure 9.6).

We shall concern ourselves solely with the component of $\mathcal{M}$ along Oz, still called the longitudinal component $\mathcal{M}_z$ (in the direction of the magnetic field):

(a) If ω and ω_0 differ considerably, so that $|\delta\omega| \gg \omega_1$, the rotation axis is practically coincident with the Oz axis, and the $\mathcal{M}_z$ component of the magnetic moment hardly varies. The orientation of the magnetic moment $\mathcal{M}$ with respect to the magnetic field $\boldsymbol{B}_0$ does not change; as we foresaw, the rotating field $\boldsymbol{B}_1$ has hardly any effect on the magnetic moment.

(b) If ω and ω_0 are nearly equal to one another, so that $|\delta\omega| \leqslant \omega_1$, the $\mathcal{M}_z$ component varies according to an equation of the form

$$\mathcal{M}_z = C_1 \cos\sqrt{(\omega_1^2 + \delta\omega^2)}\, t + C_2$$

(where C_1 and C_2 are constants) so that the orientation of the moment $\mathcal{M}$ with respect to a fixed field $\boldsymbol{B}_0$ alternates with a frequency $\sqrt{(\omega_1{}^2 + \delta\omega^2)}$. It may be pointed out that in quantum mechanical calculations, a sinusoidal variation of the transition probabilities with the same frequency $\sqrt{(\omega_1^2 + \delta\omega^2)}$ is obtained.

(c) In particular, when $\omega = \omega_0$ or $\delta\omega = 0$, the rotation axis is along OX, and the variation of $\mathscr{M}_z$ has its maximum amplitude

$$\mathscr{M}_z = C \cos \omega_1 t$$

If the time t during which the rotating field acts is restricted to half a period

$$T/2 = \pi/\omega_1$$

the magnetic moment vector describes a semi-circle around $\boldsymbol{B}_1$ and the longitudinal component $\mathscr{M}_z$ changes sign.

> To summarise, the rotating field $\boldsymbol{B}_1$ influences the atom only when its rotation frequency $\omega/2\pi$ is very close to the Larmor frequency $\omega_0/2\pi = -(\gamma/2\pi) B_0$, whence the name magnetic resonance. When the frequencies are equal, the rotating field is capable of radically altering the orientation of the atomic magnetic moment with respect to the fixed field $\boldsymbol{B}_0$.

Comment I The interaction energy of the atom with the fixed field is $W_0 = -\mathscr{M}_z B_0$ and has the same alternating variation as the component $\mathscr{M}_z$. Depending on the circumstances, the energy of the atomic system either increases, so that it absorbs energy from the electromagnetic wave (circularly polarised) which acts as the rotating field, or the energy of the atomic system decreases, thereby adding energy to the electromagnetic wave; this is induced emission (see section 9.5.6). Each time that energy exchanges between the atomic system and the electromagnetic wave are caused by the action of the magnetic field of the wave on the magnetic dipole moment of the atom, it is described as a *magnetic dipole transition.* Magnetic resonance is a particular example of a magnetic dipole transition.

Comment II If the atom is also subjected to random interactions due to the surrounding medium (relaxation processes), the motion that we have just described is strongly perturbed after a time which is random but of the order of the relaxation time τ. The rotating field will then act coherently and steadily on the magnetic moments only for a time of the order of τ, and its effect on the magnetic moments will be significant only if $\mathscr{M}_z$ has changed by an appreciable amount during the time τ, that is if $\omega_1 \tau > 1$. *The shorter the relaxation time τ, the larger the required amplitude of the rotating field B_1.*

The two following sections describe the results obtained when relaxation is taken into account.

9.5.2 The Calculation taking Relaxation into account—The Bloch Equations

Experimentally, it is difficult to carry out measurements on isolated atoms; usually one has a collection of atoms forming a material medium whose intensity of magnetisation $\boldsymbol{M} = \sum \mathscr{M}$ may be measured, the sum being carried out over a unit volume. This magnetisation $\boldsymbol{M}$ varies both under the direct influence of the magnetic field (gyroscopic effect, section 9.1) and under the action of relaxation processes (paramagnetic effect, section 9.2).

(a) In the preceding section we wrote the vectorial equation for the evolution of a magnetic moment $\mathcal{M}$ under the influence of fields $\boldsymbol{B}_0$ and $\boldsymbol{B}_1$. By adding the two terms vectorially for all the atoms in a unit volume, one obtains

$$\left(\frac{\mathrm{d}\boldsymbol{M}}{\mathrm{d}t}\right)_{\text{lab}} = (\boldsymbol{\omega}_0 + \boldsymbol{\omega}_1) \times \boldsymbol{M}$$

(b) To this direct action of the magnetic fields must be added that of the relaxation processes described in section 9.2. Each component M of the instantaneous magnetisation vector $\boldsymbol{M}$ tends exponentially towards its equilibrium value M_e, as for random phenomena

$$\frac{\mathrm{d}(M - M_e)}{M - M_e} = -\frac{1}{\tau}\mathrm{d}t$$

(where τ is the relaxation time).

The equilibrium magnetisation $\boldsymbol{M}_e$ depends upon the total magnetic field applied to the sample, which in the present case is $\boldsymbol{B}_0 + \boldsymbol{B}_1$. However, in the determination of the equilibrium state, the effect of the rotating field $\boldsymbol{B}_1$ may be disregarded for two reasons that are valid in most experiments: (i) the field $\boldsymbol{B}_1$ is much smaller than the field $\boldsymbol{B}_0$, that is $|B_1| \ll |B_0|$; (ii) the relaxation processes are too slow for the instantaneous magnetisation to be able to follow the very rapid variation of the field $\boldsymbol{B}_1$; the field $\boldsymbol{B}_1$ rotates with the period

$$T = 2\pi/\omega \ll \tau$$

It may therefore be assumed that the equilibrium magnetisation depends only on the fixed field $\boldsymbol{B}_0$ and as a reminder we call it $\boldsymbol{M}_0$ instead of $\boldsymbol{M}_e$.

Rewriting the above equation (one equation for each component) we obtain the rate of change of magnetisation under the action of relaxation processes

$$\frac{\mathrm{d}\boldsymbol{M}}{\mathrm{d}t} = -\frac{\boldsymbol{M} - \boldsymbol{M}_0}{\tau} = \frac{\boldsymbol{M}_0 - \boldsymbol{M}}{\tau}$$

We have calculated separately the rate of change of magnetisation $\boldsymbol{M}$ under the influence of each of the processes (a) and (b); but the processes (a) and (b) have independent origins, and when they occur simultaneously, the total change of $\boldsymbol{M}$ is obtained by adding the change due to (a) and the change due to (b). Finally therefore, the rate of change of magnetisation $\boldsymbol{M}$ is equal to the sum of the two rates calculated in paragraphs (a) and (b) above

$$\left(\frac{\mathrm{d}\boldsymbol{M}}{\mathrm{d}t}\right)_{\text{lab}} = (\boldsymbol{\omega}_0 + \boldsymbol{\omega}_1) \times \boldsymbol{M} + \frac{\boldsymbol{M}_0 - \boldsymbol{M}}{\tau}$$

The solution of the problem may be simplified by the same procedure as in the preceding section, that is, by determining the vector $\boldsymbol{M}$ in a frame rotating with the angular velocity ω of the field $\boldsymbol{B}_1$. The transformation from the laboratory frame to the rotating frame may be carried out exactly as before which allows us to write

$$\boxed{\left(\frac{\mathrm{d}\boldsymbol{M}}{\mathrm{d}t}\right)_{\substack{\text{rotating}\\ \text{frame}}} = (\boldsymbol{\omega}_1 - \delta\boldsymbol{\omega}) \times \boldsymbol{M} + \frac{\boldsymbol{M}_0 - \boldsymbol{M}}{\tau}}$$

with $\boldsymbol{\omega}_1 = -\gamma\boldsymbol{B}_1$ and $\delta\boldsymbol{\omega} = \boldsymbol{\omega} - \boldsymbol{\omega}_0 = \boldsymbol{\omega} + \gamma\boldsymbol{B}_0$. Following Bloch's notation (he carried out this calculation for the first time in 1946), we let u and v be the components of the vector $\boldsymbol{M}$ along the OX and OY axes respectively of the rotating frame. By projecting the vectorial equation on to the axes, a system of differential equations is obtained

$$\frac{\mathrm{d}u}{\mathrm{d}t} = -\frac{u}{\tau} + \delta\omega v$$

$$\frac{\mathrm{d}v}{\mathrm{d}t} = -\delta\omega u - \frac{v}{\tau} - \omega_1 M_z$$

$$\frac{\mathrm{d}M_z}{\mathrm{d}t} = +\omega_1 v - \frac{M_z}{\tau} + \frac{M_0}{\tau}$$

The solution of this differential system is the sum of two functions.

(i) A transient solution that dies away after a time equal to several τ. This solution is similar to the solution in the preceding section; it is sinusoidal with a frequency $\sqrt{(\omega_1{}^2 + \delta\omega^2)}$, but now its amplitude decreases exponentially with a time constant of τ. In certain circumstances this transient solution is experimentally observable, and even provides a method of measuring relaxation times. However, we shall not calculate it here for the sake of brevity; we assume that the measurements are carried out after a delay of several τ following the application of the rotating field.

(ii) A steady-state solution, existing only after a time of several τ, with which we are concerned here.

9.5.3 Steady-state Solution of the Bloch Equations

Because the calculation of the magnetisation is carried out in the rotating frame, the steady-state solution reduces to a constant. This steady-state solution may be obtained by equating the derivatives of u, v and M_z to zero, which transforms the system of differential equations into a system of three linear equations with three unknowns

$$u - \delta\omega\tau v = 0$$

$$\delta\omega\tau u + v + \omega_1 \tau \mathcal{M}_z = 0$$

$$-\omega_1 \tau v + \mathcal{M}_z = M_0$$

The system may be easily solved by gradual substitution, and the components of the magnetisation in the rotating frame are obtained

$$(\boldsymbol{M})_{\substack{\text{rotating}\\ \text{frame}}} \begin{cases} u = \delta\omega\tau v & = M_0\left(\dfrac{\omega_1\delta\omega\tau^2}{1+\delta\omega^2\tau^2+\omega_1{}^2\tau^2}\right) \\ v = -M_0\dfrac{\omega_1\tau}{1+\delta\omega^2\tau^2+\omega_1{}^2\tau^2} & \\ M_z = M_0 + \omega_1\tau v & = M_0\left(1-\dfrac{\omega_1{}^2\tau^2}{1+\delta\omega^2\tau^2+\omega_1{}^2\tau^2}\right) \end{cases}$$

The magnetisation vector is fixed in the rotating frame, so that with respect to the laboratory frame it turns with the angular velocity ω of the rotating field, and its components are

$$(\boldsymbol{M})_{\text{lab}} \begin{cases} M_x = u\cos\omega t - v\sin\omega t = \sqrt{(u^2+v^2)}\cos(\omega t-\phi) \\ M_y = u\sin\omega t + v\cos\omega t = \sqrt{(u^2+v^2)}\sin(\omega t-\phi) \\ M_z \end{cases}$$

The rotating field therefore has two effects:

(i) a decrease of the longitudinal component M_z (that is, parallel to the fixed field $\boldsymbol{B}_0$) of the magnetisation;

(ii) the appearance of a non-zero transverse component $\boldsymbol{M}_T$ (that is, perpendicular to the fixed field $\boldsymbol{B}_0$) of the magnetisation, which rotates at the same rate as the rotating field $\boldsymbol{B}_1$. In this rotation, $\boldsymbol{M}_T$ is retarded by a certain angle ϕ with respect to $\boldsymbol{B}_1$; hence the rotating field does positive work and the corresponding energy is absorbed by the medium (see section 9.5.6).

Let us discuss the magnitude of these effects as a function of the two dimensionless parameters $\delta\omega\tau$ and $\omega_1\tau$.

(a) If the difference $\delta\omega$ between the Larmor rotation velocity ω_0 and that of the rotating field ω is very large, so that $|\delta\omega| \gg 1/\tau$ and $|\delta\omega| \gg \omega_1$ (simultaneously) then $u \simeq v \simeq 0$ and $M_z \simeq M_0$; thus the rotating field does not produce any change in the medium. Once again, we find that for any changes to occur the frequencies ω and ω_0 must be nearly equal and this justifies the name magnetic resonance.

(b) If the amplitude of the rotating field is very large, so that $\omega_1 \gg 1/\tau$ and $\omega_1 \gg |\delta\omega|$ (simultaneously) then $u \simeq v \simeq M_z \simeq 0$; thus hardly any overall magnetisation exists in the medium. This is described as saturation of the magnetic resonance.

(c) Avoiding these extreme conditions, we may choose $\omega_1\tau \approx 1$, and vary $\delta\omega$ on either side of zero. Curves representing the changes of u, v and M_z as a function of $\delta\omega$ may then be drawn (see figure 9.7).

When $\delta\omega = 0$ (centre of the resonance), u is zero, v passes through a maximum v_M, and M_z passes through a minimum. When

$$|\delta\omega| = \sqrt{(1/\tau^2+\omega_1{}^2)} = \frac{\Delta\omega}{2}$$

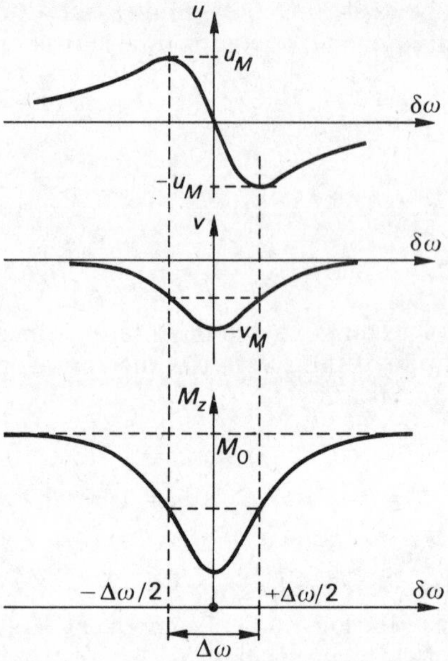

Figure 9.7 Solution of the Bloch equations when $\omega_1 \tau \approx 1$

u passes through a maximum u_M, v is reduced to half its value v_M at the centre, and the difference $M_0 - M_z$ is also reduced by half.

Twice this difference, $\Delta\omega$, is called the width at half height of the resonance curves; it is a measure of the allowable magnitude of the difference $\delta\omega$ such that a large effect may be observed. It should be noted that

$$\Delta\omega \to 2/\tau \quad \text{as} \quad \omega_1 \to 0$$

Comment I The height of these curves depends on ω_1. Thus

$$u_M = M_0 \frac{\omega_1 \tau}{2\sqrt{(1 + \omega_1{}^2 \tau^2)}}$$

is a steadily increasing function of that tends towards $M_0/2$ as ω_1 increases indefinitely. However,

$$v_M = M_0 \frac{\omega_1 \tau}{1 + \omega_1{}^2 \tau^2}$$

on the other hand is not a monotonic function of ω_1; it passes through a maximum equal to $M_0/2$ when $\omega_1 \tau = 1$, and then tends towards zero as ω_1 increases indefinitely.

On the other hand, the angle ϕ between $\boldsymbol{M}_T$ and $\boldsymbol{B}_1$ given by $\tan\phi = -v/u = -1/\delta\omega\tau$ depends only on $\delta\omega$, and not on ω_1.

Comment II To obtain the Bloch equations, we have introduced the relaxation in the form of a vectorial equation $d\boldsymbol{M}/dt = (\boldsymbol{M}_0 - \boldsymbol{M})/\tau$; by doing this, we have assumed implicitly that the three components of the vector $\boldsymbol{M}$ have the same relaxation time τ. In most cases it is necessary to distinguish between the longitudinal relaxation time τ_1 of the longitudinal component M_z and the transverse relaxation time τ_2 of the two transverse components u and v. This complicates the calculation slightly without changing its general form or the main conclusions.

Comment III The curve $v(\omega)$ has the same shape as the curve representing the absorption coefficient $K(\omega)$ in the theory of the elastically bound oscillator (see appendix 2). Physicists often call this a Lorentzian curve. It is shown in section 9.5.6 that the energy absorbed by the medium at the expense of the rotating field that produces the magnetic resonance is accurately proportional to the single component v. The variation of the power absorbed as a function of ω therefore has the same shape in both problems.

9.5.4 Experimental Verification by Radiofrequency Detection

(*a*) *It may be verified* by the use of a rotating field that magnetic resonance occurs only for one sense of rotation; hence the sign of γ may be deduced.

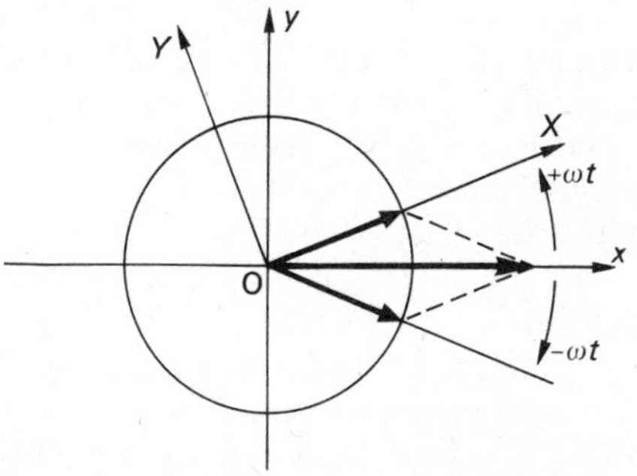

Figure 9.8 Combination of two fields rotating with equal and opposite angular velocities ω to form a linearly polarised alternating field

However, *to produce the phenomenon of magnetic resonance* it is not necessary to use a rotating magnetic field. It is sufficient to apply to the atoms under study a linear alternating magnetic field of magnitude $2B_1 \cos \omega t$ and of fixed direction perpendicular to the constant field $\boldsymbol{B}_0$ (the Ox direction in figure 9.8, drawn in the xOy plane).

It will be recalled that such a linear alternating field may be considered as the geometric sum of two fields of constant magnitude B_1, rotating in the xOy plane perpendicular to $\boldsymbol{B}_0$ with equal and opposite angular velocities. The field that rotates in the opposite sense to the Larmor rotation has hardly any effect on the atoms; only the field that rotates in the same sense as the Larmor rotation is effective. Thus all that is necessary is to pass an alternating current of frequency ω through a small coil on the Ox axis surrounding the sample being studied. If a very large magnetic field B_0 is used, the Larmor frequency occurs in the region of radar waves and the electric circuits must be replaced by waveguides. The sample of matter under study is then placed in a resonant cavity wherein an intense alternating field may be produced.

(*b*) *To detect the phenomenon*, the change of the longitudinal component $\mathcal{M}_z$ of the magnetic moments is sometimes measured (as in the Rabi method on an atomic beam for example; we describe this in the next chapter because it demonstrates quantum properties which have been overlooked so far). However, it is usually much easier to use a radiofrequency detection method, in which a property that we have described above is used, taking relaxation into account: *the appearance of a transverse component of the magnetisation* rotating at the same angular velocity ω as the rotating field. A rotating magnetic moment causes an alternating magnetic flux $\Phi = \Phi_0 \cos \omega t$ to pass through a coil, and therefore induces in it an alternating electromotive force $V = -\mathrm{d}\Phi/\mathrm{d}t = \omega\Phi_0 \sin \omega t$. Alternating voltages at a particular frequency can be selectively amplified electronically, and this allows the voltage V to be detected with great sensitivity.

The effect due to the rotating magnetisation of the medium must be distinguished from that due to the rotating field itself, but certain techniques enable this difficulty to be overcome.

(i) In the Bloch method, the detecting coil is placed in such a way that its Oy axis is exactly perpendicular to the Ox axis of the transmitting coil (which creates the alternating field $2B_1 \cos \omega t$); thus the flux created by the alternating field and passing through the receiving coil is zero.

(ii) In other methods, the effect of the flux induced by the rotating magnetisation on the transmitting circuit is observed; that is, only one coil is used and it acts simultaneously as a transmitter and a receiver. For example following Zavoisky or Purcell, the coil may be placed in a circuit similar to the usual bridge circuits used to measure coefficients of self-induction L and coefficients of voltage magnification or quality Q. Magnetic resonance gives rise to an apparent change of the self-inductance L (change of the flux $\Phi = Li$) and of the quality factor Q (loss of energy in the medium) manifested in the imbalance of the bridge. The same method may be employed at high frequencies by using waveguides: the characteristics of the resonant cavity are altered when the sample of material inside it undergoes magnetic resonance.

Comment I To achieve maximum sensitivity the detecting coil is always wound around the sample under investigation. If its axis is in the Ox or Oy direction, the sample induces a flux Φ, through the coil of cross-section S, proportional to one or other of the two components M_x and M_y of the magnetisation in the laboratory frame. The induced electromotive force is

$$-\frac{\mathrm{d}\Phi}{\mathrm{d}t} = \mu_0 S \frac{\mathrm{d}M_x}{\mathrm{d}t} = \mu_0 S\omega(u \sin \omega t + v \cos \omega t)$$

or

$$-\frac{\mathrm{d}\Phi}{\mathrm{d}t} = -\mu_0 S \frac{\mathrm{d}M_y}{\mathrm{d}t} = \mu_0 S\omega(-u \cos \omega t + v \sin \omega t)$$

In either case, the two transverse components u and v of the magnetisation in the rotating frame induce sinusoidal e.m.f.s, in quadrature with one another: one is in phase with the transmitter current (which creates the alternating field $2B_1 \cos \omega t$), the other is in quadrature with this current. A total e.m.f. is observed in the receiver coil out of phase by an angle ϕ and of amplitude $S\omega\sqrt{(u^2+v^2)}$; however, such a total electromotive force can be analysed electronically. By means of synchronous or phase-sensitive detection, one may measure the amplitudes $S\omega u$ and $S\omega v$ of the two components in phase or in quadrature with respect to the reference current. The two components u and v of the magnetisation in the rotating frame can thus be measured, and by varying the difference, $\delta\omega$, the curves drawn in section 9.5.3 may be verified experimentally.

Comment II The sensitivity of the radiofrequency detection method improves as the magnetic field B_0 increases. The amplitude of the induced e.m.f. V is equal to the product $\omega\Phi_0$, where

$$\omega \simeq \omega_0 = |\gamma| B_0$$

and Φ_0 is proportional to the transverse magnetisation M_T which is itself proportional to M_0, the equilibrium magnetisation under the influence of a field B_0. In paramagnetism, the equilibrium magnetisation M_0 is proportional to the applied field B_0, and it follows that Φ_0 is proportional to B_0.

The induced e.m.f. V is therefore proportional to $B_0{}^2$ for a coil of given area S (but at lower frequencies S may be increased by increasing the number of turns).

(*c*) Whatever method of detection is used, a magnetic resonance experiment may be carried out by

(i) keeping the field B_0 fixed and slowly varying the frequency $\nu = \omega/2\pi$ of the alternating field; this is called '*a frequency sweep*';

(ii) keeping the frequency ν fixed and slowly varying the field B_0; this is called '*a field sweep*'. In both cases, the amplitude of the effect is measured as a function of ν or of B_0; the curve of this function is called a magnetic resonance line (see figure 9.7 for examples of theoretically calculated resonance lines).

At the centre of a magnetic resonance line, the frequency ν of the alternating field equals the Larmor frequency $\nu_0 = |\gamma B_0|/2\pi$ corresponding to a magnetic field B_0. Hence the magnitude of the gyromagnetic ratio may be deduced

$$|\gamma| = 2\pi\nu/B_0$$

Conversely, if the gyromagnetic ratio γ has already been measured, the experiment may be used to measure the field B_0. The accuracy of the measurement depends on the width of the resonance line. In general the width at half height is defined as in section 9.5.3.

In section 9.5.3 we saw that on the frequency scale ω, the width $\Delta\omega$ is of the order of $2/\tau$, twice the reciprocal of the relaxation time. In other words, on the frequency scale, the width is of the order of $1/\pi\tau$, and on the magnetic field scale, the width is of the order of $2/\gamma\tau$. Measurements are therefore more accurate when relaxation times are long (so that the magnetic moments are less perturbed). Often a relative error of less than a part per million can be obtained.

9.5.5 Applications of Magnetic Resonance (E.P.R. and N.M.R.)

Two principal types of magnetic resonance experiments may be distinguished depending on whether they make use of electronic magnetic moments or nuclear magnetic moments.

(*a*) *Electron paramagnetic resonance* (*abbreviated to e.p.r.*; also known as electron spin resonance, e.s.r.). This is observed in paramagnetic media, those whose elementary constituents (atoms, ions, molecules or free radicals) possess a magnetic moment of electronic origin. Gyromagnetic ratios of the order of magnitude of the calculated value $q/2m\kappa$ are then measured. In many cases, twice this value, $q/\kappa m$, is measured as in the Einstein–de Haas or Barnett experiments but other values may also be measured. Their origin will be understood better in later chapters.

Because the method of radiofrequency detection becomes more sensitive as the magnetic field B_0 increases (see section 9.5.4), one works with high fields. From the normal value of the gyromagnetic ratio $\gamma/2\pi = 14$ GHz/tesla, a field $B_0 = 0{\cdot}7$ tesla leads to a frequency $\nu \simeq 10$ GHz which is in the microwave region, used in radar. This is why the experimental technique usually employs resonant cavities and waveguides.

E.P.R. is used extensively in research laboratories to obtain two different kinds of information: (i) the measurement of the gyromagnetic ratio of an atom, ion or molecule provides information about its internal structure (especially in the chemistry of free radicals which are always strongly paramagnetic); (ii) the measurement of relaxation times (either from linewidths or from the observation of transient phenomena) provides information about the interactions between neighbouring atoms in the medium studied. The variation of relaxation times may be studied as a function of various parameters (temperature, for example) and compared with the variation calculated from theoretical models describing these unknown interactions; these theoretical models can thus be tested.

E.P.R. is also utilised in the construction of magnetometers capable of measuring small magnetic fields (optical detection of e.p.r. in rubidium or caesium vapour), but more often nuclear magnetic resonance is employed to measure magnetic fields.

(*b*) *Nuclear magnetic resonance* (*abbreviated to n.m.r.*). The nucleus of an atom can also have a magnetic moment and an angular momentum related by a gyromagnetic ratio. Since the mass m of the protons that form nuclei is about two thousand times greater than that of electrons, it is to be expected that the gyromagnetic ratio γ of a nucleus is of the order of two thousand times smaller than that of the atomic electrons. For theoretical reasons, the angular momenta of nuclei and of electrons must be of the same order of magnitude. Hence magnetic moments of nuclei are also of the order of two thousand times smaller than those of electrons. The effects that they produce are therefore much smaller, and in particular the magnetisation corresponding to the partial orientation of nuclei in a fixed magnetic field (nuclear paramagnetism) is completely unobservable; it is always masked by the magnetisation of electronic origin, whether paramagnetism or diamagnetism. This is why in current terminology the description of substances as paramagnetic is reserved for substances having a paramagnetism of electronic origin.

However, the great sensitivity of magnetic resonance techniques and the frequency selectivity of this phenomenon allow changes of this nuclear magnetism under the influence of a rotating field of appropriate frequency to be observed; that is, they allow observation of the phenomenon of nuclear paramagnetic resonance or, more simply, nuclear magnetic resonance (n.m.r.).

The resonance frequencies thus measured are of the expected order of magnitude; for a given field B_0, the frequency ν is about two thousand times smaller in n.m.r. than in e.p.r., that is $\nu/B_0 = |\gamma|/2\pi \approx 10$ MHz/tesla.

Comment The expression given in the previous section for the e.m.f. $V = -\mathrm{d}\Phi/\mathrm{d}t$ induced in a coil of area S allows its order of magnitude to be evaluated for an n.m.r. experiment

$$V_{\text{eff}} = \frac{1}{\sqrt{2}} S\omega\mu_0 v_{\max} = \frac{1}{\sqrt{2}} S 2\pi\nu\mu_0 \frac{M_0}{2} = \frac{\pi}{\sqrt{2}} S\nu(\mu_r - 1) B_0$$

For paramagnetism of electronic origin, one usually measures

$$\mu_r - 1 \approx 10^{-4}$$

For paramagnetism of nuclear origin, a value of

$$\mu_r - 1 \approx 10^{-7}$$

may be expected. With an area $S \approx 1$ cm^2, in a field $B_0 = 1$ tesla, in other words a frequency $\nu \approx 10$ MHz, one calculates $V_{\text{eff}} = 10^{-4}$ V $= 100$ μV. This is the observed order of magnitude.

When atoms lose or gain some electrons to form an ion or when they combine to form a molecule, they usually do this in such a way that the total magnetic moment of the electrons of the ion or molecule is zero. This explains why most substances are diamagnetic and cannot be used for e.p.r. experiments. On the other hand nearly all substances contain atoms whose nuclei possess a magnetic moment, and they can be used for n.m.r. experiments. This explains the widespread use of n.m.r. in chemistry laboratories.

(i) The observation of a line corresponding to the characteristic gyromagnetic ratio of a nucleus allows the presence of that element in the product being studied to be identified. Analyses can thus be carried out, and in some instances concentrations may be found.

(ii) The characteristic n.m.r. line of an element or group of atoms (for example, the methyl group) is often very slightly displaced in frequency. This frequency 'shift' expresses the fact that the corresponding nucleus is subjected to a very small magnetic field by neighbouring atoms which adds to the field imposed by the experimenter. Measurement of this very small additional field provides information about the environment of the nucleus. It is often a characteristic of the particular molecular group in which the nucleus is embedded.

N.M.R. has a number of other applications and these are increasing in both the scientific and technological fields. The most common technological application is undoubtedly in the construction of very accurate magnetometers, by simply using hydrogen nuclei (protons) contained in a small volume of water. The importance of these developments justifies the space given in this chapter to the phenomenon of magnetic resonance. Furthermore we return to it in the following chapter in order to develop quantum aspects of the phenomenon.

9.5.6 Postscript—Calculation of the Energy Changes

(*a*) *Consideration of the steady-state solution, taking into account relaxation* (see section 9.5.3). The appearance of a transverse component of the total magnetisation leads to an *absorption of energy* by the medium, as we now show.

(i) The rotating field $\boldsymbol{B}_1$ exerts on the magnetisation $\boldsymbol{M}$ a couple whose resultant moment $\boldsymbol{\Gamma}_1 = \boldsymbol{M} \times \boldsymbol{B}_1$ has a longitudinal component $\boldsymbol{\Gamma}_{1z}$ along the Oz axis. The origin of this component is the transverse component $\boldsymbol{M}_T$ of the magnetisation: $\boldsymbol{\Gamma}_{1z} = \boldsymbol{M}_T \times \boldsymbol{B}_1$. The moment of the forces about the Oz axis does work in rotation about the Oz axis.

(ii) The sign of this work depends on the sense of $\boldsymbol{\Gamma}_{1z}$, that is on the relative alignment of $\boldsymbol{M}_T$ and $\boldsymbol{B}_1$: when γ is negative, ω_0 and ω_1 are positive, so the component v of the magnetisation along the OY axis is negative. On the other hand ω must be close to ω_0 in order for an effect to be observed, so that ω is also positive. Thus it may be seen that in the rotation around Oz, $\boldsymbol{M}_T$ is delayed with respect to $\boldsymbol{B}_1$ and that $\boldsymbol{\Gamma}_{1z}$ is positive like ω. For nuclei of atoms where γ is positive, all the signs are reversed and the conclusion is the same, in other words $\boldsymbol{M}_T$ is delayed with respect to $\boldsymbol{B}_1$ and $\boldsymbol{\Gamma}_{1z}$ has the same sign as ω.

Hence the work done by the forces exerted by the magnetic field is positive. Therefore the magnetic moments continually absorb energy from the electromagnetic wave that provides the rotating field. But magnetic moments in steady-state motion cannot accumulate this energy; they continually transfer it to the medium in the form of heat, through the relaxation processes.

(iii) The power P corresponding to the amount of work done in unit time may be calculated

$$P = \boldsymbol{\Gamma}_1 \cdot \boldsymbol{\omega} = \Gamma_{1z}\,\omega = -vB_1\,\omega$$

(P is positive since v and ω have opposite signs). It is proportional to the sole rotating component v perpendicular to the field B_1. (This is described by saying that during the rotation the component v is in quadrature with respect to the field B_1, whereas the component u is in phase with B_1.)

(*b*) *Consideration of an oscillating solution in the absence of relaxation* (see section 9.5.1) (this may also be applied to the transient solution in the presence of relaxation). We consider an isolated atom of magnetic moment $\mathcal{M}$ in the absence of relaxation (see section 9.5.1). The rotating field $\boldsymbol{B}_1$ exerts on the magnetic moment $\mathcal{M}$ a couple of resultant moment $\boldsymbol{\Gamma}_1 = \mathcal{M} \times \boldsymbol{B}_1$; we calculate the work done by this couple during the motion of the magnetic moment.

In the laboratory frame, we wrote

$$\left(\frac{\mathrm{d}\mathcal{M}}{\mathrm{d}t}\right)_{\mathrm{lab}} = (\omega_0 + \omega_1) \times \mathcal{M}$$

so that the vector $\omega_0 + \omega_1$ (a function of time) represents the instantaneous rotation vector of the moment $\mathcal{M}$. In these circumstances, the work done by the couple in a short time $\mathrm{d}t$ is

$$\mathcal{T}_1 = \Gamma_1 \cdot (\omega_0 + \omega_1)\,\mathrm{d}t = \Gamma_1 \cdot \omega_0\,\mathrm{d}t = \Gamma_{1z}\,\omega_0\,\mathrm{d}t$$

(since ω_1 is perpendicular to Γ_1 it makes no contribution to the scalar product). Γ_{1z} is expressed solely in terms of the component $\mathcal{M}_Y$ of the moment along the rotating axis OY perpendicular to B_1

$$\mathcal{T}_1 = \Gamma_{1z}\,\omega_0\,\mathrm{d}t = -\mathcal{M}_Y B_1\,\omega_0\,\mathrm{d}t = -\mathcal{M}_Y\,\omega_1 B_0\,\mathrm{d}t$$

This formula is analogous to that obtained in the case of the steady-state solution if ω is replaced by ω_0 and v, a component of the total magnetisation, by $\mathcal{M}_Y$, a component of the magnetic moment of a single atom. However, in the steady-state solution v always has the opposite sign to ω_0 because the total transverse magnetisation rotates with the same velocity as the field B_1 and with a constant delay with respect to it.

In the present case, on the other hand, the individual atomic moment has a different motion from that of the rotating field B_1 and the component $\mathcal{M}_Y$ changes sign in the course of time.

We have studied the motion of the magnetic moment $\mathcal{M}$ in the rotating frame (see section 9.5.1), and thus we deduce that

$$\mathcal{M}_Y = \mathcal{M}_\perp \sin\sqrt{(\omega_1{}^2 + \delta\omega^2)}\,t$$

$\mathcal{M}_\perp$ and $\mathcal{M}_\parallel$ are the components of $\mathcal{M}$, perpendicular and parallel respectively to the rotation axis in the rotating frame.

$\mathcal{M}_Y$ keeps the same sign during half a period $T/2$ of this rotation. Let us calculate the work done during this half period

$$\mathcal{T}_1 = \int_0^{T/2} - \mathcal{M}_Y B_0\,\omega_1 \sin\sqrt{(\omega_1{}^2 + \delta\omega^2)}\,t\,\mathrm{d}t$$

$$= \left[-\mathcal{M}_Y B_0 \frac{\omega_1}{\sqrt{(\omega_1{}^2 + \delta\omega^2)}} \cos\sqrt{(\omega_1 + \delta\omega^2)}\,t\right]_0^{T/2}$$

$$\mathcal{T}_1 = +2\mathcal{M}_\perp B_0 \frac{\omega_1}{\sqrt{(\omega_1{}^2 + \delta\omega^2)}} = 2\mathcal{M}_\perp B_0 \sin\theta$$

introducing the angle θ between Oz and the rotation axis in the rotating frame. During the same half period, the component $\mathcal{M}_z$ of the magnetic moment passes from one extreme value to another (see figure 9.9, drawn in the rotating plane XOZ)

$$\mathcal{M}_z(0) = \mathcal{M}_\parallel \cos\theta + \mathcal{M}_\perp \sin\theta$$

$$\mathcal{M}_z(T/2) = \mathcal{M}_\parallel \cos\theta - \mathcal{M}_\perp \sin\theta$$

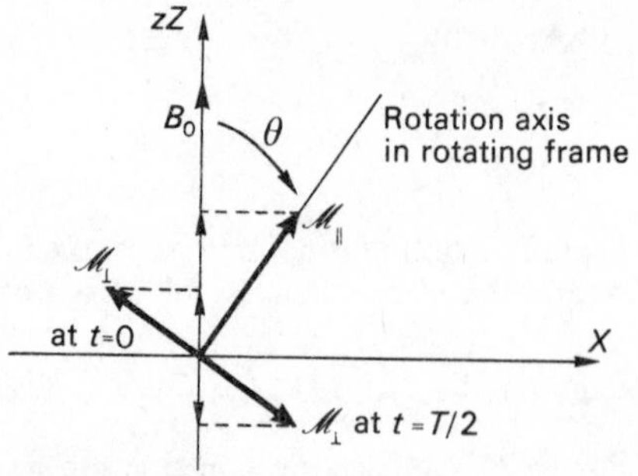

Figure 9.9 Components of $\mathcal{M}$ in the rotating frame

The energy of the atomic system $W_0 = -\mathcal{M}_z B_0$ has therefore undergone a change $\delta W_0 = 2\mathcal{M}_\perp B_0 \sin\theta$. Thus $\delta W_0 = \mathcal{T}_1$; this is the work done by the rotating field which alows the energy of the system to change. During this half period, the rotating field has done positive work; the electromagnetic wave has lost energy which it has given to the atomic system. During the following half period, the signs are reversed; the work done by the rotating field is negative; and the electromagnetic wave receives energy from the atomic system which loses it.

This calculation gives a more detailed account of the alternation between induced emission and absorption, already discussed in Comment I of section 9.5.1.

9.6 General Conclusions

Consideration of the orbital motion of electrons has given rise to the idea of a gyromagnetic ratio, γ, a constant of proportionality between the magnetic moment $\mathcal{M}$ and the angular momentum $\boldsymbol{\sigma}$ of an atom such that $\mathcal{M} = \gamma\boldsymbol{\sigma}$. The experiments that we have described in this chapter confirm both the usefulness and validity of this idea since they have allowed a number of phenomena to be explained; they also re-open the question of the origin of this gyromagnetic ratio since the measured numerical values are rarely equal to the value calculated from electronic orbits $\gamma = (1/\kappa)(q/2m)$. It will only be by the use of quantum theory results in volume 2, chapter 5 that the origin of the gyromagnetic ratio will be explained and its exact numerical value calculated.

All the experiments described in this chapter rely also on the phenomenon of paramagnetism, namely, the existence of a magnetic moment $\mathcal{M}$ associated with an atom and to which statistical laws apply. Henceforward it should be remembered that the existence of this magnetic moment $\mathcal{M}$ is in fact a quantum property. If the laws of classical statistics were applied to the motion of electrons, taking into consideration all their degrees of freedom, it would be concluded that no magnetisation could occur (Van Leeuwen's theorem); in other words, the magnetisation corresponding to the diamagnetic effect would balance the magnetisation in the opposite direction, corresponding to the paramagnetic effect.

In this chapter we have argued in classical terms; but by postulating at the beginning of this chapter the existence of an atomic magnetic moment, we have implicitly made a quantum hypothesis. These quantum properties will be clarified in the following chapter.

10

Spatial Quantisation

10.1 The Stern and Gerlach Experiment

10.1.1 Principle of the Experiment

The experiment carried out on atomic magnetic moments by Stern and Gerlach in 1921 is one of the most fundamental of atomic physics: it demonstrates irrefutably the limitations of classical theories and constitutes one of the pillars of quantum mechanics and of the theory of measurement. In addition it has important applications in the laboratory and in technology.

In the experiments described in the preceding chapter, we have assumed that the magnetic fields $\boldsymbol{B}$ used were uniform (or 'homogeneous'), at least on the atomic scale, and in their influence on atomic magnetic moments, we have taken into account only the resultant moment

$$\boldsymbol{\Gamma} = \mathcal{M} \times \boldsymbol{B}$$

Under these circumstances, we saw that all the experiments result in the measurement of the same atomic parameter—the gyromagnetic ratio γ—but none of them allow measurement of the magnetic moment $\mathcal{M}$ of the atom itself (the only quantity that can be measured is the intensity of magnetisation $\boldsymbol{M}$, that is, the magnetic moment in a unit volume containing a large number of atoms).

On the other hand, if an inhomogeneous magnetic field $\boldsymbol{B}$ is used, variable from one point of the atom to another, the magnetic forces have a non-zero resultant $\boldsymbol{F}$ which tends to displace the atom; we shall see how this may result in a direct measurement of the magnetic moment $\mathcal{M}$ of the atom.

In section 8.1 we discussed the detailed expression for the resultant $\boldsymbol{F}$. It depends on the components of the magnetic moment and on the gradient of the magnetic field, that is on the derivatives with respect to the co-ordinates of the components of the field $\boldsymbol{B}$. Thus, for the z component of the force

$$F_z = \mathcal{M} \cdot \frac{\partial \boldsymbol{B}}{\partial z} = \mathcal{M}_x \frac{\partial B_x}{\partial z} + \mathcal{M}_y \frac{\partial B_y}{\partial z} + \mathcal{M}_z \frac{\partial B_z}{\partial z}$$

Let us assume that the magnetic field $\boldsymbol{B}$ is parallel to the axis Oz. In the preceding chapter (section 9.1, Larmor precession) we saw that the effect of the resultant moment

$$\boldsymbol{\Gamma} = \mathcal{M} \times \boldsymbol{B}$$

induces a rotation around the direction of the field $\boldsymbol{B}$, that is the Oz axis. So the component $\mathcal{M}_z$ is constant whereas the $\mathcal{M}_x$ and $\mathcal{M}_y$ components undergo a very rapid sinusoidal variation, and their mean value in time is zero. The effect of the resultant $\boldsymbol{F}$ is experimentally observable only if it acts for a long enough time, and consequently only the mean value of F_z in time is important

$$\bar{F}_z = \mathcal{M}_z \frac{\partial B_z}{\partial z}$$

Observation of the displacement of an atom under the influence of an inhomogeneous magnetic field can therefore provide a measurement of the longitudinal component $\mathcal{M}_z$ (parallel to the magnetic field) of its magnetic moment. This is the principle of the experiment carried out by Stern and Gerlach, which we shall now describe.

10.1.2 Description of the Experimental Arrangement

The experiment uses the atomic beam technique. The latter was developed in 1911 by the French physicist Dunoyer and was used to carry out a direct confirmation of the kinetic theory of gases. The German physicist Stern had the idea of passing an atomic beam between the pole pieces of a magnet. The experimental arrangement is shown in figure 10.1 (for greater clarity the scale is not consistent; dimensions perpendicular to the atomic beam are strongly exaggerated).

The atoms under study are introduced into a container which is connected by a small hole to a large highly evacuated enclosure (the residual pressure is less than 10^{-5} torr; this is low enough for the mean free path of atoms to be much larger than the dimensions of the enclosure). When the container is heated, it fills with vapour at the saturated pressure corresponding to the temperature T. Some atoms escape continually through the hole in the wall,

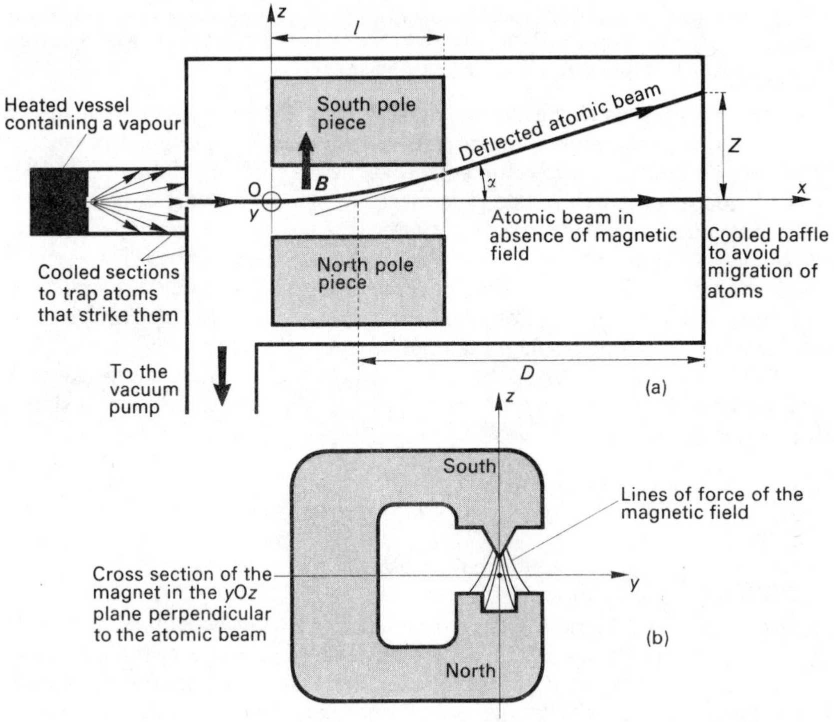

Figure 10.1 Diagram of the Stern and Gerlach experiment

and since these atoms no longer meet any obstacle, they have straight-line trajectories. However, in order to enter the large enclosure, the atoms must pass through a second hole pierced in a diaphragm. Of the atoms that escape from the vapour, those whose velocity is directed along a line joining the two holes are thus selected; they cross the large enclosure in a straight line forming what is called an atomic beam.

Comment An atomic beam can be observed in several different ways.

(a) By optical resonance in which the beam is illuminated with the light from a spectral lamp containing the same atoms. The atoms in the beam absorb photons corresponding to a resonance line, and re-emit them. The beam then appears as a luminous region within the evacuated container.

(b) After a period of time, the spot formed by a large number of atoms at the point of impact of the beam and the walls of the container may be observed. It is often necessary to cool the walls in order to prevent atoms from migrating after impact (otherwise they can spread out over the surface of the walls).

(c) In order to investigate more precisely the position of an atomic beam of an alkali metal, a metal wire heated by the passage of a current may be placed within the evacuated enclosure. When the atoms contact the heated wire, they become ionised and may be detected as a current of positive ions received by an electrode negative with respect to the wire. This method operates best with alkali metals. In general, the atoms may be transformed into positive ions by bombarding them with a narrow electron beam, but this method is far less efficient.

A magnet is placed in the evacuated enclosure. It creates a magnetic field $\boldsymbol{B}$ perpendicular to the atomic beam, and its pole pieces are aligned parallel to the direction of the atomic beam such that the magnetic field may be applied over an adequate length l. The pole pieces of the magnet are constructed in such a way that the magnetic field gradient is very large. Their shape is that of a cylinder whose generators are parallel to the beam and whose cross-section is shown in figure 10.1(b): the lines of force of the magnetic field are indicated, and the law of flux conservation in a tube of force implies that the field is more intense near the pointed pole piece than it is near the other pole.

Let Ox be the direction of the atomic beam and Oz that of the magnetic field. In these conditions, an atom in the beam is subjected to a *mean* translation force with components

$$\bar{F}_x = \mathscr{M}_z\,\partial B_z/\partial x$$
$$\bar{F}_y = \mathscr{M}_z\,\partial B_z/\partial y$$
$$\bar{F}_z = \mathscr{M}_z\,\partial B_z/\partial z$$

However, because the pole pieces are cylindrical with generators parallel to Ox direction of the beam, $\partial B_z/\partial x = 0$ (disregarding edge effects at the entrance and exit of the air gap of the magnet). Furthermore, if the beam travels in the plane of symmetry xOz of the pole pieces, then $\partial B_z/\partial y = 0$. The average resultant force therefore reduces to its component $\bar{F}_z$ and the atomic beam is deflected in the xOz plane.

Thus an atomic beam moving initially with velocity v parallel to Ox and subjected to the influence of a constant force, follows a parabolic trajectory within the air gap, as in motion under gravity (it may be shown that here the gravitational force is negligible). By taking the origin of the axes and the origin of time at the atom's entry to the air gap, its motion may be written

$$x = vt$$

$$z = \frac{1}{2}\frac{\bar{F}_z}{m}t^2$$

The atom leaves the air gap after a time $t = l/v$. It then continues in a straight line, but it has been deflected by an angle α such that

$$\tan\alpha = \frac{\mathrm{d}z}{\mathrm{d}x} = \frac{\bar{F}_z t}{mv} = \frac{\bar{F}_z l}{mv^2}$$

It may be shown that the tangent to the parabola at the point of abscissa l cuts the tangent at the origin at the point of abscissa $l/2$. If D is the distance between the middle of the air gap and the plate on which the atoms are collected, their point of impact is found to be displaced by the distance

$$Z = \bar{F}_z \frac{lD}{mv^2}$$

We have carried out the calculation for a particular atom of velocity v. However, different atoms in the beam have different velocities and, for the same force $\bar{F}_z$, they are displaced by different amounts Z. Their points of impact are therefore distributed over a slightly extended spot, the maximum density of the spot corresponding to the most probable velocity v in the atomic beam. The distribution of velocities v in an atomic beam is different from that existing in the vapour from which it came, due to the fact that velocities parallel to a particular direction have been selected. It may be shown from the kinetic theory of gases that the most probable velocity v in the beam is equal to the root-mean-square velocity in the vapour from which it came (see section 10.7 at the end of this chapter); therefore $mv^2 = 3kT$, where T is the absolute temperature of the vapour, and k is Boltzmann's constant. Hence the most probable displacement at the point of impact of the atoms (maximum density of the spot for a given $\bar{F}_z$) may be deduced

$$\boxed{Z = \bar{F}_z \frac{lD}{3kT} = \mathcal{M}_z \frac{\partial B_z}{\partial z} \times \frac{lD}{3kT}}$$

All the quantities appearing in this expression are known except $\mathcal{M}_z$. Measurement of the displacement Z of the point of impact of the atomic beam therefore allows the determination of the longitudinal component $\mathcal{M}_z$ of the atomic magnetic moment.

Comment An *a priori* order of magnitude of atomic magnetic moments may be calculated from the interpretation of ferromagnetism. It is assumed that saturation of magnetisation occurs when all the atomic magnetic moments are parallel and oriented in the same direction, the intensity of magnetisation then being $M = n\mathcal{M}$, where n is the number of atoms per unit volume and can be evaluated from the formula $n = \mathcal{N}\rho\mathcal{A}$ ($\mathcal{N}$ is Avogadro's number, ρ the density, $\mathcal{A}$ the atomic mass, see section 8.3).

Knowing the order of magnitude of the magnetic induction inside saturated soft iron $B_{sat} \simeq 1{\cdot}5$ tesla, then

$$\frac{B_{sat}}{\mu_0} \simeq M = n\mathcal{M} = \mathcal{N}\mathcal{M}\frac{\rho}{\mathcal{A}} \quad \text{or alternatively} \quad \mathcal{M} \simeq \frac{B_{sat}}{\mu_0}\frac{\mathcal{A}}{\mathcal{N}\rho}$$

By taking care to use the correct units ($\mathcal{A} = 56$ c.g.s. $= 56/1000$ SI; $\rho \approx 8$ CGS $=$ 8000 SI) one obtains $\mathcal{M} \approx 10^{-20}$ c.g.s. $= 10^{-23}$ SI. If $\mathcal{M}_z$ is of this order of magnitude, then using attainable field gradients $\partial B_z/\partial z \approx 10^2$ to 10^3 tesla/m, and with reasonable dimensions for l and D and an absolute temperature $T \sim 500$ K a displacement Z may be calculated of the order of 1 cm; this is an entirely measurable effect.

As an exercise, the order of magnitude given above may be used to carry out the following calculations:

(i) to verify that the gravitational force is indeed negligible compared with the magnetic force;

(ii) to verify that the magnetic force exerts an influence for a time t long compared with the period of Larmor rotation; this justifies the approximation made in considering only the average force $\overline{F}_z$;

(ii) to calculate the gradient of the field for the soft iron pole pieces whose cross-sections are two concentric circles (see figure 10.2).

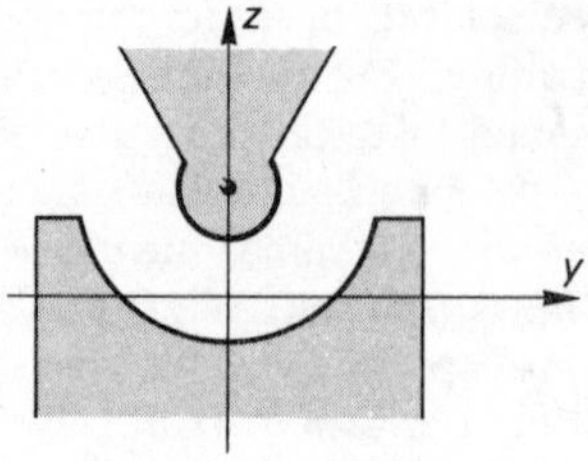

Figure 10.2 Concentric pole pieces

10.1.3 Experimental Results

We have at our disposal an apparatus for measuring the longitudinal component $\mathcal{M}_z$ of an atomic magnetic moment. What results may be expected from it? In the preceding comment we calculated the order of magnitude to be expected for the magnitude of the magnetic moment vector of an atom; but its $\mathcal{M}_z$ component depends on its orientation with respect to the Oz axis, and can be positive, negative or zero.

Statistical laws may be applied to the magnetic moments of atoms in a vapour because the many collisions which they undergo enable statistical equilibrium to be established. When they leave the vapour in the form of a beam, the atoms cease to undergo collisions and remain in the state they assumed on leaving the vapour. When the vapour is not subjected to a magnetic field, classical statistics predicts that the moment vectors $\mathcal{M}$ are equally distributed in all directions, so that the number of vectors $\mathcal{M}$ contained within a given solid angle is proportional to that solid angle; hence all values of $\mathcal{M}_z$ are equally probable. Accordingly, during a Stern and Gerlach experiment one would expect to see the atoms distributed uniformly over the screen in a region extending parallel to the direction Oz of the magnetic field, symmetric about the point of normal impact of the beam in the absence of a magnetic field. The two extremities of this region must correspond to the maximum value of $\mathcal{M}_z$.

In fact, when Stern and Gerlach carried out their experiment with atoms of silver, they did not obtain a uniform region, but instead two spots which were symmetric in relation to the point of normal impact (see figure 10.3). Contrary to the prediction of classical statistics, they observed only two values of $\mathcal{M}_z$ but these were of the expected order of magnitude (it should be noted that they did not even observe the value $\mathcal{M}_z = 0$). If the experiment is carried out on other atoms, separated spots are always observed, corresponding to discrete values of the $\mathcal{M}_z$ component; the spots are always distributed symmetrically

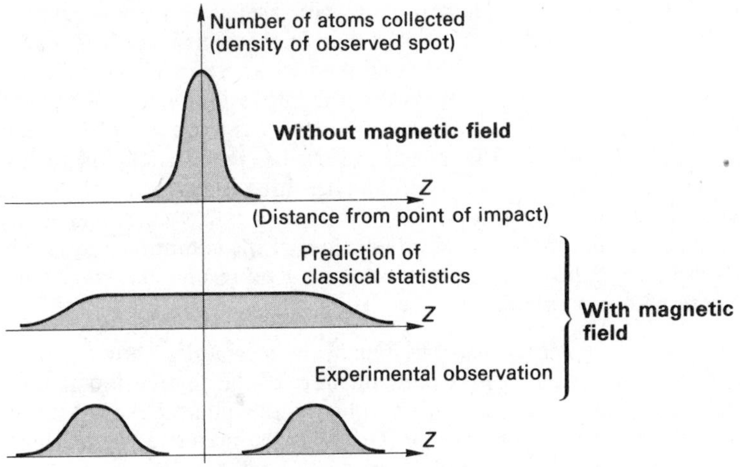

Figure 10.3 Results of the Stern and Gerlach experiment

in relation to the point of normal impact, so that the possible values are equal and opposite in pairs. This means that the magnetic moment vectors may have only certain orientations in space, and not every orientation. This is called spatial quantisation.

To summarise, the Stern and Gerlach experiment is a method of measuring directly the longitudinal component $\mathcal{M}_z$ of atomic magnetic moments (the component parallel to the magnetic field). It shows that:

(a) the order of magnitude of atomic magnetic moments is indeed that calculated from the interpretation of measurements of the intensity of magnetisation;

(b) contrary to classical theory, $\mathcal{M}_z$ can take only discrete values, equal and opposite in pairs. This phenomenon of spatial quantisation is one of the fundamental phenomena that must be explained by quantum mechanics.

10.2 The Rules of Quantisation

10.2.1 The Definition of the Quantum Numbers for Angular Momenta

We saw throughout the preceding chapter that the magnetic moment vector $\mathcal{M}$ is proportional to the angular momentum vector σ; spatial quantisation of the magnetic moments therefore involves that of the angular momenta and *vice versa.* The German physicist Sommerfeld had already made the hypothesis of spatial quantisation of angular momenta when he generalised the Bohr theory of the atom by taking into account elliptic orbits and their three-dimensional orientation in space. In this way he was led to generalise the Bohr quantisation condition (see chapter 6) by writing separately a new quantisation condition for the component σ_z of the angular momentum in a particular direction. However, Sommerfeld continued to argue in classical terms and his efforts served mainly to demonstrate the impossibility of making classical mechanics consistent with experimentally observed quantum phenomena.

Quantum mechanics on the other hand has been constructed in such a way as to overcome this difficulty. In defining the three components of the angular momentum vector as operators, then seeking the eigenvalues of these operators, and so calculating the observable values of the component σ_z (see books on quantum mechanics), quantum mechanics gives results that are in complete agreement with experiment.

(*a*) *The magnetic quantum number*. Quantum mechanics shows that (i) the various observable values of the component σ_z of the angular momentum in a particular direction differ from one another by the quantity $\hbar$ or by a multiple of $\hbar$; (ii) if one value is observable then so is the opposite value. Hence the observable values are given by the formula

$$\boxed{\sigma_z = m\hbar}$$

in which the *dimensionless number m, called the magnetic quantum number*, can be either a whole number, positive, negative or zero in which case the value $\sigma_z = 0$ is then observable (see figure 10.4(a)); or a half-integral number, that is the sum of a whole number and the fraction 1/2 in which case the value $\sigma_z = 0$ is impossible (see figure 10.4(b)).

Note: The magnetic quantum number should not be confused with the mass of the electron, though following international convention they are represented by the same letter m (the magnetic quantum number is often denoted by m_J with the subscript J).

(*b*) *The angular-momentum quantum number*. For each particular atomic system, there exists a maximum value of the component σ_z that is, a maximum value of the magnetic quantum number m. *This maximum value of m is designated by the letter J, and is called the angular-momentum quantum number*. This quantum number J characterises the magnitude of the angular-momentum vector. Knowledge of the number J characteristic of an atomic state suffices for a complete determination of all the observable values of the component $\sigma_z = m\hbar$ of its angular momentum: either (i) J is integral; the values of m are integers and the number of them is odd (the value zero is unique); or (ii) J is half-integral; the values of m are half-integral and the number of them is even. In both cases, the values of m are between $-J$ and $+J$, so that *the number of observable values of* σ_z *is* $2J+1$. Conversely, knowledge of the number of values of σ_z allows the angular-momentum quantum number J to be found. An atom whose angular momentum is zero is characterised by the quantum number $J = 0$.

It is often useful to introduce into the discussion a dimensionless vector proportional to $\boldsymbol{\sigma}$. For this reason one puts

$$\boxed{\boldsymbol{\sigma} = \hbar \mathbf{J}}$$

If $\hbar$ is chosen as the unit for measuring angular momenta, $\boldsymbol{J}$ is obtained directly from $\boldsymbol{\sigma}$. The components of $\boldsymbol{J}$ are equal to the values of the magnetic quantum number m; the maximum component of $\boldsymbol{J}$ is equal to the quantum number J.

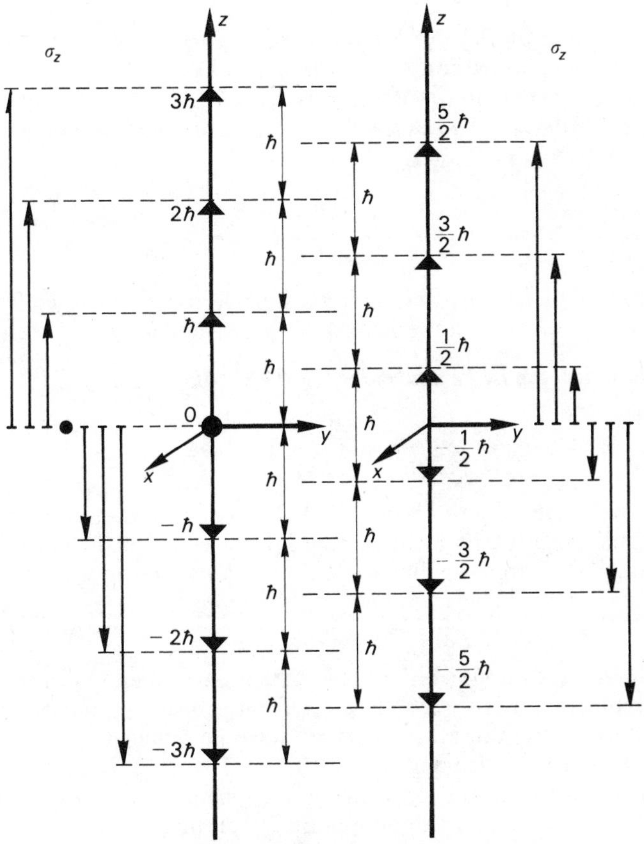

Figure 10.4 The law of spatial quantisation

Note: Care should be taken not to confuse the quantum number J with the magnitude $|\boldsymbol{J}|$ of the vector $\boldsymbol{J}$. The confusion should be readily avoided since in this section, as in the previous one, we have never discussed the magnitudes of the vectors $\mathcal{M}$ and $\boldsymbol{\sigma}$. *We have strictly discussed only the components $\mathcal{M}_z$ and σ_z of these vectors, which are the only measurable quantities in the experiments described.* Quantum mechanics shows that the introduction of the magnitudes of these vectors poses several problems (see volume 2, chapters 3 and 4).

10.2.2 Application to Magnetic Moments—the Bohr Magneton and the Landé Factor

We derive magnetic moments from angular momenta by multiplying by the gyromagnetic ratio γ

$$\mathcal{M} = \gamma \boldsymbol{\sigma} = \gamma \hbar \boldsymbol{J}$$

The normal value of the gyromagnetic ratio corresponding to an orbital motion can be substituted for γ; $\gamma = (1/\kappa)(q/2m)$.

(*Note*: in this expression m is the mass of the electron; the simultaneous use of the quantum number m in a formula or figure will be avoided.)

Thus

$$\mathcal{M} = \frac{1}{\kappa} \times \frac{q}{2m} \hbar \boldsymbol{J} = -\beta \boldsymbol{J}$$

where the negative quantity that multiplies the vector $\boldsymbol{J}$ is denoted by a single letter (since the charge q on the electron is negative: $q = -e$).

The Bohr magneton is the name given to the positive quantity

$$\beta = -\frac{1}{\kappa}\frac{q}{2m}\hbar = +\frac{1}{\kappa}\frac{e}{2m}\hbar = +\frac{1}{\kappa}\frac{eh}{4\pi m}$$

(often it is designated by the symbol μ_B). The Bohr magneton is the natural unit of atomic magnetic moments, in the same way as $\hbar$ is the natural unit of angular momenta. From the fundamental constants, one may calculate

$$\beta = 0{\cdot}9274 \times 10^{-20} \text{ erg G}^{-1}\text{(CGS)} = 0{\cdot}9274 \times 10^{-23} \text{ JT}^{-1}\text{(SI)}$$

This is in agreement with the order of magnitude derived previously for atomic magnetic moments. It is sometimes convenient to use the magnetic moment corresponding to a gram-atom—that is the Bohr magneton multiplied by Avogadro's number $\mathcal{N}$—so that $\mathcal{N}\beta = 5585$ erg G^{-1} mole^{-1} or JT^{-1} kmole^{-1}.

In general, it is useful to express the gyromagnetic ratio as a function of its normal value calculated for orbital motion, by putting

$$\boxed{\gamma = g\frac{1}{\kappa} \times \frac{q}{2m}}$$

The dimensionless number g is called the Landé factor. It is the ratio between the gyromagnetic ratio of an atom and the normal value of the orbital gyromagnetic ratio. Experiment shows that the Landé factor is usually a simple fractional number, which may be explained theoretically (see volume 2, chapter 5). With this notation, the magnetic moment vector may be written

$$\boxed{\mathcal{M} = \gamma \hbar \boldsymbol{J} = -g\beta \boldsymbol{J}}$$

and the observable values of its components may be written

$$\boxed{\mathcal{M}_z = \gamma\hbar m = -g\beta m}$$

This discussion enables us to re-examine the Stern and Gerlach experiment in greater detail. With silver atoms, only two spots are observed, that is to say, two values of $\mathcal{M}_z$ and of σ_z. Hence $2J+1=2$ and therefore the angular momentum quantum number $J=1/2$.

Furthermore if the two values of $\mathcal{M}_z$ are measured exactly, they are found to be more or less equal to the Bohr magneton; thus $\mathcal{M}_z = \pm\beta = \pm Jg\beta$. Since J is known, one may deduce that $g=2$. Exactly the same results are obtained with atoms of copper or gold, of sodium or potassium (Taylor's measurements, 1927) as well as with those of hydrogen.

With thallium atoms, two spots are also obtained so that again $J=1/2$ but this time one measures $\mathcal{M}_z = \pm\frac{1}{3}\beta$. Hence $g=2/3$.

To summarise, the Stern and Gerlach experiment permits the measurement of the angular momentum quantum number J and the Landé factor g of the atoms under study.

10.3 Zeeman Sublevels

We have already had to make use of the formula for the magnetic interaction energy W between a magnetic moment $\mathcal{M}$ and other charges in motion that create a magnetic induction $\boldsymbol{B}$.

This formula may be expressed in terms of the longitudinal component $\mathcal{M}_z$ of the magnetic moment (we always choose the Oz axis parallel to the direction of the field). When applied to an atom, $\mathcal{M}_z$ can take only the observable values indicated in the preceding section

$$\boxed{W = -\mathcal{M}\cdot\boldsymbol{B} = -\mathcal{M}_z B = -m\gamma\hbar B = +mg\beta B}$$

Since the values of $\mathcal{M}_z$ are quantised, then so are the values of the magnetic energy W: for a given magnetic induction $\boldsymbol{B}$, only certain discrete values of the magnetic energy determined by the magnetic quantum number m are observable.

Let E_0 be the energy corresponding to the normal state of the atom when it is not subjected to any external interaction; when the induction B is applied, the energy of the atom becomes $E_0 + W = E(m)$. The initial value of the energy E_0 is thus replaced by many close but distinct values $E(m)$; this is described by saying that the energy level E_0 is split into several sublevels. These allow the Zeeman effect to be explained correctly (see chapter 11) and this is why they are called Zeeman sublevels; each corresponds to a particular value of the magnetic quantum number m.

From the possible values of the quantum number m, it may be deduced that the Zeeman sublevels are equally spaced; the difference between two neighbouring sublevels

$$E(m+1) - E(m) = g\beta B$$

is independent of m. This difference is proportional to the applied magnetic induction B. Figure 10.5 shows the variation as a function of B of the energies corresponding to Zeeman sublevels for two different cases.

The number of Zeeman sublevels corresponding to the same initial level E_0 is equal to the number of possible values of m, that is to $2J+1$, where J is the angular-momentum quantum number corresponding to the energy level E_0.

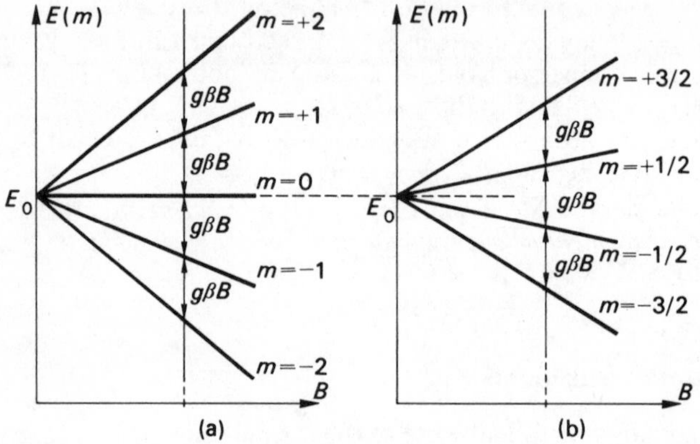

Figure 10.5 Zeeman sublevels; (a) J and m integral; (b) J and m half-integral

The $2J+1$ Zeeman sublevels correspond to different states of the atom; they differ in the orientation of the magnetic moment and angular-momentum vectors, in other words, in the different values of the longitudinal components $\mathcal{M}_z$ and σ_z. When the magnetic field is zero ($B=0$) these $2J+1$ states of the atom have exactly the same energy E_0. In quantum mechanics, the energy level E_0 corresponding to $2J+1$ different states is said to be $2J+1$ *times degenerate*. On the other hand, the application of the magnetic field makes the energy of these $2J+1$ states different; *the magnetic field is said to have lifted the degeneracy of the energy level* E_0.

Many experiments, described at the end of this chapter and in the next chapter, may be explained by the existence of Zeeman sublevels.

10.4 Application to the Calculation of the Intensity of Paramagnetic Magnetisation

In section 9.2 we explained paramagnetic magnetisation as the partial orientation of atomic magnetic moments in the direction of the applied field $\boldsymbol{B}$. In 1905, the French physicist Langevin calculated the intensity of magnetisation M as a function of B from the statistical equation of this partial orientation. However, Langevin assumed that the $\mathcal{M}_z$ component of the magnetisation could have any value between the minimum $-\mathcal{M}$ and the maximum $+\mathcal{M}$ with equal probability. The Stern and Gerlach experiment showed that Langevin's hypotheses were incorrect, and it was necessary to carry out the calculation again, taking into account the rules for spatial quantisation; this was done by the French physicist Brillouin in 1927.

10.4.1. Brillouin's Calculation

The intensity of magnetisation vector $\boldsymbol{M}$ may be obtained by summing vectorially all the atomic magnetic moments $\mathcal{M}$ contained within a unit volume. If we assume that the magnetic field $\boldsymbol{B}$ is always parallel to the Oz axis, the magnetic energy $W = -\mathcal{M}_z B$ is independent of the components $\mathcal{M}_x$ and $\mathcal{M}_y$; the latter are distributed at random and $\sum \mathcal{M}_x = \sum \mathcal{M}_y = 0$. The magnetisation vector $\boldsymbol{M}$ therefore reduces to its component M_z, simply called M.

According to Boltzmann's statistical law, the atoms have a distribution of different values of $\mathcal{M}_z$, proportional to the function $e^{-W/kT} = e^{+\mathcal{M}_z B/kT}$.

Comment By writing Boltzmann's law in this way, we are assuming that the interaction energies between the magnetic moments are negligible. This is true only if the atoms with magnetic moments are not too close to one another. This occurs in vapours or in solutions and crystal hydrates where the magnetic ions are separated by a large number of water molecules. The laws of paramagnetism are then obeyed. If the interactions between neighbouring magnetic moments are large, very different effects may be obtained (ferromagnetism or antiferromagnetism).

The observable values of $\mathcal{M}_z$ in a general case have been expressed in section 10.2 as a function of the magnetic quantum number m as $\mathcal{M}_z = -g\beta m$. It will be recalled that m varies in steps of unity from $-J$ to $+J$ and that there are $2J + 1$ values of m.

The number of atoms having the same component $\mathcal{M}_z$ (or the same quantum number m) must be calculated; this number of atoms is often called the population of the Zeeman sublevel m, and we shall denote it by $p(m)$. From the laws we have already mentioned

$$\boxed{p(m) = \frac{n}{Z} e^{-W/kT} = \frac{n}{Z} e^{+\mathcal{M}_z B/kT} = \frac{n}{Z} e^{-m\, g\beta B/kT} = \frac{n}{Z} e^{-mx}}$$

where $x = \dfrac{g\beta B}{kT}$

The coefficient of proportionality, independent of m, has been written as n/Z conforming to the conventions of statistical mechanics: n is the total number of atoms per unit volume, and the dimensionless quantity Z is called the partition function. The latter may be determined by equating the sum of the populations $p(m)$ to n

$$n = \sum_{m=-J}^{+J} p(m) = \frac{n}{Z} \sum_{m=-J}^{+J} e^{-mx} \quad \text{so} \quad Z = \sum_{m=-J}^{+J} e^{-mx} = \sum_{m=-J}^{+J} (e^{-x})^m$$

Since m varies in steps of unity, Z is the sum of the first $2J+1$ terms of a geometrical progression with a ratio $q = e^{-x}$ and a first term e^{+Jx} (for $m = -J$). Hence

$$Z = e^{+Jx}\,\frac{1-q^{2J+1}}{1-q} = \frac{e^{+Jx} - e^{-(J+1)x}}{1-e^{-x}}$$

Multiplying top and bottom by $e^{x/2}$, one obtains

$$Z = \frac{e^{(J+1/2)x} - e^{-(J+1/2)x}}{e^{x/2} - e^{-x/2}} = \frac{\sinh(J+\frac{1}{2})x}{\sinh x/2}$$

Having determined the populations of the Zeeman sublevels, we can now sum the components $\mathcal{M}_z$ for all the atoms contained in a unit volume

$$M = \sum \mathcal{M}_z = \sum_m (-g\beta m)\,p(m) = \frac{ng\beta}{Z}\sum_m (-m e^{-mx}) = \frac{ng\beta}{Z}\frac{dZ}{dx}$$

noting that the sum $\sum$ is equal to the derivative of Z with respect to x. It remains only to calculate the logarithmic derivative of Z

$$\frac{1}{Z}\frac{dZ}{dx} = \frac{(J+\frac{1}{2})\cosh(J+\frac{1}{2})x}{\sinh(J+\frac{1}{2})x} - \frac{\frac{1}{2}\cosh x/2}{\sinh x/2}$$

$$= (J+\tfrac{1}{2})\coth(J+\tfrac{1}{2})x - \tfrac{1}{2}\coth\frac{x}{2}$$

$$\boxed{M = ng\beta\left[(J+\tfrac{1}{2})\coth\,(J+\tfrac{1}{2})x - \tfrac{1}{2}\coth\frac{x}{2}\right]}$$

where $x = \dfrac{g\beta B}{kT}$.

Comment I This formula is not as complicated as it appears; the second term within the bracket may be found very easily—it ensures that the bracket and therefore the magnetisation M are zero for $J = 0$ (when atoms do not have a magnetic moment).
Comment II For the case of $J = 1/2$, often occurring in practice, the formula simplifies considerably. The bracket reduces to

$$\coth x - \coth\frac{x}{2} = \frac{1+\tanh^2 x/2}{2\tanh x/2} - \frac{1}{2\tanh x/2} = \tfrac{1}{2}\tanh\frac{x}{2}$$

so

$$\boxed{M = \frac{ng\beta}{2}\tanh\frac{x}{2}}$$

The magnetisation M is an increasing function of the parameter x. As in Langevin's calculation, it is an increasing function of the applied magnetic field B but a decreasing function of the absolute temperature T. It is interesting to study the extreme cases where x is either very large or very small.

(a) As $x \to \infty$, the two hyperbolic cotangent functions tend towards unity so that $M \to ng\beta J$. This is the value obtained when the longitudinal component of the magnetic moment has the maximum value $\mathcal{M}_z = g\beta J$ for each of the n atoms; this is called the saturated magnetisation $M_{\text{sat}} = ng\beta J$. To obtain saturation, it is necessary that $x \gg 1$, so that by ignoring the number g which is of the order of unity

$$\frac{B}{T} \gg \frac{k}{\beta} = \frac{R}{\mathcal{N}\beta} \approx 1 \text{ tesla per degree}$$

(where R is the ideal-gas constant, and $\mathcal{N}$ is Avogadro's number). With magnetic fields obtainable in the laboratory of $B \approx 1$ tesla saturation can be obtained only by lowering the temperature below 1 K, which requires rather special techniques. In fact, temperatures between 1 K and 2 K are commonly achieved in the laboratory (by rapid pumping on liquid helium in order to cool it further); it is then necessary to produce very large fields of the order of 10 tesla.

(b) If $x \ll 1$, that is for ordinary temperatures, a limited expansion can be made

$$\coth \varepsilon = \frac{1}{\varepsilon}\left(1 + \frac{\varepsilon^2}{3} + \text{etc.}\right) = \frac{1}{\varepsilon} + \frac{\varepsilon}{3} + \ldots$$

hence

$$(J + \tfrac{1}{2}) \coth (J + \tfrac{1}{2})x - \tfrac{1}{2} \coth \frac{x}{2} = \frac{x}{3}[(J + \tfrac{1}{2})^2 - \tfrac{1}{4}] = \frac{x}{3} J(J + 1)$$

$$\boxed{M = ng\beta J(J+1)\frac{x}{3} = M_{\text{sat}}(J+1)\frac{x}{3} = \frac{ng^2\beta^2 J(J+1)}{3kT} B}$$

Thus the well-known laws of paramagnetism at ordinary temperatures are established: the magnetisation is proportional to the field B and inversely proportional to the absolute temperature T, so $M = C(B/T)$ where the constant C is called Curie's constant.

10.4.2 Comparison with Langevin's Classical Theory

In classical statistics, the longitudinal component $\mathcal{M}_z$ can take any value between the minimum $-\mathcal{M}$ and the maximum $+\mathcal{M}$ with equal probability. The calculation involves the same steps as those set out above except that the discontinuous summations must be replaced by integrals. By using the parameter $y = \mathcal{M}B/kT$, Langevin's formulae may be obtained:

(i) the general formula

$$M = n\mathcal{M}\left(\coth y - \frac{1}{y}\right) = M_{\text{sat}}\left(\coth y - \frac{1}{y}\right)$$

(ii) for $y \ll 1$

$$M \simeq M_{\text{sat}}\frac{y}{3} = \frac{n\mathcal{M}^2}{3kT}B$$

where we have introduced the saturation magnetisation $M_{\text{sat}} = n\mathcal{M}$.

For comparative purposes, it is more useful to rewrite Brillouin's formulae by transforming them with the following additional notation: the maximum $\mathcal{M}_z$ component is

$$\mathcal{M} = g\beta J = \frac{M_{\text{sat}}}{n}$$

and the parameter y is given by

$$y = Jx = \frac{g\beta J}{kT}B = \frac{\mathcal{M}B}{kT}$$

Brillouin's formulae then become

(i) the general formula

$$M = n\mathcal{M}\left[\left(1 + \frac{1}{2J}\right)\coth\left(1 + \frac{1}{2J}\right)y - \frac{1}{2J}\coth\frac{y}{2J}\right]$$

(ii) for $y \ll 1$

$$M \simeq M_{\text{sat}}\left(1 + \frac{1}{J}\right)\frac{y}{3} = n\frac{\mathcal{M}^2}{3kT}\left(1 + \frac{1}{J}\right)B$$

When J becomes very large, these formulae become nearly the same as Langevin's formulae. Here we come across a general principle: the laws of quantum physics become the same as the laws of classical physics when the quantum numbers have large values.

Thus when magnetisation measurements are carried out on gadolinium sulphate which contains the ions Gd^{3+} with quantum number $J = 7/2$, it is very difficult to distinguish between the Brillouin theory and the Langevin theory.

On the other hand, when J is small, the difference between the two theories is considerable. It is particularly significant for $J = 1/2$. This is illustrated in figure 10.6; for a given saturation magnetisation M_{sat} and number of atoms per unit volume n, the variations of the magnetisation M as a function of the parameter y according to Langevin and according to Brillouin are compared. For $J = 1/2$, the Brillouin curve approaches saturation three times faster than the Langevin curve; but for $J = 2$, for example, it approaches it only 1·5 times faster.

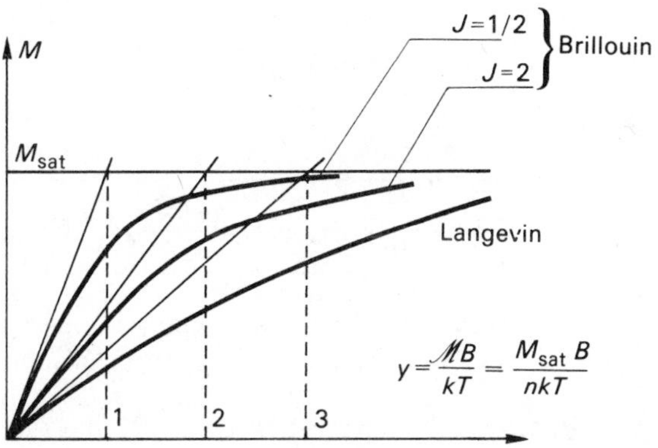

Figure 10.6 Illustration of the difference between the results of the Langevin (classical) and Brillouin (quantum) theories for two values of J

10.4.3 Experimental Verification of Brillouin's Formulae

Brillouin's formulae may be verified from two points of view.

(a) It may be confirmed that the shape of the curve $M(y)$ differs from that of Langevin and is in good agreement with that calculated by Brillouin, if the quantum number J of the atoms concerned is known from other experiments. This shape is mainly determined by the relation existing between the saturation magnetisation and the slope at the origin (see figure 10.6). This method therefore assumes that saturation may be approached by lowering the temperature sufficiently. One such investigation was carried out in 1933 by the Dutch physicists Gorter, de Haas and Van den Haendel carrying out measurements on a crystal of chrome alum down to a temperature of 1·3 K; they obtained the curve corresponding to the quantum number $J = 3/2$ characteristic of the Cr^{3+} ion in the alum.

Conversely, the curve of the magnetisation as a function of temperature (on condition the latter is taken sufficiently low) allows the angular momentum quantum number J to be estimated.

(b) It may be confirmed that the slope of the curve at the origin has the particular value calculated from the quantum number J and the Landé factor g.

Experimentally, the Curie constant C is measured such that $M = C(B/T)$; this must be compared with the theoretical value $C = n[g^2\beta^2 J(J+1)]/3k$.

The Curie constant depends on the number of atoms per unit volume n, that is to say on the pressure if the atoms are in the vapour state, or on concentration if ions are combined with other constituents of a liquid or solid medium. In order to express the results in a manner that is independent of n and that is characteristic of the atom or ion studied, it is convenient to relate the results obtained to a gram-atom. In other words the atomic Curie constant C_A is defined from the magnetic moment corresponding to a gram-atom, that is,to a collection of $\mathcal{N}$ atoms ($\mathcal{N}$ is Avogadro's number)

$$\frac{\mathcal{N}}{n} M = \mathcal{M}\,(\text{gram-atom}) = C_A \frac{B}{T}$$

where

$$\boxed{C_A = \frac{\mathcal{N}}{n} C = \mathcal{N} \frac{\beta^2}{3k} g^2 J(J+1) = \frac{\mathcal{N}\beta^2}{3R} g^2 J(J+1)}$$

(R is the gas constant). In SI units, one obtains numerically $C_A = 1{\cdot}25$ $g^2 J(J+1)$ (expressed in CGS units the atomic Curie constant would be ten times smaller).

In 1927 this formula was confirmed in detail by the German physicist Hund, using ions of the rare earths. He used the results of magnetic susceptibility measurements carried out two years previously and the values of J and g deduced from the study of spark spectra (the spectra of ions) of the rare earth elements.

We shall draw attention to two other pieces of evidence:

(i) Measurements of the magnetic susceptibilities of alkali metal vapours (sodium, potassium and so on) performed by the German Roth, gave an atomic Curie constant $C_A = 3{\cdot}7$. This is in perfect agreement with the values $J = 1/2$ and $g = 2$ found by the Stern and Gerlach method (see section 10.2), from which one finds $g^2 J(J+1) = 3$.

(ii) The magnetic susceptibility of chrome alum allows a Curie constant $C_A = 18{\cdot}7$ to be calculated for the ion Cr^{3+}, in agreement with the values $J = 3/2$ and $g = 2$ which lead to $g^2 J(J+1) = 15$.

> To summarise, measurements of the intensity of magnetisation carried out on paramagnetic substances confirm the validity of the rules for spatial quantisation and the existence of Zeeman sublevels. They are in agreement with the values of the quantum number J and of the Landé factor g measured by other methods.

The method most often used for these measurements of J and g is based on the observation of the Zeeman effect, which will be studied in the following chapter. Magnetic resonance experiments are also used to measure Landé factors.

10.5 Application to Magnetic Resonance

10.5.1 The Bohr Hypothesis applied to Adjacent Zeeman Sublevels

The quantum interpretation of magnetic resonance follows immediately from the difference between the Zeeman sublevels calculated above, if the Bohr hypothesis between two neighbouring Zeeman sublevels is applied: transitions between these two sublevels must be caused by a wave of frequency ν or of angular frequency ω such that

$$h\nu = \hbar\omega = |E(m+1) - E(m)| = |g\beta B| = |\gamma\hbar B|$$

in other words a wave of frequency $\omega = |\gamma|B$. This is the condition for magnetic resonance that we have already mentioned in section 9.5.

The wave of frequency $\omega = |\gamma|B$ can change the orientation of the atomic magnetic moments because its photons $\hbar\omega$ are exactly equal to the energy difference $\gamma\hbar B$ between two possible orientations of the atomic magnetic moment. At this stage the quantum explanation is still incomplete; it does not tell us why the electromagnetic wave must be circularly polarised nor in which sense. This will be explained in the following chapter.

It would seem that the quantum explanation is based on a different principle from the classical explanation given in the preceding chapter. In actual fact the classical calculations carried out in chapter 9 would be meaningless if applied literally to an isolated atom, but in terms of the total magnetisation $\boldsymbol{M}$ corresponding to a very large number of atoms, these classical calculations, though not of absolutely general validity, have a real meaning. In the frequently occurring situation where the angular-momentum quantum number J is 1/2, so that there are only two Zeeman sublevels, quantum mechanical calculations result in exactly the same differential equations for the components of the magnetisation vector $\boldsymbol{M}$. This is why these calculations were carried out in detail in the preceding chapter. When J has higher values, the equations become more complicated without altering the essential features of the phenomenon.

10.5.2 Experiments on an Atomic Beam by the Rabi Method

Transition probability. Whether the quantum or classical explanation is used, magnetic resonance always consists of changing the longitudinal component $\mathscr{M}_z$ of the atomic magnetic moments. Since the Stern and Gerlach experiment measures $\mathscr{M}_z$, it must enable detection of the phenomenon of magnetic resonance. This is the principle of the method used by Rabi to detect magnetic resonance, which we shall now describe.

To simplify this description, we shall confine ourselves to the case where the angular momentum quantum number J is 1/2, so that only two Zeeman sublevels exist, but this does not affect the generality of the method. In Rabi's apparatus, the atomic beam crosses the air gaps of the three different magnets in succession which create three parallel magnetic fields (see figure 10.7).

(1) The first field $\boldsymbol{B}_1$ is inhomogeneous, and deflects the atoms in parabolic paths. It separates the initial beam into two distinct beams each corresponding to a single value of $\mathscr{M}_z$. One of these two beams hits a wall and only the other continues on its path; thus atoms are selected all of whose magnetic moments have the same longitudinal component $\mathscr{M}_z$.

(2) The second field $\boldsymbol{B}_2$ is uniform and the atoms cross it in a straight line. Here the magnetic-resonance phenomenon may be produced by means of a radiofrequency electromagnetic wave of frequency $\omega = |\gamma| B_2$.

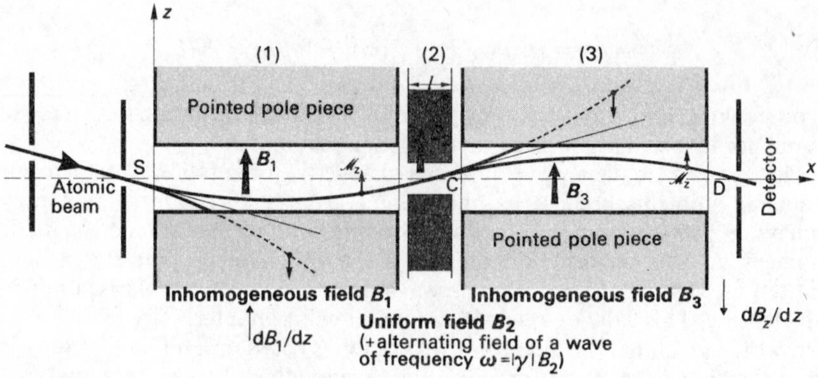

Figure 10.7 Rabi's apparatus (magnetic resonance on an atomic beam)

(3) The third field $\boldsymbol{B}_3$ is inhomogeneous like the first; in the same way, it selects atoms whose magnetic moments have a certain longitudinal component $\mathscr{M}_z$, and only these may reach the detector (where they are ionised by collision with a hot wire or by electron bombardment).

If changes of $\mathscr{M}_z$ are produced by magnetic resonance in magnet (2), the number of atoms received by the detector is similarly changed, which allows the resonance to be observed. More precisely, atoms whose initial velocity vector makes a certain angle with the Ox axis of the apparatus and whose component is positive, describe an arc of a parabola, symmetric with respect to its turning point, between the point source S and the centre C of magnet (2). (S and C are aligned along the Ox axis of the apparatus, see figure 10.7.) The velocity of the atoms at C is symmetric to the initial velocity at S with respect to the Ox axis. If magnet (3) is identical to magnet (1) but is placed the other way round (symmetric in relation to C), the same atoms describe an arc of a parabola in the field $\boldsymbol{B}_3$, symmetric in relation to C to the previous arc, and they all reach the slit D of the detector. When magnetic resonance is produced in the field $\boldsymbol{B}_2$, some of the atoms selected by magnet (1) make a transition from the state '$\mathscr{M}_z$ positive' to the state '$\mathscr{M}_z$ negative'; they are no longer able to reach the detector, and a decrease in the number of atoms received by the detector is measured.

An experiment can be carried out by irradiating region (2) of the atomic beam with an electromagnetic wave of fixed frequency ω and by varying the uniform field B_2 so as to alter the energy difference between the two Zeeman sublevels (see figure 10.8(a)). The number of atoms N recorded by the detector in a certain time is observed as a function of the field B_2, and a curve is obtained similar to that of figure 10.8(b). The number of atoms N is normally equal to N_0, but it decreases when B_2 approaches its resonance value, and passes through a minimum N_m when B_2 passes through the value $B_2 = \omega/|\gamma|$.

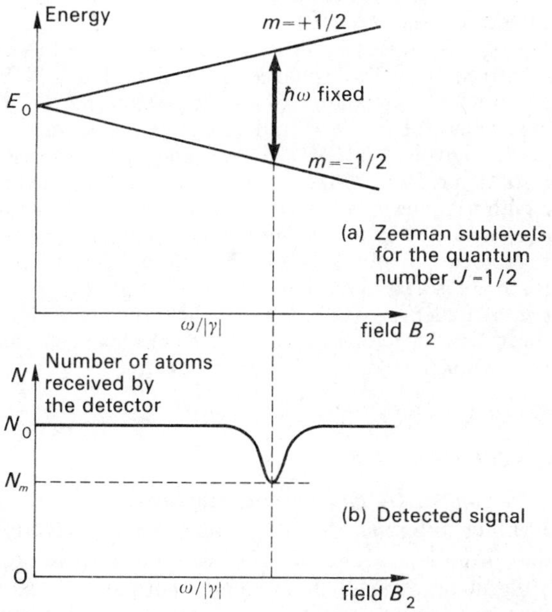

Figure 10.8 A magnetic resonance experiment performed by sweeping the magnetic field at fixed frequency

Let us now consider the centre of the resonance. The difference $N_0 - N_m$ is equal to the number of atoms that have made the transition between the two states corresponding to opposite values $m = \pm 1/2$ of the magnetic quantum number. The atoms that leave the first inhomogeneous field are all in the same quantum state $m = +1/2$, and they are subjected to the same electromagnetic wave over a path of length l, for a time t approximately the same for all of them. Only atoms whose velocities v are very close to the most probable velocity are focused on to the slit D of the detector, and their transit times $t = l/v$ across the field $\boldsymbol{B}_2$ are therefore very nearly equal. However, some remain in the initial state whilst others make a transition from the initial state to the quantum state $m = -1/2$. Here we have a direct illustration of the concept of a transition probability. The relative decrease in the number of atoms received $(N_0 - N_m)/N_0$ is equal to the probability of inducing a transition between the two quantum states after a time t by means of an electromagnetic wave. The transition probability calculated quantum mechanically may be verified experimentally.

Comment I In Rabi's experiment, the atoms leave the source S with divergent initial velocities, and the apparatus selects simultaneously (a) those whose $\mathscr{M}_z$ component is positive if their initial velocity vector makes a certain angle with Ox; (b) those whose component $\mathscr{M}_z$ is negative if their initial direction is appropriate (symmetric in relation to Ox). However, the detector can be reached only by atoms whose $\mathscr{M}_z$ component has not been changed on the way; magnetic resonance is detected simultaneously for both categories of atoms. (This may be generalised immediately to the case where $J > 1/2$.)

Comment II Historically, it was by this atomic beam method that the American physicist Rabi observed magnetic resonance for the first time in 1938; it resulted in immense progress in the understanding of atomic phenomena. The radiofrequency detection method, described in the preceding chapter, was employed only much later, by the Russian Zavoisky in 1945 for e.p.r. and by the Americans Bloch and Purcell in 1946 for n.m.r. Each of these two methods has its advantages and disadvantages and both continue to be used in the laboratory and in industry. The beam method is more complicated to set up, but for the study and measurement of individual magnetic moments of atoms, it is usually more accurate because the atoms in the atomic beam are isolated and thus the perturbations existing between neighbouring atoms in a condensed medium are avoided. Other methods of detection have recently been developed including optical detection (see chapter 11) and double-resonance methods.

10.5.3 Steady-state Experiments

Equalisation of the populations and power absorption of the wave. Experimental techniques are rather different when other methods of detection are used to study atoms that do not separate out in a beam but instead are in a material sample (solid, liquid or vapour) that remains permanently subjected to the influence of the radiofrequency wave. A dynamic equilibrium between the radiofrequency wave and the medium studied is then rapidly established in which the populations n_1 and n_2 of the sublevels $m = -1/2$ and $m = +1/2$ are stationary, despite transitions that continue to occur between them.

(a) We have seen that the spontaneous emission probability A_{21} is practically zero for radiofrequency transitions (see section 3.3.3). If the phenomenon of magnetic resonance is all that occurs, induced-emission transitions must therefore balance absorption transitions (see section 3.3.2)

$$B_{12}n_1 = B_{21}n_2$$

However, the statistical weights of the Zeeman sublevels are equal

$$G_1 = G_2 \rightarrow B_{12} = B_{21} \rightarrow n_1 = n_2$$

Therefore magnetic resonance, acting alone, equalises the populations of the two Zeeman sublevels; we have illustrated this diagrammatically in figure 10.9(a).

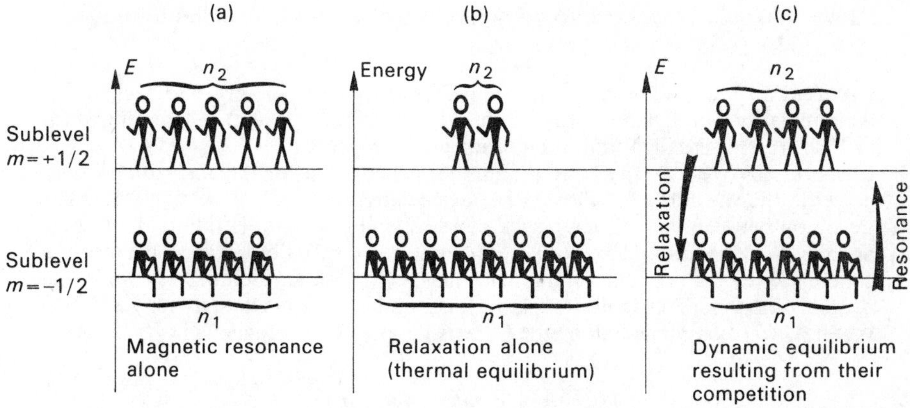

Figure 10.9 Schematic representation of the populations of the Zeeman sublevels

(b) However, magnetic resonance never acts alone in a material sample; relaxation processes must be taken into account (see section 9.2). The various random interactions between neighbouring atoms tend to establish or maintain thermal equilibrium in the medium, that is to say a distribution of atoms amongst the various Zeeman sublevels conforming to Boltzmann's law. In section 10.4 we calculated a distribution in which the Zeeman sublevels of higher energy are less populated than the sublevels of lower energy. In the absence of a direct external interaction (the radiofrequency wave being suppressed), this thermal equilibrium distribution is obtained, as shown in figure 10.9(b).

(c) In an actual experiment, a dynamic equilibrium is established resulting from the competition between these two phenomena. The populations of the Zeeman sublevels of higher energy are greater than in thermal equilibrium and the populations of sublevels of lower energy are less than in thermal equilibrium (see figure 10.9(c)). Relaxation processes tend to re-establish thermal equilibrium and therefore continually cause transitions from sublevels of higher energy to sublevels of lower energy in greater numbers than transitions in the opposite direction.

Conversely, to maintain dynamic equilibrium, the phenomenon of magnetic resonance continually produces transitions from sublevels of lower energy to sublevels of higher energy in greater numbers than transitions in the opposite direction. In other words, transitions involving absorption of photons are more numerous than transitions involving induced emission of photons ($B_{12}\, n_1 > B_{21}\, n_2$ since $n_1 > n_2$).

Transitions requiring a contribution of energy from the radiofrequency wave are therefore more numerous than the others. A certain fraction of the power in the electromagnetic wave is continually being absorbed by atoms undergoing the resonance, and the latter continually restore it to the medium in the form of heat by means of the relaxation processes. This absorption of power at the expense of the electromagnetic wave is easily measurable and allows magnetic resonance to be detected by the methods of radiofrequency detection.

Comment I If the intensity of the radiofrequency wave is very high ($\gamma B_1 \gg 1/\tau$) the populations in dynamic equilibrium are equal (figure 10.9(c) becomes identical to 10.9(a)). This is called saturation of magnetic resonance.

Comment II We may thus give a quantum explanation for the absorption of the power in the wave that has already been calculated classically in section 9.5.6. The transition between the two languages, classical and quantum, is simple when there are only two Zeeman sublevels (quantum number $J = 1/2$). For instance, if a medium contains n_1 atoms per unit volume with magnetic quantum number $m = -1/2$ and n_2 atoms corresponding to $m = +1/2$, the longitudinal component of its magnetisation vector $\boldsymbol{M}$ (the component along the Oz axis parallel to the magnetic field) is

$$M_z = (n_2 - n_1)g\beta J = (n_2 - n_1)\frac{g\beta}{2}$$

Comment III We have confined ourselves here to a qualitative analysis of the equation for the populations. For a quantitative study, the relaxation time τ would have to be introduced. Its reciprocal $1/\tau$ measures the probability of non-radiative transitions per unit time, causing the return to thermal equilibrium (to a certain extent it thus takes the place of the probability for a spontaneous radiative transition A_{21} which, as stated, is negligible). Thus the third Bloch equation, for the z component (see section 9.5.2) may be interpreted as an equation for the change of populations. However, it should be remembered that the exact calculation involves transverse quantities and that a description of the phenomenon in terms of populations alone would not be complete.

10.6 General Conclusions

The Stern and Gerlach experiment is one of the most fundamental of atomic physics because it allows direct measurement of an atomic quantity other than energy: the *longitudinal component of the magnetic moment* (directed 'along' the magnetic field) and therefore also, through the gyromagnetic ratio, the *longitudinal component of the angular momentum.*

It demonstrates that the notion of quantum values is not a characteristic of energy alone, but extends also to other atomic quantities. It is thus one of the main pillars of quantum mechanics in that it involves: (i) the idea of eigenvalues associated with each atomic quantity, and supports the concept of an operator associated with each quantity; and (ii) the experimental meaning that must be given to these eigenvalues. It is nearly always to the Stern and Gerlach experiment that one refers in considering the interpretation of quantum mechanics, in other words the theory of measurement.

When the Stern and Gerlach experiment was modified by Rabi to allow the magnetic resonance phenomenon to be observed, it also became one of the best illustrations of the concept of a transition probability.

The quantum theory of angular momentum involves the most fundamental ideas of quantum mechanics, and it is very readily proved by experiment. The importance attached to it in books on quantum mechanics is therefore not surprising.

10.7 Addendum: The Velocity Distribution Law in an Atomic Beam

10.7.1 Review of the Velocity Distribution Law in a Vapour

(a) Each velocity component has a separate gaussian probability distribution. For example the probability density for the component v_x is

$$f(v_x) = \frac{1}{\sigma\sqrt{(2\pi)}} \mathrm{e}^{-v_x^2/2\sigma^2}$$

where f is normalised such that

$$\int_{-\infty}^{+\infty} f(v_x)\,\mathrm{d}v_x = 1$$

and σ is the mean-square deviation defined by

$$\overline{v_x^2} = \int_{-\infty}^{+\infty} v_x^2 f(v_x)\,\mathrm{d}v_x = \sigma^2 = kT/m$$

This last equality may be derived from the kinetic theory explanation for the pressure P of the gas due to collisions between atoms and the walls containing them

$$P = nmv_x^2 = nm\sigma^2$$

where the number of molecules per unit volume $n = \mathcal{N}/V$, $\mathcal{N}$ being Avogadro's number and V the molar volume. Hence

$$PV = RT = \mathcal{N}m\sigma^2$$

where

$$\sigma^2 = \frac{R}{\mathcal{N}}\frac{T}{m} = \frac{kT}{m}$$

(b) When the three components are considered simultaneously, the product of the three probability densities must be calculated relative to the three independent variables v_x, v_y, v_z

$$F(v_x, v_y, v_z) = f(v_x)f(v_y)f(v_z) = \frac{1}{2\pi\sigma^3\sqrt{(2\pi)}} \mathrm{e}^{-v^2/2\sigma^2}$$

introducing the magnitude of the velocity v such that

$$v^2 = v_x^2 + v_y^2 + v_z^2$$

(c) If only the magnitude v of the velocity is considered, the function $F(v_x, v_y, v_z)$ must be integrated over a region of velocity space such that $v^2 \leqslant v_x^2 + v_y^2 + v_z^2 \leqslant (v + \mathrm{d}v)^2$.

The probability density for the magnitude v of the velocity may thus be obtained

$$F(v) = 4\pi v^2 F(v_x, v_y, v_z) = \frac{2}{\sigma^3\sqrt{(2\pi)}} v^2 e^{-v^2/2}$$

$F(v)$ is normalised such that

$$\int_0^\infty F(v)\,dv = 1$$

the mean-square velocity C is defined by

$$C^2 = \overline{v^2} = \int_0^\infty v^2 F(v)\,dv = 3\sigma^2 = 3kT/m$$

10.7.2 Application to the Intensity of an Atomic Beam

Let us consider the number of atoms per unit volume in the vapour. If n is the total number of atoms per unit volume, the number of atoms in a unit volume with velocity between v and $v + dv$ is $nF(v)\,dv$. The directions of their velocities are distributed uniformly within a solid angle of 4π; the number of atoms per unit volume whose velocity is v (to within dv) and directed within the solid angle Ω, is therefore

$$n\frac{\Omega}{4\pi}F(v)\,dv$$

In a beam, on the other hand, we do not want the number of atoms present per unit volume; but rather the quantity Q of atoms transported in unit time, in other words the total intensity of the beam. The number of atoms of velocity v that cross an aperture of area S in one second, are those contained within a cylinder of base area S and height v. We are interested only in those whose velocity lies within the solid angle Ω determined by the second aperture (see figure 10.10).

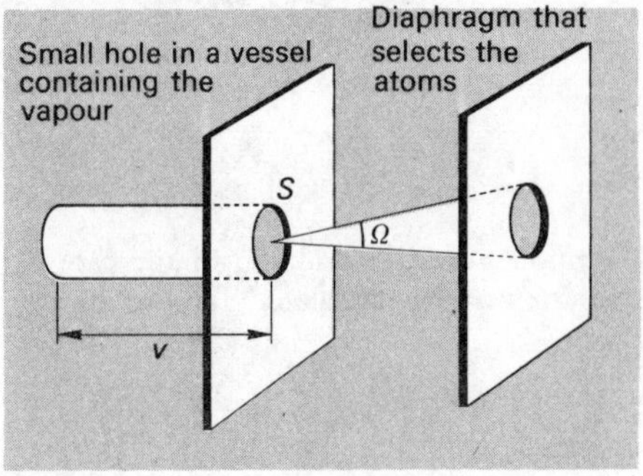

Figure 10.10 Formation of an atomic beam by two apertures

The contribution to the intensity Q of atoms with velocity v (to within dv) is therefore

$$dQ = vSn\frac{\Omega}{4\pi}F(v)\,dv$$

so

$$\frac{dQ}{dv} = nS\frac{\Omega}{4\pi}vF(v) = n\frac{S\Omega}{2\pi\sigma^3\sqrt{(2\pi)}}v^3\,e^{-v^2/2\sigma^2}$$

Atoms with high velocity transported in an atomic beam are proportionally more numerous than the high velocity atoms present in the vapour; atoms of high velocity strike the walls more frequently and therefore escape through the hole pierced in the wall in greater numbers.

The most probable velocity among the atoms transported in the beam is that for which the differential intensity dQ/dv is a maximum. Its derivative is proportional to

$$\frac{d}{dv}[v^3\,e^{-v^2/2\sigma^2}] = e^{-v^2/2\sigma^2}\left(3v^2 - \frac{v^4}{\sigma^2}\right)$$

This is zero for $v^2 = 3\sigma^2 = C^2$. Thus the most probable velocity among the atoms in the beam is equal to the mean-square velocity of the atoms in the vapour.

11

The Angular Momentum of Radiation

11.1 Classical Point of View. Rotation under the Influence of a Circularly Polarised Wave

The notion that a circularly polarised electromagnetic wave could transport angular momentum, so as to cause rotational motion, dates from the beginning of the century. Poynting concluded this in 1909 by considering a mechanical analogy. However, an exact calculation may be made by calculating the forces exerted by the wave on electric dipoles induced in an anisotropic medium (Sadowsky, 1900; Epstein, 1914).

This calculation is carried out in section 11.1.1 below, but it is not necessary to study it in detail. All that is necessary are the results, which are summarised in table 11.1. What is important is the way they are interpreted in section 11.1.2 in terms of angular momentum. It will be noted that the start of this chapter is similar to that of chapter 2.

11.1.1 Classical Calculation of the Couple Exerted on a Crystalline Foil

To explain the classical basis for the concept of the momentum of radiation (section 2.1) we calculated the radiation pressure from very general assumptions. The calculation that we carry out here in order to establish classically the concept of angular

Table 11.1 Interaction of a wave with a crystalline plate (normal incidence)

Incident wave	Thickness and orientation of crystalline plate	Emerging wave	Resultant moment applied to the crystalline plate	Change of angular momentum (projected on to direction of propagation)	
				of crystalline plate $\delta\sigma_{\text{crystal}}$	of the wave $\delta\sigma_{\text{wave}}$
Linear, oriented at 45° to principal axes ($\alpha = 0$)	$\lambda/4$ with slow axis Ox $\left(\psi = \frac{\pi}{2} + k2\pi\right)$	Right-circular	$\bar{\Gamma}_z = +P/\omega$	$+W/\omega$	$-W/\omega$
	$\lambda/4$ with slow axis Oy $\left(\psi = -\frac{\pi}{2} + k2\pi\right)$	Left-circular	$\bar{\Gamma}_z = -P/\omega$	$-W/\omega$	$+W/\omega$
Left-circular ($\alpha = +\pi/2$)	$\lambda/4 \left(\psi = \pm\frac{\pi}{2} + k\pi\right)$	Linear	$\bar{\Gamma}_z = +P/\omega$	$+W/\omega$	$-W/\omega$
Right-circular ($\alpha = -\pi/2$)	$\lambda/4 \left(\psi = \pm\frac{\pi}{2} + k\pi\right)$	Linear	$\bar{\Gamma}_z = -P/\omega$	$-W/\omega$	$+W/\omega$
Linear, oriented at 45° to principal axes ($\alpha = 0$)	$\lambda/2$ ($\psi = \pi + k2\pi$)	Linear (perpendicular to the initial direction)	$\bar{\Gamma}_z = 0$	0	0
Left-circular ($\alpha = +\pi/2$)	$\lambda/2$ (ditto)	Right-circular	$\bar{\Gamma}_z = +2P/\omega$	$+2W/\omega$	$-2W/\omega$
Right-circular ($\alpha = -\pi/2$)	$\lambda/2$ (ditto)	Left-circular	$\bar{\Gamma}_z = -2P/\omega$	$-2W/\omega$	$+2W/\omega$
Any	λ ($\psi = k2\pi$)	Unchanged	$\bar{\Gamma}_z = 0$	0	0

momentum starts with assumptions that are much less general; we study the normal propagation of a plane wave across a thin anisotropic crystalline foil.

The crystalline foil is perpendicular to the axis of propagation Oz of the electromagnetic wave. Let Ox and Oy be the directions of the principal axes of the crystalline foil; an electric field parallel to the Ox axis propagates in the foil with a refractive index n_x, so that the velocity of propagation is c/n_x, while an electric field parallel to Oy propagates with a refractive index $n_y \neq n_x$.

To study the propagation within the foil of a wave of any polarisation, the electric field $\boldsymbol{E}$ must be resolved into its two components along Ox and Oy; a wave of any polarisation can be represented by its two components

$$E_x = A\cos\omega t, \qquad E_y = B\cos(\omega t - \alpha)$$

The angle α depends on whether the polarisation of the wave is circular or elliptic; it is zero for linear polarisation, and is $\pi/2$ when the axes of the elliptic vibration coincide with the principal axes of the crystalline foil. Circular polarisation corresponds to the case where $\alpha = \pi/2$ and $A = B$ simultaneously.

The above equations represent the wave at the entrance face of the crystalline foil in the plane $z = 0$. When the wave has crossed a thickness z of the material, its components are retarded through propagation and they become

$$E_x = A\cos\omega[t - n_x(z/c)] = A\cos(\omega t - \phi_x)$$

$$E_y = B\cos\{\omega[t - n_y(z/c)] - \alpha\} = B\cos(\omega t - \phi_y)$$

Under the influence of the electric field $\boldsymbol{E}$, each part of the crystalline foil is electrically polarised with an intensity of polarisation that may be calculated with the aid of the electric induction vector $\boldsymbol{D} = \varepsilon_0 \boldsymbol{E} + \boldsymbol{P}$ whose components are (remembering the relation $\varepsilon_r = n^2$ between the refractive index n and the relative dielectric constant ε_r) $D_x = n_x^2 \varepsilon_0 E_x$ and $D_y = n_y^2 \varepsilon_0 E_y$.

Within the anisotropic medium, the induction vector $\boldsymbol{D}$ is no longer collinear with the electric field $\boldsymbol{E}$; for this reason the forces exerted by the field $\boldsymbol{E}$ on the electric dipole moments are no longer zero. Let us calculate the resultant moment of these forces per unit volume

$$\frac{\mathrm{d}\boldsymbol{\Gamma}}{\mathrm{d}\mathscr{V}} = \boldsymbol{P} \times \boldsymbol{E} = \boldsymbol{D} \times \boldsymbol{E}$$

This resultant moment reduces to its component along the Oz axis

$$\frac{\mathrm{d}\Gamma_z}{\mathrm{d}\mathscr{V}} = D_x E_y - D_y E_x = (n_x^2 - n_y^2)\varepsilon_0 E_x E_y$$

$$= (n_x^2 - n_y^2)\varepsilon_0 AB\cos(\omega t - \phi_x)\cos(\omega t - \phi_y)$$

Transforming the product into a sum of cosines, one obtains

$$\frac{\mathrm{d}\Gamma_z}{\mathrm{d}\mathscr{V}} = (n_x^2 - n_y^2)\varepsilon_0 \frac{AB}{2}[\cos(2\omega t - \phi_x - \phi_y) + \cos(\phi_x - \phi_y)]$$

All that is measurable experimentally is the mean value of this resultant moment over a time interval

$$\frac{\mathrm{d}\bar{\Gamma}_z}{\mathrm{d}\mathscr{V}} = (n_x^2 - n_y^2)\varepsilon_0 \frac{AB}{2}\cos(\phi_x - \phi_y) = (n_x^2 - n_y^2)\varepsilon_0 \frac{AB}{2}\cos\left(\omega\frac{n_x - n_y}{c}z - \alpha\right)$$

To obtain the average resultant moment for the whole crystalline foil, we must integrate over its volume. If S is its cross-sectional area and l its thickness, then one finds

$$\bar{\Gamma}_z = \int \frac{\mathrm{d}\Gamma_z}{\mathrm{d}\mathcal{V}} S\,\mathrm{d}z = S(n_x^2 - n_y^2)\,\varepsilon_0 \frac{AB}{2}\frac{c}{\omega(n_x - n_y)}\left[\sin\left(\omega\frac{n_x - n_y}{c}z - \alpha\right)\right]_0^l$$

$$\boxed{\bar{\Gamma}_z = \frac{S\varepsilon_0 c}{2\omega}(n_x + n_y)\,AB\,[\sin(\psi - \alpha) + \sin\alpha] \quad \text{where } \psi = \omega\frac{n_x - n_y}{c}l}$$

The angle ψ represents the dephasing introduced between the two components E_x and E_y in crossing the crystalline foil. If the foil is isotropic $\psi = 0$ and it may be seen that $\bar{\Gamma}_z = 0$. More generally, a moment $\bar{\Gamma}_z$ of zero is also obtained if the anisotropic crystalline foil does not change the polarisation of the wave that crossed it ($\psi = 2\pi$ or a multiple of 2π).

It is interesting to compare $\bar{\Gamma}_z$ with the average power transported by the wave across the foil

$$\bar{P} = S\frac{c}{n_x}\overline{E_x D_x} + S\frac{c}{n_y}\overline{E_y D_y} = S\varepsilon_0 c(n_x \overline{E_x^2} + n_y \overline{E_y^2}) = \frac{S\varepsilon_0 c}{2}(n_x A^2 + n_y B^2)$$

In some special cases a very simple relationship is obtained between $\bar{\Gamma}_z$ and $\bar{P}$

$$A = B \rightarrow \quad \boxed{\bar{\Gamma}_z = \frac{\bar{P}}{\omega}[\sin(\psi - \alpha) + \sin\alpha]}$$

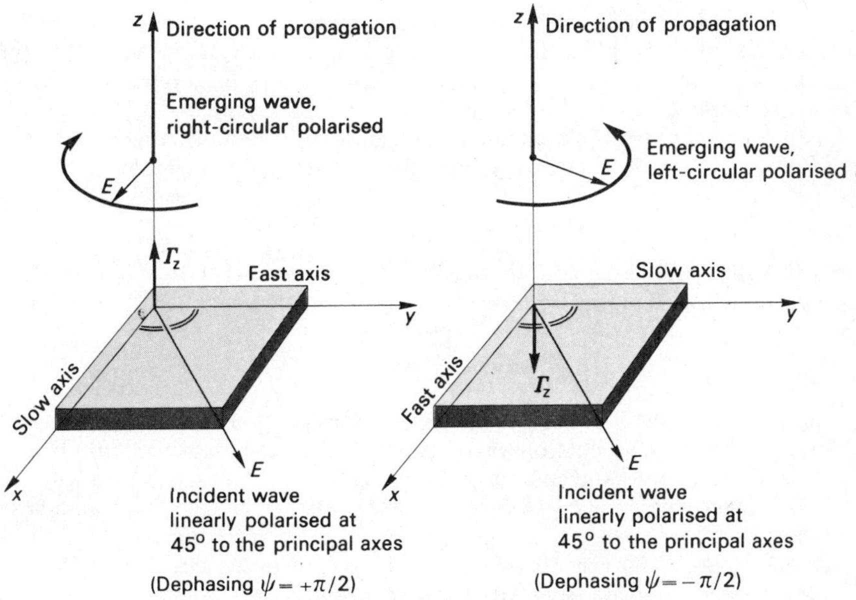

Figure 11.1 Interaction of a wave with a λ/4 plate

Any elliptically polarised wave can always be decomposed into a linear combination of linearly polarised and circularly polarised waves. In order to draw interesting conclusions from this calculation we need only consider those particular cases for which $A = B$. The results may be obtained very easily and are summarised in table 11.1.

Comment To assist the reader with the problem of signs it may be recalled that:

(i) the angle ψ is positive if Ox is the slow axis and Oy the fast axis of the crystalline foil ($n_x > n_y$); ψ is negative in the opposite situation;

(ii) if a $\lambda/4$ plate with principal axes xOy transforms linearly polarised light into circularly polarised light, the electric field vector at the exit from the plate rotates from the fast axis towards the slow axis while sweeping around the quadrant containing the direction of the incident electric field (see figure 11.1);

(iii) the light is described as left-circularly polarised if the electric field rotates in the trigonometric sense around the axis of propagation of the wave; the light is described as right-circularly polarised if the electric field rotates in the opposite sense (see figure 11.1).

11.1.2 Interpretation in Terms of Angular Momentum

If a crystalline foil alters the polarisation of the wave crossing it, it is subjected to a couple of moment Γ_z which we may determine from the theory given in the preceding section. The results obtained are set out in table 11.1. This couple, which may be positive or negative, tends to rotate the foil, that is to say, gives it a positive or negative angular momentum. The angular momentum equation relates the moment of the forces to the rate of change of angular momentum σ_{crystal} of the crystalline foil: $\mathrm{d}\sigma_{\text{crystal}}/\mathrm{d}t = \mathbf{\Gamma}$.

Disregarding the source of the electromagnetic wave, we may consider the wave and the crystalline foil as forming an isolated system, and apply the general law of conservation of angular momentum to this system. To ensure conservation of angular momentum, it must be accepted that the electromagnetic wave also has an angular momentum $\boldsymbol{\sigma}_{\text{wave}}$ and that its variation is opposed to that of the angular momentum of the crystal, such that

$$\boldsymbol{\sigma}_{\text{crystal}} + \boldsymbol{\sigma}_{\text{wave}} = \text{constant}$$

In a short time interval t, the wave transports an energy $W = Pt$ across the crystalline foil, and undergoes a change of angular momentum.

$$\delta\boldsymbol{\sigma}_{\text{wave}} = -\,\delta\boldsymbol{\sigma}_{\text{crystal}} = -\,\bar{\Gamma}_z\, t$$

that may be defined for each of the special cases previously studied (again this involves the components of wave propagation in the Oz direction). This is shown in the right-hand column of Table 11.1.

If a linearly polarised wave, carrying energy W, is changed into a circularly polarised wave, the angular momentum of this wave changes by W/ω. If a circularly polarised wave of energy W is changed into a circularly polarised wave of the opposite sense, its angular momentum changes by twice this amount $2W/\omega$.

All these results, together with their signs, may be interpreted by acknowledging that:

(i) a linearly polarised wave does not have angular momentum;
(ii) a left-circularly polarised wave of energy W has a positive angular momentum $\sigma_z = +W/\omega$;
(iii) a right-circularly polarised wave of energy W has a negative angular momentum $\sigma_z = -W/\omega$.

Thus in the latter two cases, the electric field of the wave rotates in the trigonometric sense around the corresponding angular momentum vector.

ω may be defined algebraically, as the angular velocity of the electric field vector of the wave around the reference axis Oz (thus ω is positive if the electric field rotates in the trigonometric sense with respect to the Oz axis, whether the wave is right-circular or left-circular).

It is unimportant whether the Oz axis is oriented in the direction of propagation or in the opposite sense; in every case the angular momentum of a circularly polarised wave is given by the algebraic expression

$$\boxed{\sigma_z = W/\omega}$$

The very close similarity between this problem and the problem of the momentum of radiation for normal incidence (see chapter 2) should be noted.

We have replaced:

the momentum $\boldsymbol{p}$	by the angular momentum $\boldsymbol{\sigma}$
the translation force $\boldsymbol{F}$	by the resultant couple $\boldsymbol{\Gamma}$
the propagation velocity c	by the angular velocity ω of the electric field
the equation $\mathrm{d}p_z/\mathrm{d}t = F_z = P/c$	by the equation $\mathrm{d}\sigma_z/\mathrm{d}t = \Gamma_z = P/\omega$
the equation $p_z = W/c$	by the equation $\sigma_z = W/\omega$

11.1.3 Experimental Verification

We have postulated that a quarter-wave or half-wave crystalline plate subjected to a circularly polarised wave must be forced into rotation. This effect was confirmed experimentally in 1936 by the American Beth, followed a short time later by the Englishman Holbourn. The diagram of the experiment is shown in figure 11.2.

A half-wave crystalline plate is suspended by a torsion wire. It is illuminated by a very intense light beam (power P) of left-circular polarisation, for example, and transforms it into a beam of right-circular polarisation; because of this it is subjected to a couple of moment $\Gamma_z = 2P/\omega$. The effect may be doubled by reflecting the light beam emerging from the half-wave plate back on itself by means of a mirror. If a quarter-wave plate is inserted on the outward and return journey (see figure 11.2), its polarisation is altered such that the second

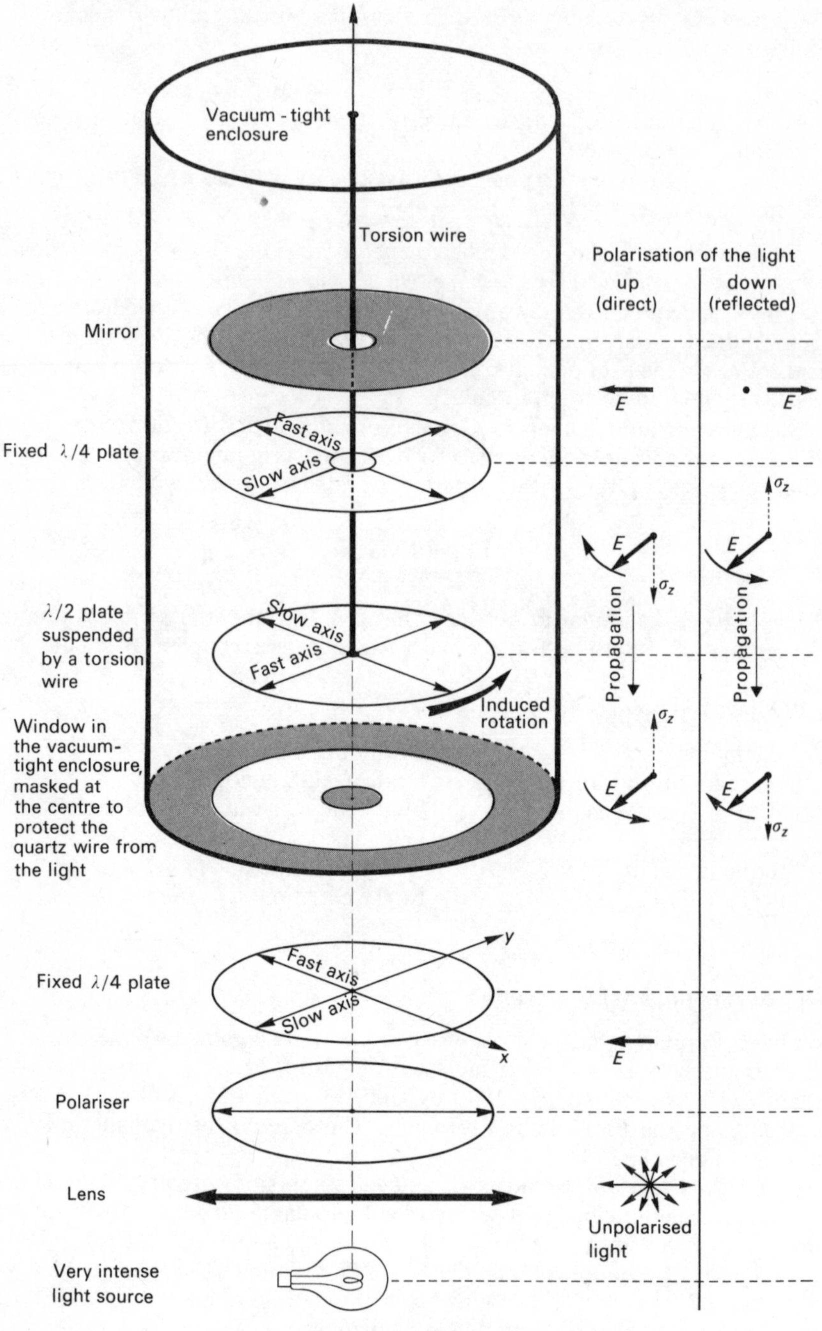

Figure 11.2 Diagram of Beth's experiment

passage through the half-wave plate in the opposite direction gives rise to an effect of the same sign as the first passage. The half-wave plate is thus subjected to a couple of moment $\Gamma_z = 4P/\omega$.

The couple is measured from the angle of rotation of the torsion wire, and if the power P emitted by the lamp is determined, this equation may be verified. As the effect studied is very weak, parasitic effects must be carefully eliminated. The half-wave plate is suspended in a vacuum, and the light must be prevented from hitting the torsion wire. If the quarter-wave plate that polarises the light from the lamp is turned through 90°, so as to invert the sense of the circular polarisation, the effect changes sign. If this sign is changed periodically with time intervals of half the natural period of the torsional pendulum thus created, a sustained oscillation of larger amplitude is produced (see the Einstein–de Haas experiment, section 9.3).

The quarter- and half-wave plates retain their properties approximately over a fairly extended range of wavelength; this allows use of a luminous power P of the order of a fraction of a watt. However, the frequency ν of the light waves is high, greater than 10^{14} Hz for the visible and near infrared, so that $\omega = 2\pi\nu$ is greater than 10^{15} s^{-1}. A resultant moment Γ of the order of 10^{-15} N m is thus calculated. With quartz wires whose torsional constant is as small as 10^{-12} N m rad^{-1}, the angle of deflection obtained is only a small fraction of a degree, but the effect is nevertheless observable.

It should be noted that the moment Γ varies inversely as the frequency of the wave. Therefore the effect must be easier to observe with electromagnetic waves of lower frequency. In this way the Italian Carrara in 1949 carried out without great difficulty an experiment similar to Beth's by using radiofrequency waves. The half-wave plate is replaced by a device for absorbing radiofrequency waves, and the rotation of the device is observed when it absorbs a circularly polarised wave. The frequency may be reduced even further: the stator of an alternating electric motor creates a magnetic field rotating at a frequency of $\nu = 50$ Hz, in other words a circularly polarised electromagnetic wave of very low frequency. A rotor that absorbs the power P of this wave is subjected to a very strong driving couple Γ. If a synchronous induction motor is used, the rotor rotates at the same angular velocity $\omega = 2\pi\nu$ as the magnetic field, and the power consumed by the motor is equal to the work done per unit time by the driving couple: $P = \Gamma\omega$.

> The relation $P = \Gamma\omega$ between the couple Γ and the power P is identical in both the problem of the synchronous motor and the problem of absorption of a circularly polarised wave.

11.2 Angular Momentum of Circularly Polarised Photons—Application to Magnetic Resonance Experiments

We have shown that a circularly polarised wave of frequency ω carries, together with an energy W, an angular momentum parallel to the direction of propagation, whose magnitude is equal to W/ω. A photon of this circular wave transports an energy $W = h\nu = \hbar\omega$; therefore it carries simultaneously an angular momentum of $W/\omega = \hbar$. In contrast to the energy and momentum, *the angular momentum transported by a photon is therefore independent of the frequency of the wave.*

We re-establish in a different way one of the results of the preceding chapter, namely, that the constant $\hbar$ is the natural unit of angular momentum. We have discussed the magnitude and direction of the angular momentum vector corresponding to a photon. Its sense is determined by the rule stated in the preceding section: the electric field of the corresponding wave rotates in the trigonometric sense around the angular momentum vector of the photon (see figure 11.3). On the other hand a photon corresponding to a linearly polarised wave does not transport angular momentum.

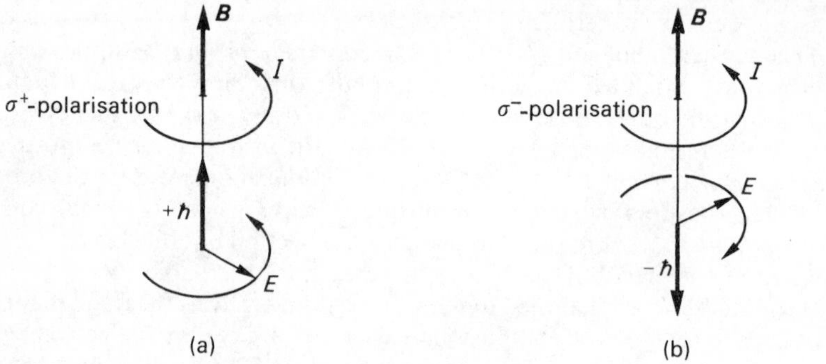

Figure 11.3 Sense of rotation of the electric field in σ^+- and σ^--polarised waves

11.2.1 Polarisation of the Wave that Produces Magnetic Resonance

Considerations of the angular momentum of a photon allow us to complete the quantum mechanical explanation of the magnetic resonance phenomenon described in section 10.5.1. We have applied the Bohr condition to transitions between Zeeman sublevels, in other words the conservation of total energy; but the conservation of total angular momentum must also be ensured during these transitions.

When an atom makes a transition between two neighbouring sublevels of quantum numbers m and $m+1$, the longitudinal component of its angular momentum changes discontinuously between the two values $\sigma_z = m\hbar$ and $\sigma_z = (m+1)\hbar$. Since the angular momentum of the atom changes by $\hbar$, the angular momentum of the electromagnetic wave must also change by $\hbar$ in the opposite sense; this occurs if an atom absorbs or emits a circularly polarised photon (whose axis of rotation is in the direction of the applied magnetic field). This is why magnetic resonance transitions (absorption or induced emission) can be brought about only by circularly polarised radiofrequency waves. We shall now discuss the signs.

(*a*) *The normal situation where the gyromagnetic ratio γ is negative and the Landé g factor is positive.* Here we saw that the energy of the Zeeman sublevels

$$E(m) = E_0 + mg\beta B$$

is an increasing function of the quantum number m (see section 10.3). Thus in transitions between Zeeman sublevels, the longitudinal component of the angular momentum σ_z and the energy $E(m)$ vary in the same manner: the absorption of a photon (transition from m to $m+1$) causes both the energy and the angular momentum σ_z of the atom to increase; the emission of a photon (transition from $m+1$ to m) causes both the energy and the angular momentum σ_z of the atom to decrease.

To ensure conservation of angular momentum in both cases, the photon absorbed or emitted must transport an angular momentum whose longitudinal component is positive: the electric field of the corresponding circular wave rotates in the trigonometric sense around the magnetic field $\boldsymbol{B}$, and therefore in the same sense as the magnetising current I (see figure 11.3(a)). The symbol σ^+ represents *the circular polarisation thus fixed in relation to the magnetic field* $\boldsymbol{B}$*, completely independent of the direction of propagation of the wave* (see section 8.4, classical Zeeman effect).

(*b*) *When the gyromagnetic ratio* γ *is positive, and therefore the Landé g factor is negative.* Here (some examples of this will be given in volume 2, chapter 6), the energy and the longitudinal component of the angular momentum of the atom vary in opposite senses during magnetic resonance transitions. This can occur only if the absorbed or emitted photon transports an angular momentum whose longitudinal component is negative. The electric field of the corresponding wave rotates around the magnetic field $\boldsymbol{B}$ in a negative sense, that is to say, in a sense opposed to the magnetising current I (see figure 11.3(b)). The symbol σ^- represents a circular polarisation in the opposite sense to that described previously.

The result obtained in both cases is in good agreement with the classical theory of magnetic resonance (section 9.5). The condition obtained classically $\omega = -\gamma B$ gives rise to a positive angular velocity (σ^+-polarisation) when the gyromagnetic ratio is negative and a negative angular velocity (σ^--polarisation) when γ is positive.

Comment A wave linearly polarised parallel to the field $\boldsymbol{B}$ (symbolised by the letter π) cannot transport any longitudinal angular momentum and can never produce magnetic resonance.†

11.2.2 Rotational Motion caused by Magnetic Resonance

The exchange of angular momentum between atom and radiation during magnetic resonance transitions has been demonstrated directly in experiments carried out since 1960 by the Italian physicist Gozzini and his students (see, for example, Alzetta, Arimondo, Ascoli and Gozzini, *U Nuovo Cimento* **52B** 379–91 (1967)).

In section 10.5 we saw that if magnetic resonance is produced in the steady state, on atoms within a material medium, this medium continually absorbs

† *Translator's note:* In this book, the term magnetic resonance is confined to transitions occurring between adjacent Zeeman sublevels, and excludes transitions between the Zeeman sublevels of two different states E_1 and E_2.

a certain power P. This power is supplied by the electromagnetic wave to the resonant atoms, and is then transmitted to the material medium through relaxation processes. However, the wave is circularly polarised and therefore transports, at the same time as an energy W, an angular momentum W/ω.

In a unit time, the medium must absorb both a power P and an angular momentum $\mathrm{d}\sigma_z/\mathrm{d}t = P/\omega$; therefore it is subjected to a couple whose resultant moment $\Gamma_z = \mathrm{d}\sigma_z/\mathrm{d}t = P/\omega$. (It should be noted that this argument does not depend on the existence of photons.) By suspending the sample in which magnetic resonance is produced at the end of a torsion wire, this moment Γ_z may then be measured. It may then be verified that Γ_z depends on the difference $\delta\omega = \omega - \omega_0$ between the frequency ω of the wave and the Larmor frequency $\omega_0 = \gamma B_0$ in the magnetic field B_0. It is zero when $\delta\omega$ is large and is a maximum when $\delta\omega = 0$. In the normal case where the gyromagnetic ratio is negative, the absorbed wave has polarisation σ^+; it confirms that the moment vector is directed in the same sense as the magnetic field, so that the sample tends to rotate in the same sense as the magnetising current.

Comment In section 9.5.6 we calculated classically the longitudinal component Γ_{1z} of the moment Γ_1 exerted by a rotating field $\boldsymbol{B}_1$ on the total magnetisation $\boldsymbol{M}$ of the atoms. Since the magnetisation $\boldsymbol{M}$ rotates steadily with a constant angular velocity ω, the resultant moment of all the forces exerted on it is zero; the third Bloch equation is obtained in effect by writing

$$\mathrm{d}M_z/\mathrm{d}t = 0; \quad \text{or} \quad \mathrm{d}\sigma_z/\mathrm{d}t = \Gamma_R + \Gamma_{1z} = 0$$

Thus by means of the relaxation processes, the material medium exerts a mean force of moment $\Gamma_R = -\Gamma_{1z}$ on the magnetisation $\boldsymbol{M}$. From the principle of action and reaction, it may be deduced that reciprocally the magnetisation $\boldsymbol{M}$ exerts a force of moment $-\Gamma_R = +\Gamma_{1z}$ on the material medium, so that forces of moment Γ_{1z} are transmitted to the material medium by means of the relaxation processes.

Since the magnetisation $\boldsymbol{M}$ rotates with the same velocity as the rotating field, the classical calculation for the power consumed is the same as the problem of the synchronous motor: $P = \Gamma_{1z}\omega$. This is the same as the equation derived above $\Gamma_{1z} = P/\omega$.

11.3 Selection Rule for the Magnetic Quantum Number. The Zeeman Effect

11.3.1 The Selection Rule

A detailed study has been made of the conservation of angular momentum in magnetic resonance transitions. However, we have failed to call attention to an important point: we have discussed only transitions between neighbouring Zeeman sublevels of quantum numbers m and $m+1$. This is because direct transitions are never observed between Zeeman sublevels that are farther apart: m and $m+2$ or m and $m+3$ and so on. (For a wave of a particular frequency ν, these would occur in a magnetic field B two or three times smaller than for a normal transition, such that $h\nu = 2g\beta B$ or $3g\beta B$ and so on.)

In such transitions, the angular momentum of the atom would change by $2\hbar$ or $3\hbar$; conservation of angular momentum could not then be ensured since a photon transports a maximum angular momentum of $\hbar$. Therefore it may be understood why all transitions are forbidden where the magnetic quantum number would change by more than one.

This prohibition can be generalised to any radiative transition. It may also be stated as follows: the experimentally observed transitions may always be interpreted by assuming that the change Δm of the magnetic quantum number of the atom takes one of three values

$$\boxed{\Delta m = m_2 - m_1 = -1, 0 \text{ or } +1}$$

Following the usual convention, we denote the higher energy state ($E_2 > E_1$) with the subscript 2 such that:

(i) if $\Delta m = +1$, the angular momentum of the atom changes in the same sense as its energy; the transition corresponds to photons of angular momentum $+\hbar$:σ^+ circular polarisation;

(ii) if $\Delta m = -1$, the angular momentum of the atom changes in the opposite sense to its energy; the transition corresponds to photons of angular momentum $-\hbar$:σ^- circular polarisation;

(iii) if $\Delta m = 0$, the angular momentum of the atom does not change, and the photons whose angular momentum is zero, have linear polarisation π.

Theoretical arguments allow the following result to be confirmed (see volume 2, section 4.1): any radiative transition is forbidden between two energy levels that do not conform to the condition framed above. This condition allows pairs of levels to be selected between which, *a priori*, a radiative transition is possible. This is described as a *selection rule*. This is the first example of a selection rule that we have met, but there are others for other quantum numbers and we shall discuss them later on (the quantum number J for example, has the same selection rule as m: $\Delta J = +1, 0$ or -1).

The selection rule for the magnetic quantum number helps in understanding the details of the Zeeman effect. Conversely, the agreement between Zeeman effect experiments and our interpretation of them is the best experimental proof of the validity of this selection rule (and also of the three possible values for the longitudinal component of the angular momentum of a photon: $-\hbar, 0, +\hbar$).

11.3.2 The Frequencies and Numbers of Zeeman Components

The Zeeman effect is observed experimentally when a magnetic field $\boldsymbol{B}$ is applied to an operating spectral lamp (see figure 11.4). Each of the characteristic spectral lines of the emitting atoms is replaced by several close but distinct lines, called Zeeman components (see figure 11.6). We have already mentioned this in section 8.4 when commenting upon the shortcomings of the classical explanation for the Zeeman effect.

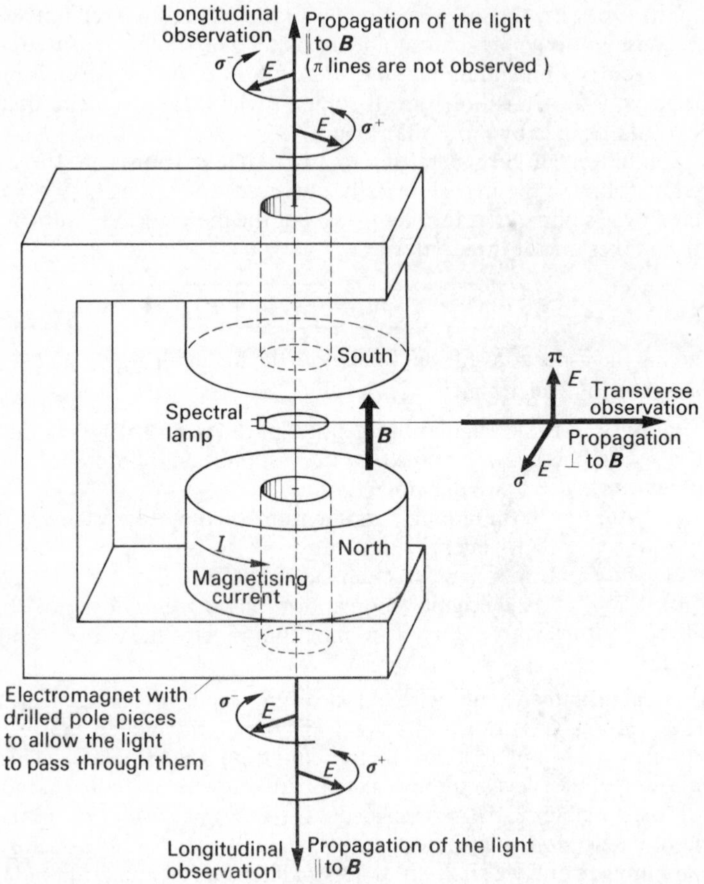

Figure 11.4 Observation of Zeeman's effect. Definition of the polarisations of the light

On the other hand, the quantum explanation is extremely simple. Let us consider a spectral transition of frequency ν_{12} between the two energy levels E_1 and E_2 of an atom: $h\nu_{12} = E_2 - E_1$.

When a magnetic field $\boldsymbol{B}$ is applied, the energy levels E_1 and E_2 are split into magnetic sublevels. Let J_1, m_1 and g_1 be the two quantum numbers and the Landé factor corresponding to the lower level E_1 and let J_2, m_2 and g_2 be the same quantities corresponding to the higher level E_2. Following section 10.3, the energies of the Zeeman sublevels are

$$E(m_1) = E_1 + m_1 g_1 \beta B$$

$$E(m_2) = E_2 + m_2 g_2 \beta B$$

The transition between levels E_1 and E_2 is replaced by transitions between the sublevels m_1 of the state E_1 and the sublevels m_2 of the state E_2. As an

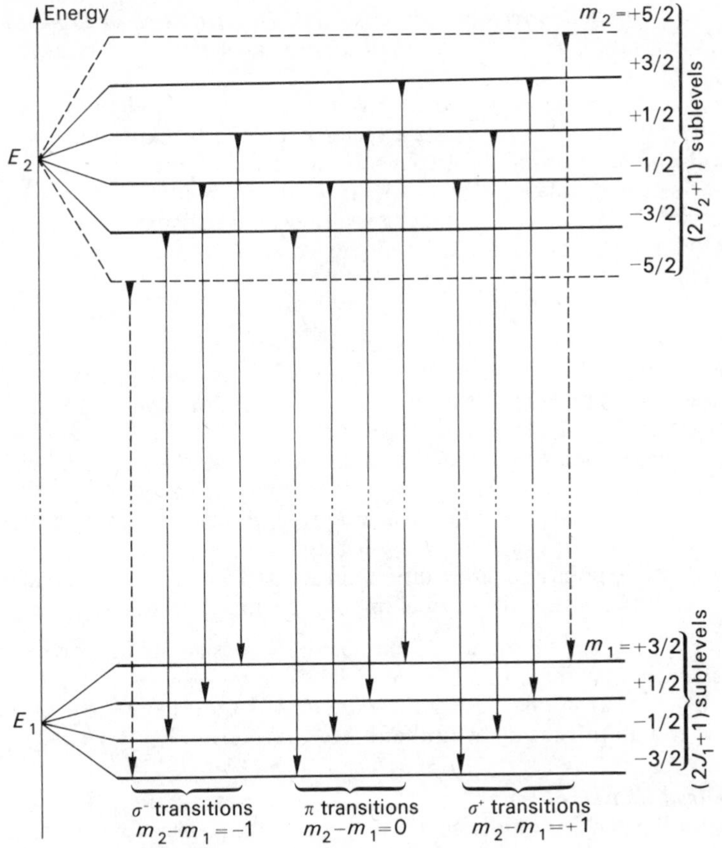

Figure 11.5 Representation of the Zeeman components on an energy diagram for one particular value of the field **B**. The figure is drawn for the particular case where $J_1 = 3/2$ and $J_2 = 5/2$ (taking the levels and transitions drawn in broken lines into account) or $J_2 = 3/2$ (ignoring the parts drawn in broken lines)

example, consider the energy level diagram in figure 11.5. In this diagram the transitions corresponding to each of the three values of the difference $m_2 - m_1$ allowed by the selection rule have been separated into three distinct groups.

The frequencies corresponding to these transitions are given by the Bohr condition

$$h\nu = E(m_2) - E(m_1) = E_2 - E_1 + m_2 g_2 \beta B - m_1 g_1 \beta B$$
$$= h\nu_{12} + (m_2 g_2 - m_1 g_1)\beta B$$

and so

$$\nu = \nu_{12} + (m_2 g_2 - m_1 g_1)\delta\nu_0 \quad \text{where } \delta\nu_0 = \frac{\beta}{h} B = \frac{1}{\kappa}\frac{e}{4\pi m} B$$

$\delta\nu_0$ is the frequency difference calculated classically; it is equal to the Larmor frequency of an atom having the classical gyromagnetic ratio (a Landé factor $g = 1$).

If m_1 and m_2 are given all possible values, this formula allows the various Zeeman components observed in a spectrograph to be explained. It should not be forgotten that of the frequencies ν thus calculated, only those that obey the selection rule $m_2 - m_1 = +1, -1$, or 0 are observable. For a detailed explanation of the spectrographic observations, the transitions corresponding to the three distinct values of the difference $m_2 - m_1$ should be considered separately.

(*a*) *The transitions $m_2 - m_1 = 0$.* These take place without any change in the angular momentum of the atom, and therefore involve photons without angular momentum, corresponding to a linearly polarised wave; this is called π-polarisation.

The number of corresponding lines depends on the number of possible values of m_1 and m_2; it is the smaller of the two numbers $2J_1 + 1$ and $2J_2 + 1$ (which are respectively the number of values of m_1 and the number of values of m_2). Their frequencies are $\nu_\pi = \nu_{12} + m_1(g_2 - g_1)\delta\nu_0$.

Thus these π-polarised lines are equidistant from one another and are symmetric in relation to the normal line ν_{12} (changing m_1 to $-m_1$).

If J_1 and J_2 are integral, the number of π lines is odd and one of them coincides with the normal line ν_{12} (the value $m_1 = 0$ exists).

If J_1 and J_2 are half-integral on the other hand, the number of π lines is even and there is no central one (see figure 11.6).

(*b*) *The transitions $m_2 - m_1 = +1$.* The atom increases its longitudinal angular momentum at the same time as its energy, so that the photons must have a longitudinal angular momentum $+\hbar$; they belong to a circularly polarised wave rotating around the magnetic field $\boldsymbol{B}$ in the same sense as the magnetising current; they are described as having σ^+-polarisation (see figure 11.3(a) or figure 11.4).

The number of σ^+ lines is equal to the number of π lines if J_1 and J_2 are different (see figure 11.5, including the levels drawn in dotted lines). On the other hand if $J_1 = J_2$ the number of σ^+ lines is $2J_1 = 2J_2$; the number of σ^+ lines is one less than the π lines (see figure 11.5, excluding the dotted lines). Their frequencies are $\nu_{\sigma+} = \nu_{12} - g_2\delta\nu_0 + m_1(g_2 - g_1)\delta\nu_0$. Thus the σ^+ lines are also equidistant from one another and the distance between them is equal to the distance between the π lines; however, they are displaced in relation to the π lines by a constant amount $g_2\delta\nu_0$.

(*c*) *The transitions $m_2 - m_1 = -1$.* The atom loses angular momentum when its energy increases, so that the photons must have a longitudinal angular momentum $-\hbar$; they belong to a circularly polarised wave rotating around the magnetic field $\boldsymbol{B}$ in a sense opposed to that of the magnetising current; they are described as having σ^--polarisation (see figure 11.3(b) or figure 11.4). The

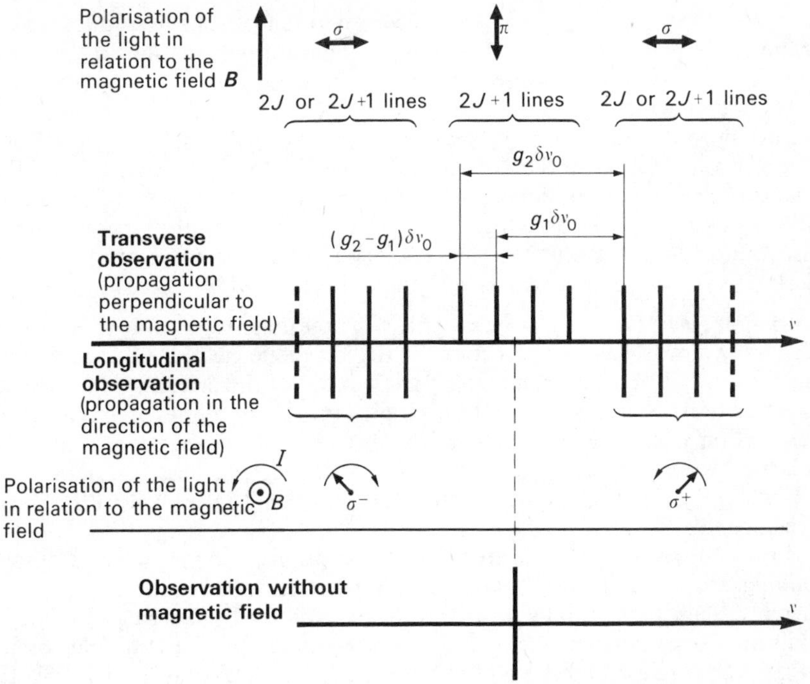

Figure 11.6 Observation of the Zeeman components in a spectrograph. J is the smaller of the two quantum numbers J_1 and J_2. This figure corresponds to the energy diagram of figure 11.5. It has been assumed that g_1 and g_2 are nearly equal. When g_1 and g_2 are very different, the three groups of lines overlap

frequencies of the lines are

$$\nu_{\sigma^-} = \nu_{12} - g_2\delta\nu_0 + m_1(g_2 - g_1)\delta\nu_0$$

To each value of m_1 there is a corresponding negative value, and so the σ^- lines are symmetric with the σ^+ lines in relation to the normal line of frequency ν_{12}.

11.3.3 Experimental Observation and Polarisations

All the results obtained in the preceding section may be verified from figure 11.6, where the lines observed in a spectrograph together with their polarisation are given for the case corresponding to the energy level diagram of figure 11.5 (and with the same convention for those parts drawn in dotted lines). A distinction has been made in the upper and lower halves of figure 11.6, between observations carried out with light received longitudinally (parallel to $\boldsymbol{B}$) and transversely (perpendicular to $\boldsymbol{B}$; see figure 11.4).

The predicted number of components is found in transverse observation as well as their calculated frequency differences. In longitudinal observation, however, the π lines ($m_2 - m_1 = 0$) are not observed. This requires us to examine the precise meaning of the three preferred polarisations σ^+, σ^- and π, and their conditions of observation.

The way in which the spectral lines called σ^+, σ^- or π appear to be polarised according to the direction of observation agrees exactly with the predictions of the classical theory of the Zeeman effect (see section 8.4). The results of the quantum theory of radiation enable the association of each type of transition with a classical dipole; thus the radiation emitted by an atom has the same distribution in space and the same polarisation as the wave calculated by classical theory (see appendix 2) with the following conditions.

(a) It is necessary to associate with π transitions, $m_2 - m_1 = 0$, a dipole oscillating linearly in the direction of the magnetic field $\boldsymbol{B}$. The radiation diagram is that of a linear antenna, which does not transmit any wave in the direction of its own axis; hence the disappearance of the π lines in longitudinal observation is explained. Whatever the direction of propagation, a linearly polarised wave is observed. The direction of the electric field is obtained by projecting the dipole in the direction of the field $\boldsymbol{B}$ on to the plane of the wave (see appendix 2).

(b) It is necessary to associate with σ^+ transitions, $m_2 - m_1 = +1$, a dipole rotating around $\boldsymbol{B}$ in the trigonometric sense (the sense of the magnetising current). Such a dipole may be analysed classically by considering its components in two perpendicular directions, representing two linear dipoles oscillating with a phase difference of $\pi/2$ (see section 8.4, classical Zeeman effect). In this way the circular polarisation observed when the propagation is longitudinal is explained, as well as the linear polarisation, perpendicular to $\boldsymbol{B}$, observed for transverse propagation (see figure 11.4). For an oblique direction of propagation, an elliptically polarised wave is observed; the shape of the ellipse is obtained by projecting the circle described by the dipole on to the plane of the wave.

(c) It is necessary to associate with σ^- transitions, $m_2 - m_1 = -1$, a dipole rotating in the opposite sense, and what has been said for σ^+ transitions may be repeated with a change only in the signs.

Polarisation is an attribute of the electromagnetic wave, and it is impossible to discuss it fully in the corpuscular language of photons. Once again there appears a difficulty related to wave–corpuscle duality. On the other hand, the frequencies of the spectral lines are related to the quantisation of energy and only photon language can explain the fact that the number of Zeeman components varies from one spectral line to another, and can sometimes be very large. Whereas the classical explanation always leads to three Zeeman components, one for each polarisation, the classical triplet is only a particular case of the more general Zeeman effect; in the particular case where the Landé factors of the two levels W_1 and W_2 are equal, one observes

$$g_2 = g_1 = g \rightarrow \left\{ \begin{array}{l} \text{(i) } \nu_\pi = \nu_{12} \\ \text{(ii) } \nu_{\sigma^+} = \nu_{12} + g\delta\nu_0 \\ \text{(iii) } \nu_{\sigma^-} = \nu_{12} - g\delta\nu_0 \end{array} \right\} \text{ independent of } m_1 \text{ and } m_2$$

If the common Landé factor is equal to unity, the differences between the three components of the triplet are equal to the difference $\delta\nu_0$ calculated classically.

We have already calculated the numerical value of the classical difference

$$\delta\nu_0/B = 14 \text{ GHz/tesla}$$

It indicates the general order of magnitude of the differences observed. Since the Doppler width of spectral lines is usually greater than 1000 MHz, it is evident that strong magnetic fields B must be used in order to separate the Zeeman components experimentally.

When well-separated Zeeman components are observed, an accurate measurement of their differences enables the Landé factors g_1 and g_2 of the two levels concerned to be calculated. Moreover, the number of components provides information about the two angular momentum quantum numbers J_1 and J_2. This is why the Zeeman effect has been in the past, and remains today, one of the most powerful methods available for investigating atomic structures.

Comment: Grotrian diagrams. The representation of transitions between Zeeman sublevels on a normal energy level diagram often gives rise to confusion. For this reason diagrams designed by the German Grotrian are often used to represent the Zeeman components schematically. Instead of representing the Zeeman sublevels of a particular level by horizontal lines corresponding to the various values of energy (figure 11.5), they are represented by points on a horizontal axis, as a function of the magnetic quantum number m (figure 11.7). The two axes corresponding to the lower energy level (below) and the higher energy level (above) are placed one above the other, in such a way that equal values of m are situated on the same vertical.

The transitions $m_2 - m_1 = +1$, of polarisation σ^+, are then represented by arrows slanting towards the left. The transitions $m_2 - m_1 = 0$, of π-polarisation, are represented by vertical arrows. The transitions $m_2 - m_1 = -1$, of σ^--polarisation, are represented by arrows slanting towards the right. Transitions that would be represented by arrows of greater inclination are forbidden.

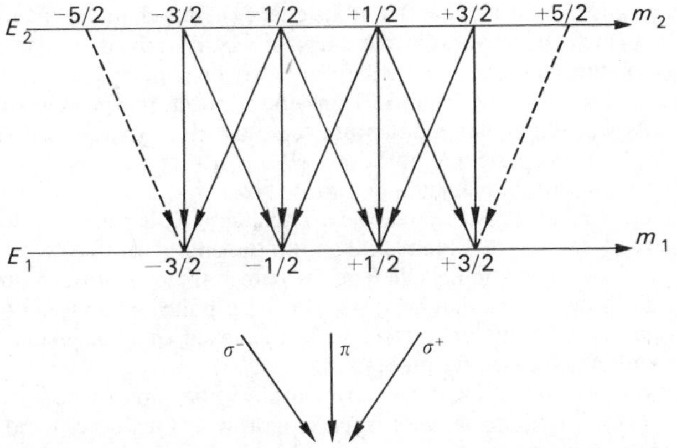

Figure 11.7 Schematic representation of the Zeeman components (as in figures 11.5 and 11.6 the transitions drawn in broken lines exist if $J_2 = 5/2$ but do not exist if $J_2 = 3/2$)

11.4 Application to the Optical Detection of Radiofrequency Resonances

Radiofrequency resonances refer to all phenomena involving radiative transitions whose frequency falls in the region of radio waves (magnetic resonance transitions are special cases of radiofrequency transitions). Optical methods of detecting radiofrequency resonances have been developed since 1950, and have led to a wide variety of experiments. In order to understand their principle of operation, it is best to describe an especially simple example in detail, and we choose the first experiment of this kind carried out by the French physicist Brossel in 1949.

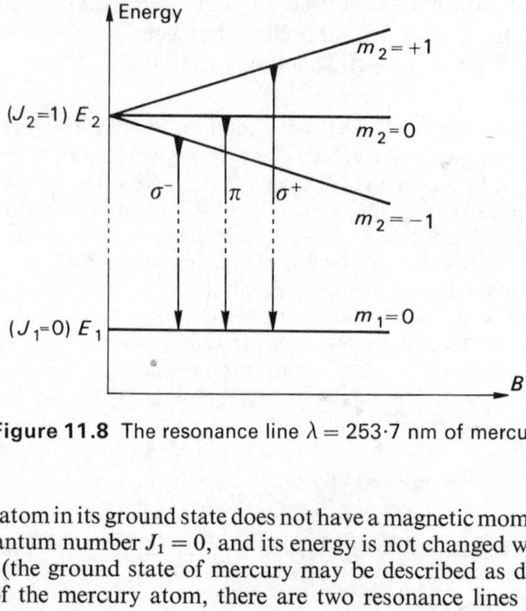

Figure 11.8 The resonance line $\lambda = 253{\cdot}7$ nm of mercury

The mercury atom in its ground state does not have a magnetic moment: its angular momentum quantum number $J_1 = 0$, and its energy is not changed when a magnetic field is applied (the ground state of mercury may be described as diamagnetic). In the spectrum of the mercury atom, there are two resonance lines of wavelengths $\lambda = 185$ nm and $\lambda = 253{\cdot}7$ nm. The 253·7 nm line connects the ground state E_1 to an excited state E_2 whose angular momentum quantum number is $J_2 = 1$ (this state is denoted by 6^3P_1 in spectroscopic notation). (We consider only the even isotopes of mercury whose nuclear angular momentum is zero.)

When a magnetic field $\boldsymbol{B}$ is applied, the excited state E_2 is split into three Zeeman sublevels $m_2 = -1$, 0 and +1. Figure 11.8 shows the energy of the Zeeman sublevels as a function of the magnetic field B. The resonance line is split into three Zeeman components as shown in the diagram; one line of π-polarisation and of unchanged frequency, one line of σ^+-polarisation and of increased frequency, one line of σ^--polarisation and of decreased frequency.

If the magnetic field $\boldsymbol{B}$ is weak, it is experimentally impossible to separate these three components by their frequencies with a spectrograph. On the other hand it is always easy to separate them by their polarisation. If an optical resonance experiment is carried out on mercury vapour, using a linearly polarised incident wave, of π-polarisation for example, only one of the three sublevels of the excited state ($m_2 = 0$) is involved. Thus the fluorescent light re-emitted by the mercury vapour also has π-polarisation.

Let us now suppose that magnetic resonance occurs between the Zeeman sublevels of the excited state. During their stay in the excited state (lifetime $\tau = 1{\cdot}1 \times 10^{-7}$ s) a fraction of the atoms will undergo a transition from the sublevel $m_2 = 0$ to the sublevels

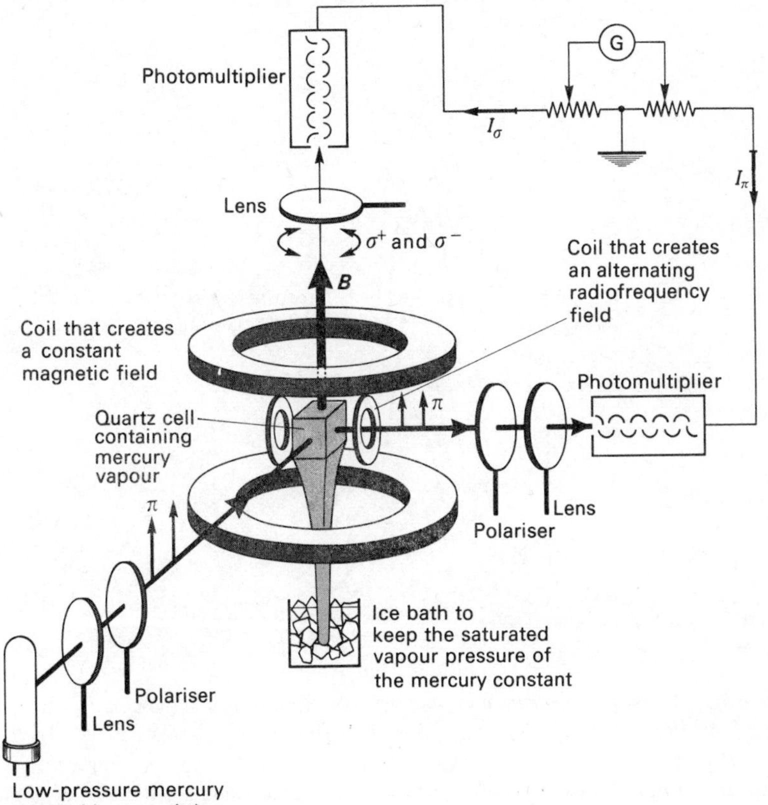

Figure 11.9 A magnetic resonance experiment on an excited state of mercury

$m_2 = -1$ and $m_2 = +1$, and from these latter sublevels the atoms will emit light of σ-polarisation. The appearance of a certain fraction of σ-polarisation in the fluorescent light is a very sensitive indication of the effectiveness of the magnetic resonance.

A diagram of Brossel's experiment is shown in figure 11.9. To obtain the best possible signal, two detectors are used, one for measuring the intensity I_π re-emitted with π-polarisation parallel to the magnetic field $\boldsymbol{B}$, the other for measuring the intensity I_σ re-emitted with σ^+ or σ^- circular polarisation. The two detectors are connected in opposition to a galvanometer whose deflection is thus proportional to changes of the quantity $I_\sigma - I_\pi$. This arrangement has two advantages: during the resonance, since the changes in I_π and I_σ are in opposite senses, the signal is increased in amplitude; in addition, fluctuations in the intensity of the exciting light are eliminated. More modern versions of the detection circuitry may be constructed, by employing various electronic techniques. The experiment is carried out at a constant frequency and for a constant amplitude of the alternating radiofrequency field, by slowly sweeping the magnetic field B. Figure 11.10 shows a family of curves obtained

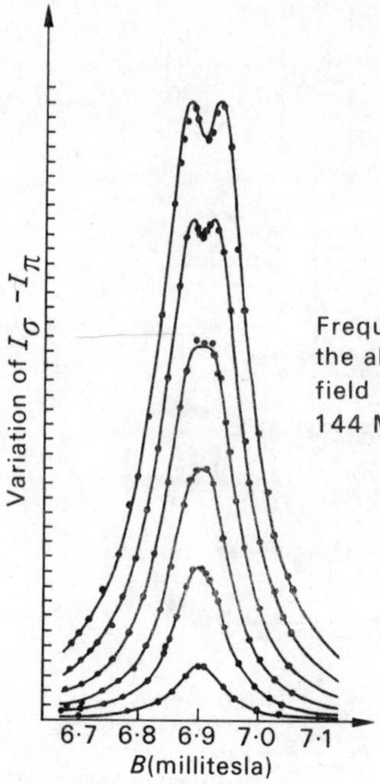

Figure 11.10 Magnetic resonance curves for Brossel's experiment (each curve corresponds to a constant amplitude of the alternating field)

for various amplitudes of radiofrequency field. As an exercise, the reader should verify, using the values given in this diagram, that the Landé factor of the 6^3P_1 level of mercury has a value very close to 1·48. (The complicated shape of the curves with a central dip is related to the value of the quantum number $J = 1$; the shapes of the curves calculated in chapter 9 are strictly valid only for the case of $J = 1/2$.)

In spite of its rather different appearance, this experiment is not without similarity to Rabi's experiment (section 10.5).

(a) Irradiation of the vapour by linearly polarised light replaces the first inhomogeneous field; it enables preparation of atoms that are all in the same Zeeman sublevel.

(b) The lifetime τ of the excited level E_2 replaces the transit time t of the atoms across the second uniform field; the atoms undergo magnetic resonance for a limited time τ or t.

(c) The detection of σ- or π-polarised light replaces the third, inhomogeneous, field and allows the number of atoms that have made a transition to be measured.

The case that we have described is obviously simple. However, generally speaking, this method is based on the fact that *the polarisation or intensity of the light emitted by the atoms depends upon their distribution between the different Zeeman sublevels. Magnetic resonance, by altering the populations of the Zeeman sublevels, also changes the polarisation or intensity of the light.*

The same idea may also be applied to ground states; instead of the light emitted, it is the polarisation or intensity of the light absorbed by the atoms that depends on their distribution between the Zeeman sublevels. All this may be generalised to radiofrequency resonances between closely spaced levels other than Zeeman sublevels. These optical methods in radiofrequency resonance spectroscopy form part of a more general class of experiments known as *double resonance*. Under this heading are denoted all experiments in which radiative transitions at two distinct frequencies are produced simultaneously in the same sample.

Comment The time during which atoms remain in the excited state can vary widely from one atom to another: τ is only the mean value. In other words, the lifetime τ is the time constant with which the magnetisation in the excited state spontaneously decays; it therefore plays a role very similar to that of a relaxation time (see chapter 9), and determines the width of the magnetic resonance lines $\delta\omega \approx 1/\tau$. This is a method of measuring lifetimes of excited states (see volume 2, chapter 7).

11.5 Application to Optical Pumping

The phenomenon of magnetic resonance is manifested mainly by changes in the populations of Zeeman sublevels. In section 10.5 we indicated the direction in which this change occurs: magnetic resonance tends to equalise the populations of the Zeeman sublevels. Therefore the greater the differences between the initial populations of these sublevels, the greater will be the effect produced.

The method of optical detection that was described in the previous section, is especially efficient when applied to excited states because the atomic system studied may then be prepared by populating preferentially a particular Zeeman sublevel (but not always 100 per cent as in the example described previously). On the other hand, when ground-state sublevels are studied, the initial populations, in the absence of magnetic resonance, are determined by the thermal equilibrium of the medium; we have calculated these populations in section 10.4.1 (Brillouin's calculation). If the absolute temperature T is very low and if the magnetic field B is strong, very large differences of population in thermal equilibrium can be obtained. However, at ordinary temperatures in a weak magnetic field, the differences between populations in thermal equilibrium are very small, and consequently magnetic resonance is difficult to detect.

Optical pumping, invented by the French physicist Kastler in 1950, is a method that allows the differences of populations between Zeeman sublevels to be created or increased. The principle of this method rests on the conservation of angular momentum in exchanges between atoms and radiation and this is why it forms a very suitable conclusion to this chapter. As in the preceding section, we shall explain the method in relation to an example chosen for its simplicity.

It will be recalled that sodium lamps provide a dominant spectral line in the visible; in fact it is a doublet formed from two very close lines, D_1 of wavelength $\lambda = 589{\cdot}6$ nm and D_2 of wavelength $\lambda = 589{\cdot}0$ nm, which may be separated by means of an interference filter. The D_1 line connects two levels having the same angular momentum quantum number $J_1 = J_2 = 1/2$; thus in a magnetic field each of them is split into two Zeeman sublevels so that the D_1 spectral line is split into four components: two of π-polarisation, one of σ^+-polarisation and one of σ^--polarisation (see figure 11.11).

An optical resonance experiment is carried out on sodium vapour using incident light of for example, σ^+ circular polarisation. Atoms in the ground sublevel $m_1 = +1/2$ cannot absorb the incident light; the latter can, however, be absorbed by atoms in the ground sublevel $m_1 = -1/2$, which are then raised into the sublevel $m_2 = +1/2$ of the excited state.

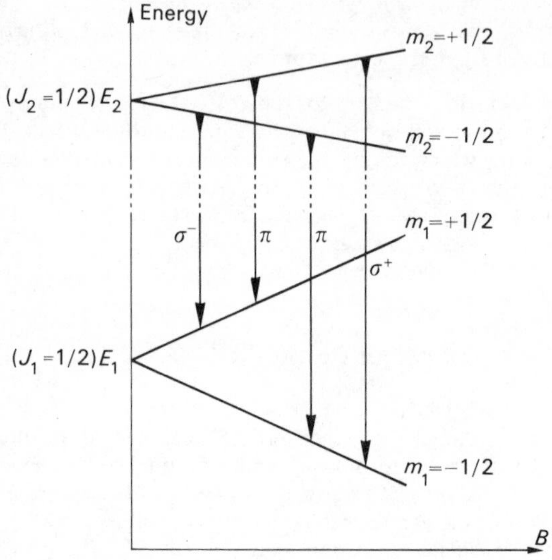

Figure 11.11 D_1 line $\lambda = 589{\cdot}6$ nm of sodium

Atoms can remain in the excited state for a very short time only (the lifetime of the excited state is about 10^{-8} s), and subsequently they make a spontaneous transition to the ground state. This spontaneous transition can occur through either the σ^+ Zeeman component or the π Zeeman component according to a probability law that may be calculated with quantum mechanics. In the latter case, the atom returns to the $m_1 = +1/2$ sublevel of the ground state, whereas it had started from the $m_1 = -1/2$ sublevel. Therefore this process has allowed a fraction of the atoms of one Zeeman sublevel to be transferred to another, so that their populations have been changed. This is described by saying that the atomic magnetic moments have been oriented. The atoms may be thought of as being taken from a reservoir constituting one of the Zeeman sublevels in order to transfer them into a second reservoir constituting the other Zeeman sublevel of the ground state, by making them pass momentarily through a pump that constitutes the excited state; hence the term optical pumping.

Relaxation processes continually try to re-establish thermal equilibrium, and the difference of populations obtained is the result of a dynamic equilibrium between optical pumping and relaxation.

(a) If the optical pumping is slow (low light intensity) and the relaxation fast (short relaxation time), the difference of populations obtained is not far removed from thermal equilibrium; the degree of orientation is described as weak.

(b) If the optical pumping is fast (high light intensity) and the relaxation slow (long relaxation time), the difference of population obtained can be far removed from thermal equilibrium; the degree of orientation is described as strong.

(In the case in which the quantum number $J = 1/2$, the degree of orientation is quantitatively defined from the respective populations p_+ and p_- of the two Zeeman sublevels $m = +1/2$ and $m = -1/2$ by the formula $\rho = (p_+ - p_-)/(p_+ + p_-.)$

Figure 11.12 shows a very simplified diagram of such an optical pumping experiment. The light from a sodium spectral lamp after passing through a polariser and a quarter-wave plate which circularly polarise it is focused into a glass bulb containing sodium vapour. The average direction of this light beam is parallel to the fixed magnetic field $\boldsymbol{B}_0$ to which the bulb of sodium is subjected. This is all that is required for optical pumping.

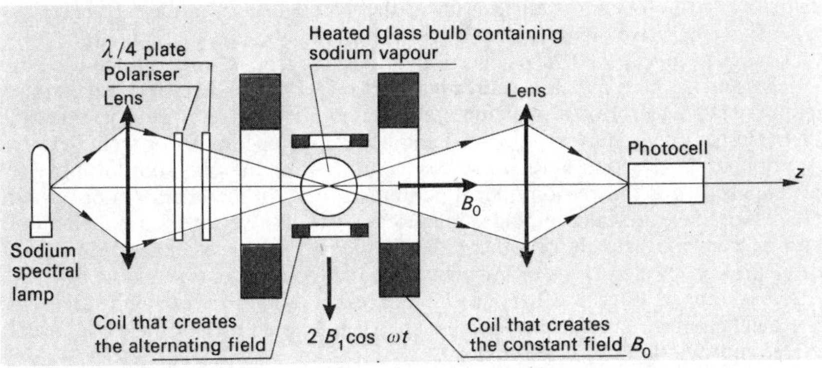

Figure 11.12 An experimental arrangement for producing and observing optical pumping

In order to detect changes of population of the Zeeman sublevels in the vapour, the light that has crossed the bulb is received by a photoelectric cell. The transmitted light intensity, measured by the cell, changes inversely as the amount of light absorbed and depends on the populations of the Zeeman sublevels. A change in intensity of the light transmitted by the vapour is observed, either when magnetic resonance takes place by means of a coil that creates an alternating field perpendicular to $\boldsymbol{B}_0$; or when the quarter-wave plate is rotated so as to change the polarisation of the light and suppress the pumping effect; or when the lamp is suddently turned on—a certain time is necessary for the optical pumping to change the populations of the Zeeman sublevels and to achieve the dynamic equilibrium that has been discussed. A transitory variation of the light transmitted over a short time is observed.

Generally speaking, the optical pumping method consists of irradiating atoms with optical resonance light, σ^+ (or σ^-) circularly polarised, which transports positive (or negative) angular momentum. The atoms absorb the angular momentum of the light as they are raised to the excited state, and when they fall back by a spontaneous transition, emitting partially depolarised light, they retain some of this angular momentum. *Overall, the assembly of atoms absorbs a fraction of the angular momentum of the circularly polarised incident light; this alters their total angular momentum, and thus their distribution between the Zeeman sublevels.*

(i) If the irradiation is carried out with σ^- light, their average angular momentum decreases, their angular momenta being oriented preferentially in the opposite direction to the magnetic field; thus they accumulate in greater numbers in the sublevels whose m quantum number is negative. In the more usual case where the Landé g factor is positive, these are also the sublevels of lower energy, and so the population differences existing in thermal equilibrium have been increased. In a way, *the atoms have been artificially cooled.*

(ii) If the irradiation is carried out with σ^+ light, the average angular momentum of the atoms increases; their angular momenta are oriented preferentially in the direction of the magnetic field, so that they accumulate in greater numbers in the sublevels whose m quantum number is positive. In this case, where g is positive, these are the sublevels of higher energy. Thus the sublevels of high energy are more populated than the sublevels of low energy. The population differences have changed sign in relation to thermal equilibrium and this is described by saying that a *population inversion* has been achieved. It should be noted that this population inversion would result from the Boltzmann statistical law if the temperature T in the formula were replaced by a negative quantity; this is why it may also be said that a *negative temperature* has been achieved. The concept of a negative temperature is often used by physicists and has been mentioned already in section 4.3. It should be recalled that the continuous passage from positive temperatures to negative temperatures goes through infinite temperature (equal populations) and not zero temperature. The achievement of a population inversion is the necessary condition for the operation of a maser (see chapter 4); for this reason optical pumping is used in the operation of certain masers. However, optical pumping is more often used as a very sensitive method of detecting magnetic resonance or other radio-frequency resonances. It has opened up a vast area of new experiments of importance in the development of atomic physics. It also has important technical applications since it is used in very accurate and sensitive magnetometers (used *inter alia* for spatial measurements of magnetic fields) and in frequency standards (see volume 2, chapter 7).

11.6 General Conclusions

We have described experiments in which rotational motion is caused by absorption or transformation of a circularly polarised electromagnetic wave. They are a striking illustration of the notion of angular momentum associated with radiation. The relation between angular momentum and energy is simple to remember since the fundamental equation defining it is identical to that of a synchronous motor

$$d\sigma_z/dt = \Gamma_z = P/\omega$$

This equation allows us to calculate the angular momentum of a photon and to confirm the constant $\hbar$ as the natural unit of angular momentum on the atomic scale.

The conservation of angular momentum in radiative transitions leads us to the extremely important concept of a selection rule. This allows us to understand all the details of the phenomenon of spectral decomposition under the action of a magnetic field, or the Zeeman effect. The observation of this phenomenon is one of the most powerful methods of investigation available in atomic physics.

12

The Magnetic Moment and Angular Momentum of the Free Electron

12.1 The Hypothesis of Spin

In chapter 8 we introduced the idea of an atomic magnetic moment due to the orbital motion of electrons around the nucleus. In chapter 9, however, we also saw that the measured values of the gyromagnetic ratio are often different from the values thus calculated.

The Landé g factor has been defined as the actual gyromagnetic ratio of an atom γ divided by its calculated value

$$\boxed{\gamma = g\frac{1}{\kappa} \times \frac{q}{2m}}$$

The existence of Landé g factors different from unity shows that the explanation of atomic magnetism cannot be accounted for entirely by the orbital motion of the electrons.

This was confirmed by Sommerfeld's work in generalising Bohr's theory to the case of elliptic orbits, oriented in the three spatial directions. He could not obtain agreement with experiment either for the number of energy levels or for the values of the angular momentum quantum number J. In particular, the existence of half-integral quantum numbers presented insurmountable problems.

To overcome these difficulties, Uhlenbeck and Goudsmit in 1925 suggested that the electron itself could rotate like a top, and thus possess its own angular momentum and its own magnetic moment even if it is not in motion as a whole. They called this intrinsic angular momentum '*spin*' (as in the motion of a top). Without taking this model too seriously, the main idea behind it, which is one of the keystones of atomic physics, is now accepted: the existence of an intrinsic angular momentum and magnetic moment of the electron.

Progress in experimental and theoretical physics has demonstrated the validity of the hypothesis that the electron itself, independent of its motion as a whole, has

(a) an intrinsic angular momentum whose components can have only the two opposite values

$$\boxed{\sigma_z = +\tfrac{1}{2}\hbar \quad \text{or} \quad \sigma_z = -\tfrac{1}{2}\hbar}$$

and so the corresponding magnetic quantum number m_s can have only the two values

$$\boxed{m_s = +\tfrac{1}{2} \quad \text{or} \quad m_s = -\tfrac{1}{2}}$$

(b) an intrinsic magnetic moment whose components are equal to the Bohr magneton

$$\boxed{\mathcal{M}_z = +\beta \quad \text{or} \quad \mathcal{M}_z = -\beta}$$

and so the gyromagnetic ratio of the electron is twice the classical value calculated for orbital motion. Thus the electron is characterised by

$$\boxed{\text{a Landé factor } g = 2}$$

The way in which magnetic moments of different origin (orbital and spin) are added in order to form the total magnetic moment of the atom may be understood more easily by using quantum mechanics. For this reason we postpone to a later chapter (volume 2, chapter 3) the study of the addition of angular momenta, which formed historically the first experimental proof of the existence of spin. Some relatively recent experiments have allowed the magnetic moments of free electrons to be observed, under conditions where there is no other cause of magnetism and their gyromagnetic ratio has been measured. These experiments now provide us with a more direct and striking proof of the existence of spin and of the intrinsic magnetic moment of the electron. The two remaining sections of this chapter are devoted to describing these experiments. (No new fundamental concept will be found in these two sections: the hurried reader who has no difficulty in accepting the ideas of Uhlenbeck and Goudsmit, could omit these two sections.)

The existence of spin was interpreted theoretically in 1928 when Dirac established the foundations of relativistic quantum mechanics. More recently (1947) hyperfine structure measurements by Rabi and his associates led Breit to make the hypothesis of an 'anomalous' magnetic moment for the electron, slightly different from the Bohr magneton. This was soon confirmed by accurate measurements of gyromagnetic ratios carried out by Kusch and Foley on various atomic states. The theory of quantum electrodynamics has resulted in the application of corrections to the Dirac theory. A limited expansion in terms of the fine structure constant $\alpha = e^2/4\pi\varepsilon_0\hbar c$ allows one to obtain to a first approximation

$$\mathcal{M}_z = \beta\left(1 + \frac{\alpha}{2\pi}\right) = \beta(1 + 1{\cdot}16 \times 10^{-3})$$

12.2 Larmor Rotation of Free-electron Spins, Polarised by Scattering

12.2.1 Polarisation of Electron Spins by Scattering

In chapter 5 we described the α-particle scattering experiment carried out by Rutherford in order to provide evidence for the nuclei of atoms. When α-particles pass close to a nucleus, they describe a hyperbolic trajectory and are deflected through an angle that can be very large. If high-energy electrons are directed on to a very thin metal foil, they are also scattered in all directions by the same interaction between electric charges.

However, if the electron itself is a magnet, this must be taken into account. The electromagnetic interaction between an electron in motion and a nucleus of an atom does not depend solely on the charge of the electron but also on its intrinsic magnetic moment. Calculation shows that the electron tends to be deflected in a plane perpendicular to its spin. Furthermore, if the electrons deflected in a particular direction are collected, those whose spins are oriented perpendicular in one sense to the scattering plane are more numerous than those whose spins are oriented in the opposite sense. This is described by saying that the electrons should be partially polarised.

In order to confirm experimentally this partial polarisation of the electrons, a double scattering experiment may be carried out, similar to that carried out to study the polarisation of X-rays: the first scattering produces the polarisation, the second detects it. The diagram of this experiment is shown in figure 12.1: a primary beam of electrons from an accelerator is directed along the Oy axis and strikes a gold foil at O. (With a weight of 0·2 mg/cm^2, its thickness would be 0·2 μm or about 1000 interatomic distances; such a foil is sufficiently thin for each electron to have a high probability of being scattered by a single nucleus only.) Electrons scattered at right angles are collected in the Oz direction and their spins are aligned preferentially in the Ox direction.

To test this property, these electrons are made to undergo a second scattering by placing another gold foil on the Oz axis at O′. The intensity of the flux of scattered particles is measured as a function of the direction in which this second scattering occurs.

The electrons whose spins are parallel to O′x' must be deflected preferentially

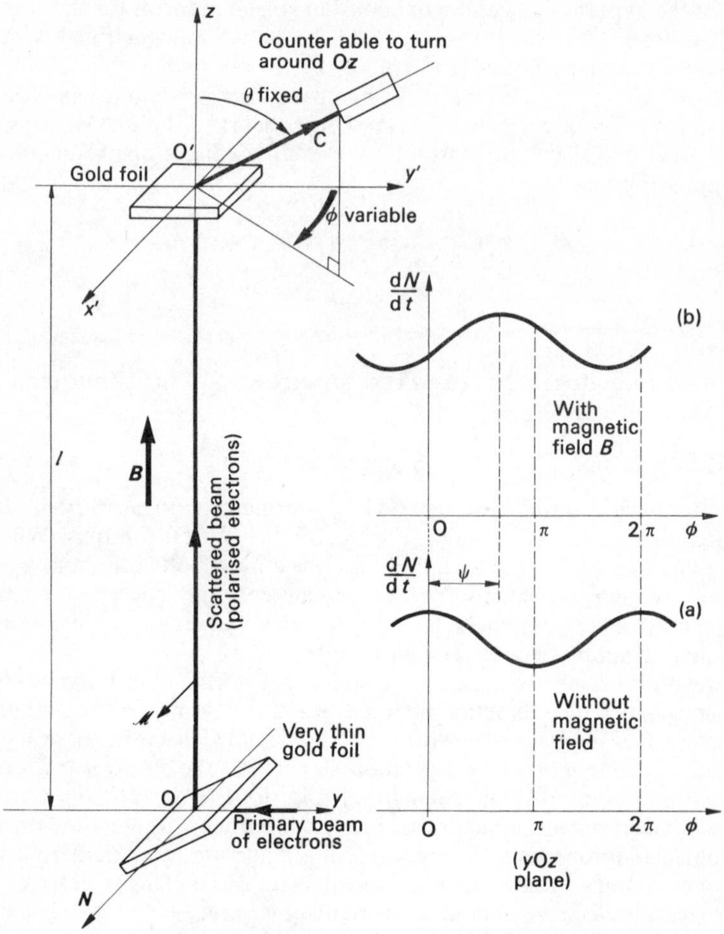

Figure 12.1 Electrons polarised by scattering

in the half-plane $y'O'z\,(y' > 0)$, perpendicular to $O'x'$; since these electrons are more numerous than the others, more electrons must be collected after scattering in the $y'O'z$ half-plane $(y' > 0)$ than in other planes.

This is confirmed by turning a particle counter C around the $O'z$ axis; it collects the scattered electrons in a direction making a fixed angle θ with the $O'z$ axis. The number of electrons scattered per unit time dN/dt is measured as a function of the angle ϕ between the $zO'C$ plane and the $O'y'$ axis, and the curve (a) of figure 12.1 is obtained: dN/dt goes through a maximum for $\phi = 0$ ($y'O'z$ half-plane) and through a minimum for $\phi = \pi$.

Comment The amplitude of the variation of dN/dt as a function of ϕ depends on the kinetic energy of the electrons and on the angle θ (on the nuclei as well, if the type of the foil is changed). It may reach about 10 per cent of the average value of dN/dt.

The scattering process disperses the electrons into a solid angle of 4π. In any one particular direction only a very small fraction of the incident electrons is collected. It is not surprising therefore that the efficiency of this double scattering process is very small. With a primary beam of 1 μA, that is 6×10^{12} electrons/s, a counting rate dN/dt of the order of 100 electrons/s is obtained.

12.2.2 Larmor Rotation of the Spins

In what follows we need not be concerned with understanding the exact mechanism of this phenomenon of spin polarisation. It is enough to know that there is a means of partially orienting the spins of free electrons and of detecting this orientation, in order to understand an experiment involving Larmor rotation carried out in the U.S.A. in 1954 by Louisell, Pidd and Crane.

Let us apply a magnetic field $\boldsymbol{B}$ parallel to Oz along the length of path OO′ described by the electrons between the two scatterings. This magnetic field is parallel to the velocity of the electrons; the Laplace force is therefore zero and the rectilinear trajectory of the electrons is unperturbed. However, in the magnetic field, the electron spins carry out a Larmor rotation with angular velocity $\omega = -\gamma B$ (see section 9.1). During their passage from O to O′, all the spins rotate by a certain angle ψ around Oz (in the same sense as the magnetising current since γ is negative).

Of the electrons starting from O, most have spins that are parallel to Ox; of the electrons arriving at O′, most therefore have spins that make an angle ψ with Ox'. If another curve is drawn giving dN/dt as a function of ϕ, the curve (b) of figure 12.1 is obtained: the minimum and maximum of the curve are displaced by an angle ψ in relation to the initial curve (a). This provides striking evidence for the Larmor rotation of the spins.

The measurement of ψ permits the direct determination of the gyromagnetic ratio of the electron spins. However, some precautions must be taken because of the use of high-energy electrons moving at relativistic velocities—the value of the gyromagnetic ratio depends on the velocity v of the electrons.

Let us calculate the Larmor rotation in a frame where the electrons are stationary, the proper frame for electrons moving with velocity v relative to the laboratory. Suppose a transformation is carried out between two frames in motion relative to one another with a velocity parallel to the Oz axis; from relativistic electromagnetism, the components along this axis E_z and B_z of the electric and magnetic fields are not altered. However, if the components of the vectors $\boldsymbol{E}$ and $\boldsymbol{B}$ perpendicular to the velocity are zero in one frame they are also zero in the other frame. Thus in our problem the magnetic field $\boldsymbol{B}$ is the same in the laboratory frame as in the proper frame of the electrons. In the proper frame therefore we find the equation for the Larmor angle of rotation: $\psi = -\gamma B t_p$ in which γ is the usual value of the gyromagnetic ratio, B is the magnetic field measured in the laboratory frame but t_p is the time between the two scatterings measured in the proper frame of the electrons.

Since the distance l between the two gold foils is known, the time interval $t = l/v$ during which the electrons travel between two scatterings is also known in the laboratory frame. From the general relationship in relativity between the proper time and the laboratory time, we may deduce that $t_p = t\sqrt{(1 - v^2/c^2)}$.

Hence we may find the angle of rotation of the spins

$$\psi = -\sqrt{(1 - v^2/c^2)}\,\gamma B t = -\sqrt{(1 - v^2/c^2)}\,\gamma B l/v$$

This formula allows γ to be calculated from the measured values of ψ; the expected value $\gamma = \dfrac{1}{\kappa}\dfrac{q}{2m}$ is obtained.

Comment I Everything takes place as if the spins in the laboratory frame had the gyromagnetic ratio

$$\gamma' = \gamma\sqrt{(1 - v^2/c^2)}\frac{1/\kappa \times q/m}{\sqrt{(1 - v^2/c^2)}} = \frac{1}{\kappa}\frac{q}{\text{relativistic mass}}$$

Comment II If the distance l is long enough and the field B of sufficiently high intensity, the angle ψ can be greater than 2π; but the displacement of the curves (a) and (b) allow ψ to be measured only within 2π. B must be increased gradually and the progressive displacement of the curve (b) must be followed in order to determine the angle ψ completely. In Louisell, Pidd and Crane's experiment where the distance l was about 10 m in a field of 10 millitesla, and the electrons had a velocity close to c, the spins turned between four and five times. The energy of the electrons may be calculated as an exercise.

Experiments using the same principle enable the spins of the short-lived particles studied in high-energy physics to be determined.

12.2.3 The Measurement of $g - 2$

In section 12.1, we indicated that the magnetic moment of the electron is slightly greater than the Bohr magneton; therefore the Landé g factor is greater than 2 and we may write

$$g = 2(1 + a)$$

The theory of quantum electrodynamics has allowed an accurate evaluation of this difference in the form of an expansion in successive terms of powers of α, the fine structure constant. It is clearly very important to carry out an experiment to measure the magnetic moment of the electron very accurately so that this may be compared with the theoretical value. The experiment discussed in section 12.2.2 requires the transit time $t = l/v$ to be known in order to evaluate γ. The former quantity is difficult to ascertain accurately, and it would be unrealistic to expect to determine γ with a relative precision of less than one part in a thousand, which corresponds to the order of magnitude of the correction to be determined.

Louisell, Pidd and Crane devised (1953) an ingenious improvement to their experiment which led to measurements by Schuff, Pidd and Crane (1961) and Wilkinson and Crane (1963). Quantitative discussion of this experiment is difficult and only a general qualitative idea can be given here.

We know that an electron, subjected to a magnetic field $\boldsymbol{B}$ perpendicular to its velocity $\boldsymbol{v}$, describes a circular path of radius R in a plane perpendicular to the field, with an angular velocity

$$\omega_c = \frac{v}{R} = \frac{1}{\kappa}\frac{qB}{m}$$

This angular velocity (the 'cyclotron' frequency) is a characteristic of the electron; it equals the precessional angular velocity of a magnetic moment of Landé factor $g = 2$.

Electrons are selected whose velocity $\boldsymbol{v}$ has a small component along the Oz axis parallel to the magnetic field $\boldsymbol{B}$; these electrons describe a tight helix around an axis parallel to Oz (figure 12.2(a)). They are emitted in a polarised state at the point O (by

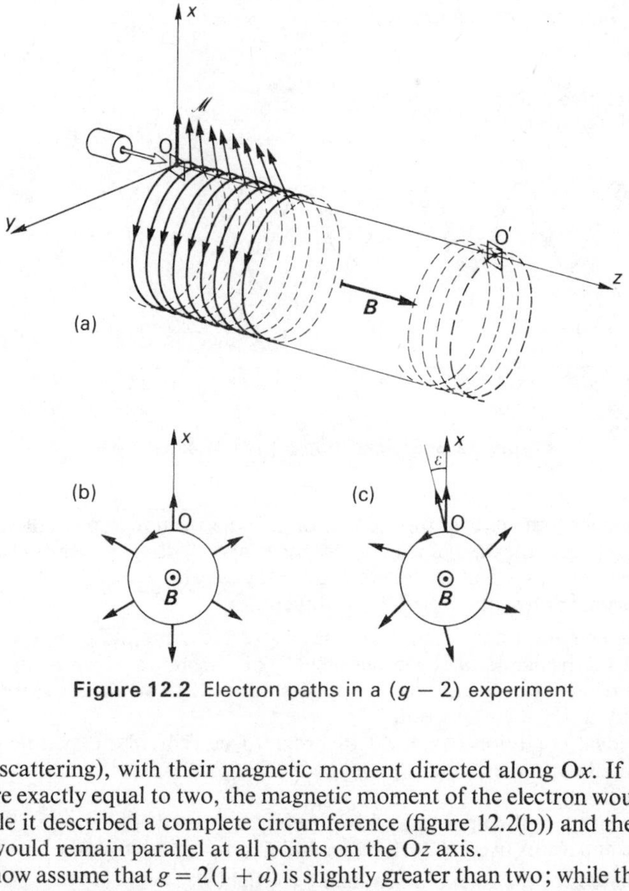

Figure 12.2 Electron paths in a $(g - 2)$ experiment

means of scattering), with their magnetic moment directed along Ox. If the Landé factor were exactly equal to two, the magnetic moment of the electron would precess by 2π while it described a complete circumference (figure 12.2(b)) and the magnetic moment would remain parallel at all points on the Oz axis.

Let us now assume that $g = 2(1 + a)$ is slightly greater than two; while the electron describes a complete circumference (figure 12.2(c)) the magnetic moment no longer turns through 2π but through

$$2\pi(1 + a) = 2\pi + \varepsilon$$

ε being the angle defined in figure 12.2(c); the magnetic moments at the various points on the Oz axis are no longer parallel to one another.

Suppose a second gold foil serving as an analyser is placed at O′, as in the experiment described in section 12.2.2. According as the electrons arrive at O′ with a magnetic moment parallel to the x axis or to the y axis, the current received by the detector will be a maximum or a minimum. By continuously varying the pitch of the helix, with the magnetic field B remaining constant, the transit time t between O and O′ changes as does the number of revolutions carried out by the electron. The curve of the detector current (figure 12.3) as a function of t has maxima and minima. The transit time t is difficult to measure absolutely because of the uncertainty in the starting point. However, the difference $t_q - t_p$ between the transit times corresponding to two maxima, denoted by p and q, can be determined to high precision. During this interval $t_q - t_p$, the number of circumferences described by the electron is $N_{qp} = \omega_c(t_c - t_p)/2\pi$. An accurate measurement of the field B enables ω_c to be calculated and hence the value of N_{pq}.

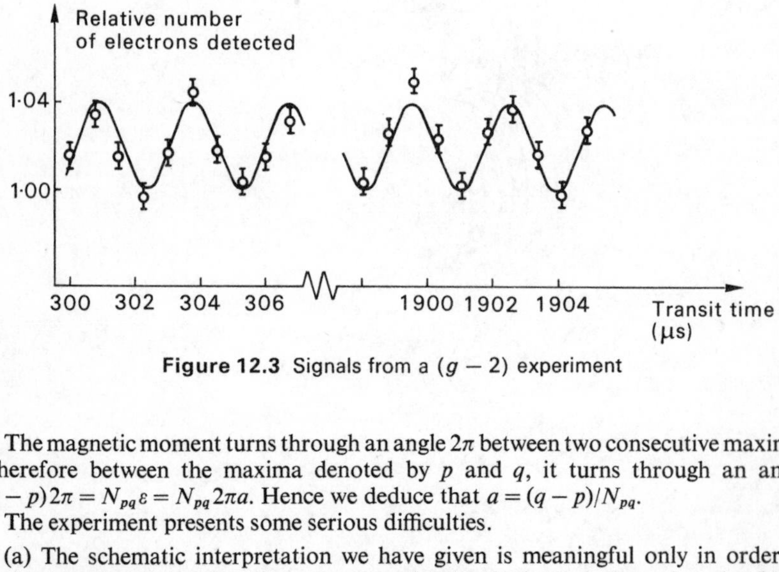

Figure 12.3 Signals from a $(g-2)$ experiment

The magnetic moment turns through an angle 2π between two consecutive maxima. Therefore between the maxima denoted by p and q, it turns through an angle $(q-p)2\pi = N_{pq}\varepsilon = N_{pq}2\pi a$. Hence we deduce that $a = (q-p)/N_{pq}$.

The experiment presents some serious difficulties.

(a) The schematic interpretation we have given is meaningful only in order to understand the principle of the experiment. The theoretical discussion should be carried out relativistically and this introduces difficulties (we shall give a reason for this difficulty in volume 2, section 3.2).

(b) The ideal conditions assumed in order to describe the experiment are not generally satisfied; in particular, the magnetic field is not exactly uniform along the path of the electron.

Many versions of the apparatus have been constructed. The essential characteristics of the apparatus used by Wilkinson and Crane were as follows:

Dimensions of the solenoid producing the field B:
length 2·40 m, diameter 0·60 m
distance OO′ = 0·60 m
Diameter of the circumferences described by the electrons: 0·15 m
Energy of the electrons: from 45 to 114 keV
Magnetic field: 9·4 to 15·3 mT
Maximum transit time: 1900 μs

The following result was obtained

$$a = 0{\cdot}001\,159\,622\ (\pm 27)$$

(the number in brackets gives the error in the last two figures). However, a complete discussion by Rich (1968) of the experimental conditions and of Wilkinson and Crane's results led to the adoption of the experimental value

$$a = 0{\cdot}001\,159\,557\ (\pm 30)$$

in slight disagreement with the theoretical value of

$$a = \alpha/2\pi - 0{\cdot}328\,\alpha^2/\pi^2 = 0{\cdot}001\,159\,614\ (\pm 3)$$

(using the value $\alpha^{-1} = 137{\cdot}038\,9\ (\pm 3)$).

Therefore theoreticians have now evaluated the corrections to third order, in α^3.

12.3 Magnetic Resonance on Free-electron Spins

In earlier chapters we studied various types of experiments that provide evidence of the gyromagnetic ratios of atoms. Among these different types of experiment, there are some that allow gyromagnetic ratios to be measured with high precision; these are the magnetic resonance experiments. If the spin of the electron has all the properties of other angular momenta and magnetic moments, it should be possible to apply the magnetic resonance method to it. Unfortunately none of the detection methods we mentioned with regard to magnetic resonance are applicable in the case of electrons.

(a) The method of radiofrequency detection (section 9.5.4) is sensitive to the magnetic flux produced by atomic systems. For this to be sufficiently intense, the number of magnetic moments per unit volume must be rather high. For this reason this method is applied in practice only to condensed media, solids or liquids. It is clearly difficult to achieve a sufficiently high concentration of free electrons, especially as it is undesirable for them to be perturbed by too frequent collisions.

(b) Rabi's method (section 10.5) derived from the Stern and Gerlach experiment, is applicable only to electrically neutral particles. With charged particles, it is practically impossible to avoid the presence of parasitic electrostatic forces greater than the magnetic force.

(c) The optical method assumes the existence of internal energy levels of the system, that do not exist for a simple free electron.

Magnetic resonance of free-electron spins has been observed, however, by Dehmelt in the U.S.A. in 1958, by using *exchange collisions*: when two atoms of a vapour enter into a collision, the total angular momentum of the assembly of the two atoms is conserved, but that of each atom may be altered separately. Thus one of the atoms may gain the angular momentum that the other loses; the two atoms would exchange angular momentum during such a collision. Let us suppose for example that two different atoms with the same quantum number $J = 1/2$ are present in the same vapour: exchange collisions tend to equalise the proportions of each species present between atoms of positive angular momentum and atoms of negative angular momentum.

Dehmelt made use of exchange collisions between free electrons and sodium atoms; he produced free electrons within the vapour of sodium, by passing an electric discharge through it. Simultaneously he carried out optical pumping on the sodium atoms as described in section 11.5: the preferential orientation given to the sodium atoms by the optical pumping is communicated by means of exchange collisions to the free electron spins. An orientation, or a polarisation, of the electron spins is thus produced, a necessary condition if magnetic resonance is to produce a change in the medium by equalising once again the number of spins aligned in the two opposite directions.

This change is detected by the effect it has on the orientation of the sodium atoms. In exchange collisions with disoriented electrons and sodium atoms lose more angular momentum than in collisions with partially oriented electrons. It follows that the rate of orientation of the sodium atoms decreases, which is manifest by the optical signal received by the photoelectric cell.

By observing magnetic resonance of the sodium atoms under the same conditions, Dehmelt could thus compare accurately the gyromagnetic ratio of the free electrons with that of the sodium atoms. He was able to confirm, with a relative precision of 10^{-5}, that the spin gyromagnetic ratio is the same whether the electrons are free or bound to an atom.

Appendix 1

Electromagnetic Formulae

adaptable to all the usual systems of units

The formulae are rationalised, but using the coefficients ε_0, μ_0 and the velocity of light c, we introduce a coefficient κ such that

$$\boxed{\varepsilon_0 \mu_0 c^2 = \kappa^2}$$

In the SI (MKSA) system

$$\kappa = 1 \qquad 4\pi\varepsilon_0 = \frac{1}{9 \times 10^9} \qquad \frac{\mu_0}{4\pi} = 10^{-7}$$

In the gaussian system (electric units of the CGS electrostatic system and magnetic units of the CGS electromagnetic system)

$$\kappa = c \qquad 4\pi\varepsilon_0 = 1 \qquad \mu_0/4\pi = 1$$

Definition of the fields

$$\boldsymbol{f} = q\boldsymbol{E} + \frac{1}{\kappa} q\boldsymbol{v} \times \boldsymbol{B}$$

Maxwell's equations

$$\text{(I)}\ \begin{cases} \operatorname{curl} \boldsymbol{E} + \dfrac{1}{\kappa}\dfrac{\partial \boldsymbol{B}}{\partial t} = 0 \\ \operatorname{div} \boldsymbol{B} = 0 \end{cases} \qquad \text{general solution} \begin{cases} \boldsymbol{E} = -\operatorname{grad} V - \dfrac{1}{\kappa}\dfrac{\partial \boldsymbol{A}}{\partial t} \\ \boldsymbol{B} = \operatorname{curl} \boldsymbol{A} \end{cases}$$

$$\text{(II)}\ \begin{cases} \operatorname{curl} \boldsymbol{B} - \dfrac{\varepsilon_0 \mu_0}{\kappa}\dfrac{\partial \boldsymbol{E}}{\partial t} = \dfrac{\mu_0}{\kappa}\boldsymbol{j} \\ \operatorname{div} \boldsymbol{E} = \rho/\varepsilon_0 \end{cases}$$

By adopting the *Lorentz gauge*

$$\operatorname{div} \boldsymbol{A} + \frac{\varepsilon_0 \mu_0}{\kappa}\frac{\partial V}{\partial t} = 0$$

Hence *equations of propagation of the potentials* are found which justify the relation between ε_0, κ and c

$$\Delta \boldsymbol{A} - \frac{\varepsilon_0 \mu_0}{\kappa^2}\frac{\partial^2 \boldsymbol{A}}{\partial t^2} = -\frac{\mu_0}{\kappa}\boldsymbol{j}$$

$$\Delta V - \frac{\varepsilon_0 \mu_0}{\kappa^2}\frac{\partial^2 V}{\partial t^2} = -\frac{\rho}{\varepsilon_0}$$

The retarded potentials (general solution of the propagation equations)

$$\boldsymbol{A}(t, \boldsymbol{r}) = \frac{\mu_0}{4\pi\kappa}\iiint \frac{\boldsymbol{j}(t - r/c)}{r}\,\mathrm{d}\mathscr{V}, \qquad V(t, \boldsymbol{r}) = \frac{1}{4\pi\varepsilon_0}\iiint \frac{\rho(t - r/c)}{r}\,\mathrm{d}\mathscr{V}$$

Hence, for the static or quasi-static case ($\boldsymbol{j}$ and ρ vary slowly with time):

$$\boldsymbol{B} = \frac{\mu_0}{4\pi\kappa}\iiint \frac{\boldsymbol{j} \times \boldsymbol{r}}{r^3}\,\mathrm{d}\mathscr{V} \quad \text{or} \quad \frac{\mu_0}{4\pi}\int \frac{I\,\mathrm{d}\boldsymbol{l} \times \boldsymbol{r}}{r^3} \qquad \textit{(Laplace's law)}$$

$$\boldsymbol{E} = \frac{1}{4\pi\varepsilon_0}\iiint \frac{\rho \boldsymbol{r}}{r^3}\,\mathrm{d}\mathscr{V} \quad \text{or} \quad \frac{1}{4\pi\varepsilon_0}\frac{q\boldsymbol{r}}{r^3} \qquad \textit{(Coulomb's law)}$$

Appendix 2

Review of the Classical Theory of Radiation

The theory of dipole radiation is of considerable importance in physics. First, it allows a simple interpretation of many phenomena related to the interactions between radiation and matter, to be given in terms of classical models. Furthermore, it is the starting point for the quantum interpretation. Some of the simpler aspects of the theory which arose in the course of the book are reviewed below.

A2.1 The Radiation from an Oscillating Dipole

A2.1.1 Assumptions

A dipole is formed from two charges $+Q$ and $-Q$ at the extremities of a small linear element l centred at the origin, and lying along the Oz axis (see figure A2.1(a)). The charges Q are assumed *to occupy fixed positions* but their magnitudes vary with time; to assure charge conservation, it must be assumed in addition that a current of magnitude $I = \partial Q/\partial t = (1/l)\,\partial p/\partial t = (1/l)p'(t)$ passes through the linear element l, where p is the algebraic value of the electric dipole moment $\boldsymbol{p} = Q\boldsymbol{l}$ along Oz and p' is its derivative with respect to time. The dipole is oscillating, that is to say the dipole moment will be assumed to vary sinusoidally with time; using imaginary notation, $p = p_0 e^{i\omega t}$.

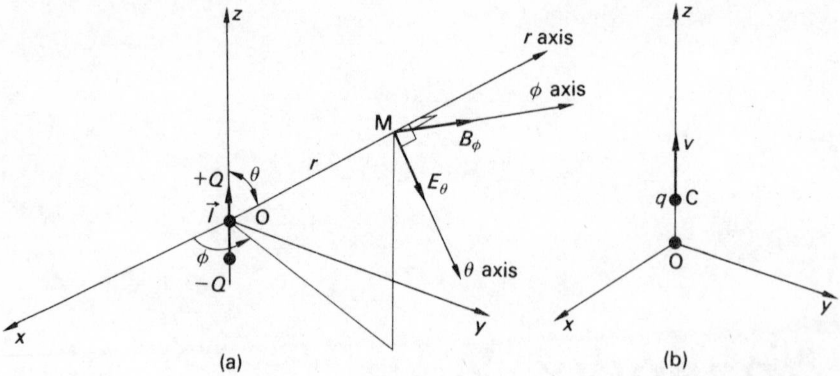

Figure A2.1

This problem is interesting from three points of view:

(1) it leads to relatively simple calculations;

(2) it allows an *exact* account of the theory of radiofrequency antennae to be given;

(3) it may be applied *as a first approximation* to an isolated charge C of constant magnitude q moving with a non-relativistic velocity $v = \mathrm{d}z/\mathrm{d}t$ (see figure A2.1(b)); it is then sufficient to put $p = qz$. The exact theory for the case of an isolated charge in motion utilises the Liénart–Wiechert potentials, but it is complicated and contains many correction terms in v/c.

A2.1.2 Retarded Potentials

We calculate retarded potentials at a point M defined by the spherical co-ordinates r, θ, ϕ (see figure A2.1(a)).

(1) The vector potential

$$\boldsymbol{A}(M, t) = \frac{\mu_0}{4\pi\kappa} \frac{I(t - r/c)}{r} \boldsymbol{l} = \frac{\mu_0}{4\pi\kappa} \frac{1}{r} \boldsymbol{p}'\left(t - \frac{r}{c}\right)$$

This may be reduced to its component along the Oz axis

$$A_z(M, t) = \frac{\mu_0}{4\pi\kappa} \frac{1}{r} p'\left(t - \frac{r}{c}\right)$$

(2) The scalar potential may be obtained most simply from the Lorentz condition

$$\frac{\partial V}{\partial t} = -\frac{\kappa}{\varepsilon_0 \mu_0} \operatorname{div} \boldsymbol{A} = -\frac{\kappa}{\varepsilon_0 \mu_0} \frac{\partial A_z}{\partial z} = -\frac{\kappa}{\varepsilon_0 \mu_0} \cos\theta \frac{\partial A_z}{\partial r}$$

By taking account of the fact that A_z depends on r both through the factor $1/r$ and through the factor $p(t-r/c)$, then integrating with respect to time, one obtains

$$V(M,t)=\frac{1}{4\pi\varepsilon_0}\cos\theta\left[\frac{1}{r^2}p\left(t-\frac{r}{c}\right)+\frac{1}{rc}p'\left(t-\frac{r}{c}\right)\right]$$

A2.1.3 The Radiation Fields

Calculation of the fields may be carried out using the two equations

$$\boldsymbol{E}=-\operatorname{grad}V-\frac{1}{\kappa}\frac{\partial A}{\partial t}$$

and

$$\boldsymbol{B}=\operatorname{curl}\boldsymbol{A}$$

In spherical co-ordinates, one obtains the components

$$E_r(M,t)=\frac{1}{4\pi\varepsilon_0}\cos\theta\left[\frac{2}{r^3}p\left(t-\frac{r}{c}\right)+\frac{2}{r^2c}p'\left(t-\frac{r}{c}\right)\right]$$

$$E_\theta(M,t)=\frac{1}{4\pi\varepsilon_0}\sin\theta\left[\frac{1}{r^3}p\left(t-\frac{r}{c}\right)+\frac{1}{r^2c}p'\left(t-\frac{r}{c}\right)+\frac{1}{rc^2}p''\left(t-\frac{r}{c}\right)\right]$$

$$E_\phi(M,t)=0$$

$$B_r(M,t)=0$$

$$B_\theta(M,t)=0$$

$$B_\phi(M,t)=\frac{\mu_0}{4\pi\kappa}\sin\theta\left[\frac{1}{r^2}p'\left(t-\frac{r}{c}\right)+\frac{1}{rc^2}p''\left(t-\frac{r}{c}\right)\right]$$

At a sufficiently great distance r, the terms in $1/r$ in the expressions for E_θ and B_ϕ dominate. These terms are proportional to p'', the second derivative of the electric dipole moment, and thus for an isolated charge, to its acceleration $a=\mathrm{d}^2z/\mathrm{d}t^2$.

For sinusoidal motion, complex quantities are introduced

$$p(t)=p_0\mathrm{e}^{\mathrm{i}\omega t},\qquad p'(t)=\mathrm{i}\omega p,\qquad p''(t)=-\omega^2 p$$

and hence the terms in $1/r$ become dominant when

$$r\gg c/\omega=\lambda/2\pi$$

(where λ is the wavelength corresponding to the frequency ω).

Making use of this condition, the radiation fields at large distances may be written

$$E_r \approx 0$$

$$E_\theta \approx \frac{1}{4\pi\varepsilon_0} \frac{\sin\theta}{rc^2} p''\left(t - \frac{r}{c}\right) = -\frac{1}{4\pi\varepsilon_0} \frac{\sin\theta}{r} \frac{\omega^2}{c^2} p_0 \, \mathrm{e}^{\mathrm{i}\omega t} \, \mathrm{e}^{-\mathrm{i}(\omega/c)r}$$

$$E_\phi = 0$$

$$B_r = 0$$

$$B_\theta = 0$$

$$B_\phi \approx \frac{\kappa}{c} E_\theta$$

It may easily be verified that in this approximation the fields at the point M are similar to those of a plane wave propagating in the direction of the radius vector $\boldsymbol{r}$, the term in $\mathrm{e}^{-\mathrm{i}(\omega/c)r}$ representing the dephasing due to propagation.

Comment It can be shown that the same formulae are applicable in the non-relativistic approximation ($v \ll c$) to an isolated charge in any type of motion (non-rectilinear and non-periodic), provided that the position of the charge is taken as the origin and the direction of its accleration vector $\boldsymbol{a}$ at the time $(t - r/c)$ is taken as the Oz axis. It is then permissible to write $p''(t - r/c) = qa(t - r/c)$. The emission of a radiation field (of amplitude in $1/r$, relatively important at large distances) occurs every time an electric charge has an acceleration vector. This happens for instance when high-energy electrons are suddenly stopped by collision in a metal (deceleration radiation or 'bremsstrahlung', see section 7.3.1) or when high-velocity electrons in a magnetic field are in uniform circular motion (synchrotron radiation).

A2.1.4 Total Radiated Power

The total power P radiated throughout space may be obtained by calculating by means of Poynting's vector the flux emerging from a sphere Σ (of large radius r)

$$\frac{\kappa}{\mu_0} \boldsymbol{E} \times \boldsymbol{B} \approx \frac{\kappa}{\mu_0} E_\theta B_\phi \frac{\boldsymbol{r}}{r} = \varepsilon_0 c E_\theta{}^2 \frac{\boldsymbol{r}}{r}$$

At every point on the sphere Σ this vector is perpendicular to an element of area

$$\mathrm{d}S = r^2 \sin\theta \, \mathrm{d}\theta \, \mathrm{d}\phi$$

and directed outwards such that

$$P = \iint_\Sigma \varepsilon_0 c E_\theta{}^2 r^2 \sin\theta \, \mathrm{d}\theta \, \mathrm{d}\phi = \frac{1}{16\pi^2 \varepsilon_0 c^3} \left[p''\left(t - \frac{r}{c}\right)\right]^2 \int_0^{2\pi} \mathrm{d}\phi \int_0^{\pi} \sin^3\theta \, \mathrm{d}\theta$$

The product of the two integrals is $8\pi/3$, and one obtains

$$\boxed{P = \frac{1}{6\pi\varepsilon_0 c^3} p''^2 = \frac{1}{6\pi\varepsilon_0 c^3} \omega^4 p^2 = \frac{1}{6\pi\varepsilon_0 c^3} q^2 a^2}$$

depending on whether p'' is replaced by $-\omega^2 p$ (the so-called oscillating dipole case) or by qa (as for an isolated charge q of acceleration a).

The power thus calculated is independent of the radius r of the sphere Σ; whatever its size, the same quantity of energy crosses its surface in a given time interval.

A2.2 Application to an Elastically Bound Electron. Damping of Free Oscillations

The classical theory of radiation is based mainly on the results for the dipole reviewed above. We shall now go on to show some elementary aspects of the theory. We assume that an atomic electron whose position is defined by the radius vector $\boldsymbol{r}$ is bound by an elastic force proportional to r (the Thomson model; see comments in sections 1.3.2 and 5.2.3)

$$\boldsymbol{f} = -k\boldsymbol{r}$$

If such an electron happens to be displaced from its equilibrium position, it carries out a spontaneous oscillatory motion, or free oscillation, with a natural frequency determined by the attractive force

$$\omega_0 = \sqrt{\left(\frac{k}{m}\right)}$$

(m is the mass of the electron)

We consider, for simplicity, linear motion in only one dimension; in imaginary notation

$$z = z_0 \, e^{i\omega_0 t}$$

Since this electron is accelerating, it generates an electromagnetic wave that is a function of its acceleration $a = d^2z/dt^2$, and that continuously carries away energy to infinity.

Using the results of the preceding section, we can calculate *the mean value* in time of the power transported by the wave

$$\bar{P} = \frac{1}{6\pi\varepsilon_0 c^3} q^2 \omega_0^4 \overline{z^2} = \frac{1}{6\pi\varepsilon_0 c^3} q^2 \omega_0^4 \frac{z_0^2}{2}$$

On the other hand, it is known that an oscillator of mass m, of natural frequency ω_0 and of amplitude z_0, stores an energy

$$W = \tfrac{1}{2} m \omega_0^2 z_0^2$$

(This result may be obtained by writing, for example, the kinetic energy of an oscillator as it passes through its equilibrium position, where the potential energy corresponding to the attractive force is zero.) To satisfy the principle of conservation of energy, the amplitude z_0 of the oscillations must gradually decrease with time in such a way that the loss of energy W is exactly counterbalanced by the energy carried away by the wave

$$\frac{\mathrm{d}W}{\mathrm{d}t} = -\bar{P} = -\frac{1}{6\pi\varepsilon_0 c^3} q^2 \omega_0^4 \frac{z_0^2}{2} = -\frac{q^2 \omega_0^2}{6\pi\varepsilon_0 c^3 m} W$$

Thus the energy W obeys the differential equation

$$\boxed{\frac{1}{W}\frac{\mathrm{d}W}{\mathrm{d}t} = -\frac{q^2 \omega_0^2}{6\pi\varepsilon_0 c^3 m} = -\frac{1}{\tau}}$$

where the constant representing the second term has been called $1/\tau$ with the dimensions of an inverse time.

Integration of this well-known differential equation shows that the energy W decreases exponentially with a time constant

$$W = W_0 \mathrm{e}^{-t/\tau}$$

and by taking the square root of the energy, the variation of amplitude may be deduced

$$z_0 = C \mathrm{e}^{-t/2\tau}$$

The amplitude decreases with twice the time constant. The oscillatory motion practically stops after a time of the order of τ, the time τ being called the lifetime of the oscillator.

The complete equation of motion of the electron, including this damping phenomenon, leads to a third-order differential equation. A simplified solution may be obtained to a good approximation by describing the damping as a viscous friction term; the differential equation for the free oscillation may then be written

$$\frac{\mathrm{d}^2 z}{\mathrm{d}t^2} + \gamma \frac{\mathrm{d}z}{\mathrm{d}t} + \omega_0^2 z = 0$$

which leads to the solutions

$$z = C \exp\left\{\left[-\frac{\gamma}{2} \pm \sqrt{\left(\frac{\gamma^2}{4} - \omega_0^2\right)}\right] t\right\} \approx C \exp\left[\left(-\frac{\gamma}{2} \pm \mathrm{i}\omega_0\right) t\right] \quad \text{if } \gamma \ll \omega_0$$

in agreement with the results found above when $\gamma = 1/\tau$.

A2.3 Forced Oscillations of an Elastically Bound Electron

A2.3.1 Steady-state Motion of the Electrons

The differential equation written at the end of the preceding section is the same for the three co-ordinates of the electron and can be applied directly to the vector $\boldsymbol{r}$ defining the position of the electron in space.

When the electron is subjected to the influence of an external sinusoidal electric field $\boldsymbol{E} = \boldsymbol{E}_0 \mathrm{e}^{\mathrm{i}\omega t}$, of fixed direction and frequency ω, different from ω_0, the differential equation describing its motion becomes

$$\frac{\mathrm{d}^2 \boldsymbol{r}}{\mathrm{d}t^2} + \gamma \frac{\mathrm{d}\boldsymbol{r}}{\mathrm{d}t} + \omega_0{}^2 \boldsymbol{r} = \frac{q}{m} \boldsymbol{E} = \frac{q}{m} \boldsymbol{E}_0 \mathrm{e}^{\mathrm{i}\omega t}$$

From this the *steady-state* solution may be found directly

$$\boldsymbol{r} = \frac{1}{\omega_0{}^2 - \omega^2 + \mathrm{i}\omega\gamma} \frac{q}{m} \boldsymbol{E}_0 \mathrm{e}^{\mathrm{i}\omega t} = \frac{q}{m} \frac{1}{\omega_0{}^2 - \omega^2 + \mathrm{i}\omega\gamma} \boldsymbol{E}$$

If an electromagnetic wave of frequency ω propagates in a material medium containing N electrons per unit volume (all elastically bound and having the same natural frequency ω_0), an oscillating current density $\boldsymbol{j}$ appears in this medium, such that

$$\boldsymbol{j} = Nq \frac{\mathrm{d}\boldsymbol{r}}{\partial t} = \frac{Nq^2}{m} \frac{1}{\omega_0{}^2 - \omega^2 + \mathrm{i}\omega\gamma} \frac{\partial \boldsymbol{E}}{\mathrm{d}t}$$

A2.3.2 The Complex Refractive Index and the Coefficient of Absorption

The Maxwell–Ampère equation

$$\mathrm{curl}\, \boldsymbol{B} = \frac{\mu_0}{\kappa} \left(\varepsilon_0 \frac{\partial \boldsymbol{E}}{\partial t} + \boldsymbol{j} \right)$$

may be written to take account of the motion of the electrons

$$\mathrm{curl}\, \boldsymbol{B} = \frac{\mu_0}{\kappa} \varepsilon_0 \left(1 + \frac{Nq^2}{m\varepsilon_0} \frac{1}{\omega_0{}^2 - \omega^2 + \mathrm{i}\omega\gamma} \right) \frac{\partial \boldsymbol{E}}{\mathrm{d}t} = \frac{\mu_0}{\kappa} \varepsilon_0 \varepsilon_r \frac{\partial \boldsymbol{E}}{\partial t}$$

where we have put

$$\varepsilon_r = 1 + \frac{Nq^2}{m\varepsilon_0} \frac{1}{\omega_0{}^2 - \omega^2 + \mathrm{i}\omega\gamma}$$

Thus we have expressed the Maxwell–Ampère equation in a form equivalent to that describing a dielectric medium with a complex relative dielectric constant ε_r. The theory of propagation of electromagnetic waves thus leads to the introduction of a complex refractive index $(n - \mathrm{i}k)$ such that

$$\varepsilon_r = (n - \mathrm{i}k)^2 = n^2 - k^2 - 2\,\mathrm{i}nk \approx n^2 - 2\,\mathrm{i}k$$

where we have made the following approximations

the real part of the refractive index $n \approx 1$
the imaginary part of the refractive index $k \ll 1$

Equating real and imaginary parts of ε_r then allows one to write

$$\text{real refractive index } n = 1 + \frac{Nq^2}{2m\varepsilon_0} \frac{\omega_0{}^2 - \omega^2}{(\omega_0{}^2 - \omega^2)^2 + \omega^2\gamma^2}$$

$$\text{imaginary refractive index} \quad k = \frac{Nq^2}{2m\varepsilon_0} \frac{\omega\gamma}{(\omega_0{}^2 - \omega^2)^2 + \omega^2\gamma^2}$$

The curves of figures A2.2(a) and A2.2(b) show the changes of n and of k respectively as a function of the frequency ω of the incident electromagnetic

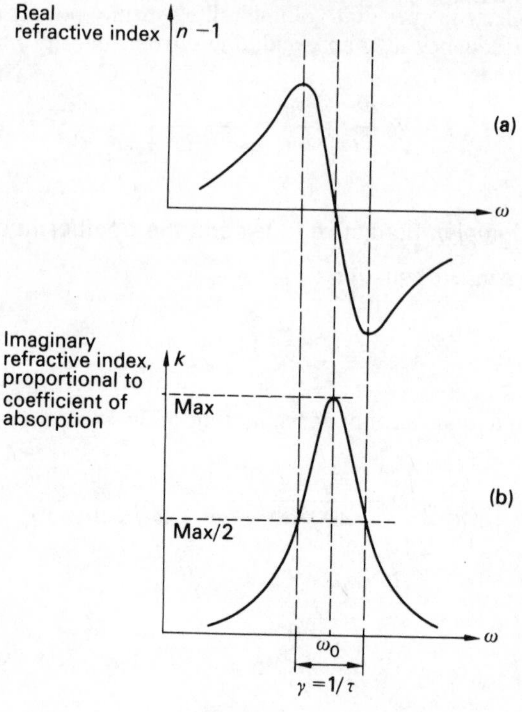

Figure A2.2

wave. The real refractive index n enables the phase velocity of the propagating waves to be calculated; it represents the normal refractive index. It should be noted that at high frequencies ($\omega > \omega_0$) the refractive index n becomes less than unity; this may be confirmed experimentally in the propagation of X-rays. The imaginary refractive index k enables the absorption coefficient of the electromagnetic wave in the medium under study to be calculated. We summarise this calculation as follows.

If one starts from the Maxwell–Ampère equation expressed in the form above and applies the theory of wave propagation, a solution may be obtained for a plane wave propagating parallel to the Ox axis

$$E(x,t) = A\,e^{i\omega[t-(n-ik)/c\,x]}$$

or, by rearranging the terms

$$E(x,t) = A\,e^{-k(\omega/c)\,x}\,e^{i\omega(t-(n/c)\,x)} = E_0\,e^{i\omega\,[t-(n/c)\,x]}$$

where

$$E_0 = A\,e^{-(\omega/c)\,kx}$$

This equation represents a wave that propagates with a phase velocity c/n and with a decreasing amplitude E_0.

Its intensity $E_0{}^2$ is damped according to the exponential law

$$E_0{}^2 = A^2\,e^{-2(\omega/c)\,kx} = A^2\,e^{-Kx}$$

with an absorption coefficient (see section 3.1.2)

$$\boxed{K = \frac{2\omega}{c}\,k = \frac{Nq^2}{m\varepsilon_0\,c}\,\frac{\omega^2\gamma}{(\omega_0{}^2-\omega^2)^2+\omega^2\gamma^2}}$$

Thus we obtain a theoretical expression for the absorption coefficient K as a function of frequency. This description must be considered only as a first approximation; it can be improved by introducing several types of electrons having different natural frequencies.

A2.3.3 The Absorption Coefficient in the Region of a Natural Frequency

The expression obtained for K can be simplified by making the following additional assumption

$$\boxed{\gamma \ll \omega_0}$$

This assumption is often true (it means that the oscillator has time to carry out many oscillations before dying away); under these conditions the absorption coefficient K is practically zero except in a narrow frequency interval around ω_0. We can then make the approximation

$$\boxed{|\omega_0-\omega| \ll \omega_0}$$

and

$$\omega_0+\omega \approx 2\omega$$

Hence

$$K = \frac{2\omega k}{c} \approx \frac{Nq^2}{m\varepsilon_0 c} \frac{\gamma}{4(\omega_0 - \omega)^2 + \gamma^2} = \frac{Nq^2}{16\pi^2 \varepsilon_0 cm} \frac{\gamma}{(\nu_0 - \nu)^2 + (\gamma/4\pi)^2}$$

If the incident electromagnetic radiation has a continuous frequency spectrum, each frequency band is absorbed with a different coefficient $K(\nu)$. Some theories (see section 3.1.4) involve what is called the total absorption, proportional to the area bounded by the curve in figure A2.2(b)

$$\int_0^\infty K(\nu)\,d\nu$$

To facilitate the evaluation of this definite integral, it should be noted that the coefficient $K(\nu)$ is practically zero when the frequency ν differs from the natural frequency ν_0; although having no physical meaning, we can write

$$\int_0^\infty K(\nu)\,d\nu \approx \int_{-\infty}^{+\infty} K(\nu)\,d\nu = \frac{Nq^2}{4\pi\varepsilon_0 cm} \int_{-\infty}^{+\infty} \frac{(\gamma/4\pi)\,d\nu}{(\nu - \nu_0)^2 + (\gamma/4\pi)^2}$$

The latter integral may be evaluated by standard techniques in terms of an arctan, and may be shown to be equal to π. Hence the *total absorption* may be found

$$\int_0^\infty K(\nu)\,d\nu \approx \frac{Nq^2}{4\varepsilon_0 cm}$$

This expression is utilised in volume 2, section 7.4 to define oscillator strengths. It should be especially noted that the damping coefficient $\gamma = 1/\tau$ does not appear; this coefficient is a measure of the frequency width at half height of the absorption curve (see figure A2.2(b)); the area under the curve $K(\nu)$ has the same value whatever the causes of damping determining its width.

A2.3.4 Thomson Scattering of X-rays

While investigating the laws for reflection of X-rays, Imbert and Bertin Sans, in 1896, demonstrated the existence of omnidirectional X-ray scattering. J. J. Thomson interpreted this effect as due to radiation from the electrons of the material, forced into oscillation by the incident radiation. In the case of X-rays, the frequency of the incident wave is far greater than the resonance frequency of the electrons, and thus the motion of the electrons can be written by simplifying the equation in section A2.3.1

$$\omega \gg \omega_0 \rightarrow \boldsymbol{r} = -\frac{q}{m\omega^2}\boldsymbol{E}$$

Each electron in motion behaves as an oscillating dipole of moment $\boldsymbol{p} = q\boldsymbol{r}$ and emits a radiation field. The electrons are distributed randomly in space and their separations are of the order of magnitude of the wavelength of the X-rays; under these conditions the phase differences between these various radiated fields are completely random and it is their intensities that are added. (These results are in complete contrast with those relating to the scattering of visible light by a condensed solid or liquid medium.)

Therefore, we obtain the total power P_s scattered by all the oscillating electrons simply by multiplying the contribution calculated for one of them, as if it were isolated, by their total number $\mathscr{V}N$ ($\mathscr{V}$ is the volume of a sample containing N electrons per unit volume)

$$P = \mathscr{V}N \frac{1}{6\pi\varepsilon_0 c^3} \omega^4 q^2 r^2 = \mathscr{V}N \frac{q^4}{6\pi\varepsilon_0 c^3 m^2} E^2$$

Let us assume that there are n atoms per unit volume in the scattering material and that each atom contains Z electrons (the atomic number), so that $N = Zn$.

Let us also assume that the incident beam of X-rays is a parallel beam of cross-sectional area S, so that the incident power transported is $P_i = S\varepsilon_0 c E^2$.

Hence the ratio of energy scattered to energy incident is

$$\boxed{\frac{P_s}{P_i} = \frac{\mathscr{V}}{S} Zn \frac{q^4}{6\pi\varepsilon_0{}^2 c^4 m^2}}$$

This ratio can be determined experimentally and, knowing n, the number of electrons per atom, Z can be deduced. Barkla in 1909 was the first to determine the value of Z by this method.

Comment I This result can be expressed in a slightly different way in terms of the absorption coefficient K. The scattered power P_s is equal and opposite to the change δP_i in the incident power as a result of passing through the scattering sample. If the latter fills the cross-section S of the incident beam over a distance δx (its volume is therefore $\mathscr{V} = S\delta x$), one finds

$$K = -\frac{1}{P_i}\frac{\delta P_i}{\delta x} = \frac{1}{\delta x}\frac{P_s}{P_i} = Zn \frac{q^4}{6\pi\varepsilon_0{}^2 c^4 m^2}$$

The same formula can be calculated from the expression for K obtained at the end of section A2.3.2, by simplifying it using the assumption $\omega \gg \omega_0$

$$K \approx \frac{Nq^2}{\varepsilon_0 cm} \frac{\gamma}{\omega^2}$$

and then replacing γ by the expression calculated in section A2.2 for the case where the radiation is the only cause of damping of the electronic motion

$$\gamma = \frac{1}{\tau} = \frac{\omega^2 q^2}{6\pi\varepsilon_0 c^3 m}$$

Comment II *Spatial distribution and polarisation of the scattered wave.* The wave radiated by each electron is not distributed isotropically; the amplitude of the field obeys the $\sin\theta$ law derived in section A2.1; thus the intensities obey a $\sin^2\theta$ law (θ being the angle between the dipole moment $\boldsymbol{p}$ and the direction of observation). Hence the scattered power P_s must be distributed among the various directions of observation according to a $(1 + \cos^2\alpha)$ law, where α is the angle between the direction of observation and the direction of propagation Ox of the incident wave. This law is in excellent agreement with experimental measurements.

Besides this intensity variation, a certain degree of polarisation of the scattered wave is also observed, depending on the angle α, which may be explained completely by the classical theory. We shall consider only a simple case where the direction of observation Oy is perpendicular to the direction of propagation Ox of the incident wave: the forced oscillations are produced in the yOz plane. Dipoles in the Oz direction radiate in the Oy direction an electric field parallel to Oz, and dipoles in the Oy direction emit no waves in the Oy direction. This explains why the wave scattered in the Oy direction is 100 per cent linearly polarised: since half the dipoles do not contribute to this emission the wave is half as intense as it is in the forward or backward directions.

Comment III The case where $\omega \ll \omega_0$, on the other hand, is called Rayleigh scattering (in contrast with Thomson scattering). An example of this is the scattering of visible light by air molecules in the atmosphere. By simplifying the equation in section A2.3.1, a radius vector $\boldsymbol{r}$ is obtained independent of ω. The scattered power is then proportional to ω^4.

Appendix 3

Non-relativistic Elastic Collisions

Two particles that interact during a collision form an isolated system to which the laws of conservation of energy and momentum may be applied. It may easily be shown that these laws are valid in all galilean frames; they are particularly simple to apply in the centre of mass frame because the total momentum there is zero. We reviewed the properties of the centre of mass C at the beginning of section 5.1.1 and we retain the same notation.

	Radius vectors of the two particles		Velocities of the two particles	
In the laboratory	OM_1	OM_2	$V_1 = v_1 + V$	$V_2 = v_2 + V$
In the centre of mass frame	$r_1 = -\frac{m_2}{m_1 + m_2} r$;	$r_2 = \frac{m_1}{m_1 + m_2} r$	$v_1 = -\frac{m_2}{m_1 + m_2} v$;	$v_2 = \frac{m_1}{m_1 + m_2} v$

where $\boldsymbol{r} = \boldsymbol{r}_2 - \boldsymbol{r}_1$ and $\boldsymbol{v} = \mathrm{d}\boldsymbol{r}/\mathrm{d}t = \boldsymbol{v}_2 - \boldsymbol{v}_1 = \boldsymbol{V}_2 - \boldsymbol{V}_1$ (relative velocity), and where $\boldsymbol{V}$ is the velocity of the centre of mass in the laboratory frame.

From conservation of kinetic energy (the collisions are *elastic*)

$$m_1 v_1^2 + m_2 v_2^2 = \text{constant}$$

From conservation of momentum

$$m_1 \mathbf{v}_1 + m_2 \mathbf{v}_2 = 0 \quad \begin{cases} \text{in magnitude } m_1 v_1 = m_2 v_2 \\ \text{and the direction of } \mathbf{v}_1 \text{ is} \\ \text{always opposite to } \mathbf{v}_2 \end{cases}$$

$$\left.\right\} \begin{cases} v_1 = \text{constant} \\ v_2 = \text{constant} \end{cases}$$

Hence the magnitude v of the relative velocity also remains constant. Figure A3.1 summarises the results obtained in the centre of mass frame. In this figure, quantities after the collision are labelled with a 'prime': $v_1 = v_1'$; $v_2 = v_2'$ and $v = v'$.

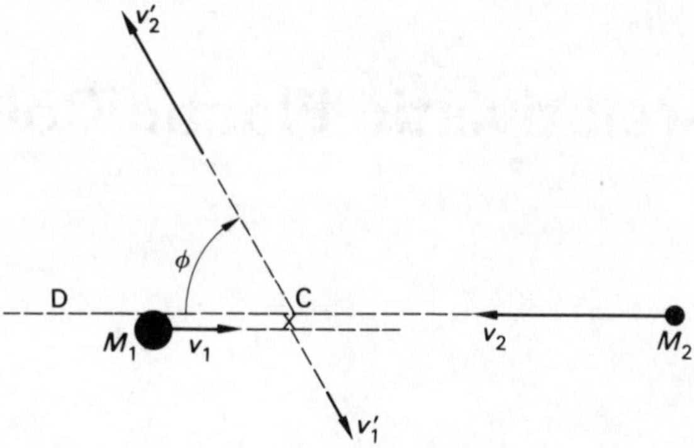

Figure A3.1

The only quantity that remains undetermined is the angle of deviation ϕ. This may be calculated as a function of the impact parameter (the distance between the two lines collinear with the vectors $\mathbf{v}_1$ and $\mathbf{v}_2$) if the interaction law between the two particles is known (see chapter 5, Rutherford's experiment).

Special case where a projectile meets a stationary target. The initially stationary target is particle number 1. Since $\mathbf{V}_1 = 0$ one finds

$$\mathbf{v} = \mathbf{v}_2$$

$$\mathbf{V} = \frac{m_2}{m_1 + m_2} \mathbf{V}_2 = \frac{m_2}{m_1 + m_2} \mathbf{v}$$

After the collision, the velocities in the laboratory are

for the target

$$\mathbf{V}_1' = \mathbf{v}_1' + \mathbf{V} = \frac{m_2}{m_1 + m_2} (\mathbf{v} - \mathbf{v}')$$

for the projectile

$$V_2' = v_2' + V = \frac{m_1}{m_1 + m_2} v' + \frac{m_2}{m_1 + m_2} v$$

It should be remembered that the two vectors v' and $v = V_2$ have the same magnitude.

(1) If the target is much heavier than the projectile

$$m_1 \gg m_2 \rightarrow \begin{cases} V_1' \approx 0 \\ V_2' \approx v', \text{ with a relative error of the order of } m_2/m_1 \end{cases}$$

the projectile keeps almost all of its kinetic energy and its angle of deviation in the laboratory frame is nearly equal to ϕ.

(2) If the masses of the target and of the projectile are of the same order of magnitude, the angle of deviation χ observed in the laboratory frame must be determined. To do this, we project the velocity vectors on to a line D parallel to the initial velocity V_2 and on to a perpendicular line (see figure A3.2).

	Initial relative velocity v	Final relative velocity v'	Final velocity of projectile V_2'
Projection on D (parallel to V_2)	v	$v \cos\phi$	$\frac{m_1}{m_1 + m_2} v \cos\phi + \frac{m_2}{m_1 + m_2} v$
Projection on a perpendicular line	0	$v \sin\phi$	$\frac{m_1}{m_1 + m_2} v \sin\phi$

Hence the angle of deviation χ in the laboratory frame is given by

$$\tan\chi = \frac{m_1 \sin\phi}{m_1 \cos\phi + m_2} = \frac{\sin\phi}{\cos\phi + m_2/m_1}$$

It is confirmed that $\chi \approx \phi$ when $m_2/m_1 \ll 1$.

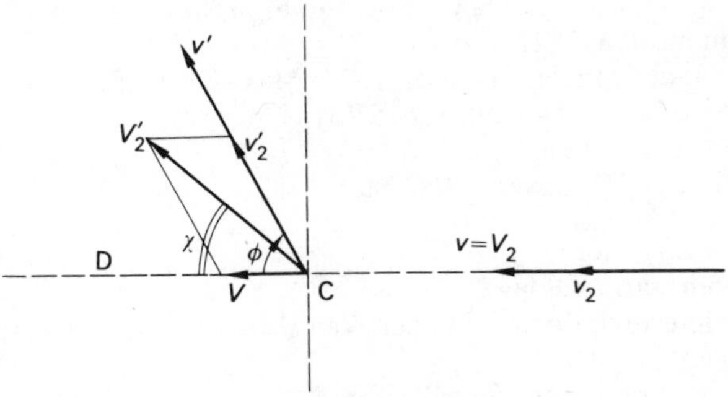

Figure A3.2

Appendix 4

Historical Summary of Atomic Physics

A4.1 The Existence of the Atom and Avogadro's Number

Proust's law of definite proportions, 1801.
Dalton's law of multiple proportions, 1807.
Gay-Lussac's law of combination by volume, 1808.
Avogadro's hypothesis, 1811.
The Avogadro–Ampère law, 1814

Interpretation of Boyle's law by Bernoulli, 1738.
Brownian motion, 1827.
Clausius' kinetic theory of gases, 1857; Maxwell, 1860.
Scattering of light: Tyndall, 1868; Rayleigh, 1871.
Boltzmann's law, 1896.
Atomic beams: Dunoyer, 1911; Stern, 1920.

Measurements of $\mathcal{N}$

(1) From statistical laws

The interpretation of Van der Waals' equation and the viscosity of gases, 1875.

Thermal radiation: Planck, 1900.

Brownian motion: Jean Perrin, 1907.

Critical opalescence: Smoluchowski, 1908.
Scattering of light by a gas: Cabannes, 1913.

(2) From the charge of the electron $-e$ compared with 1 faraday: Millikan, 1908.

(3) By counting α-particles and measuring simultaneously
the volume of helium liberated } Rutherford, Geiger, Regener, Mme.
the period of disintegration } Curie, about 1910.

(4) From atomic dimensions
Diffraction of X-rays: Roentgen, 1895; Laue, 1912; Bragg, 1914; Debye-Scherrer, 1915; Siegbahn, Compton, 1925.
Monomolecular surface layers: Devaux, Langmuir, 1917; Marcellin.

A4.2. Identification of the Electron

Volta's battery, 1800.
Laws of electrolysis: Faraday, 1833.
(Maxwell's equation, 1855.)
Cathode rays: Hittorf, 1869; canal rays: Goldstein, 1886.
The theory of ions: Arrhenius, 1887.
Transport of negative charge by cathode rays: Jean Perrin, 1895.
The Zeeman effect ($\Delta\omega = eB/2m$), 1896.
Direct measurement of e/m for cathode rays: J. J. Thomson, 1897.
Lorentz's theory of the electron, 1897.
Direct measurement of e: Millikan, 1908.
Thermionic diode: Fleming, 1904. (Triode: Lee de Forest, 1907; improved by Langmuir, 1915.)
Measurement of e from fluctuations in thermionic emission: Hull and Williams, 1925 (from Schottky's formula, 1918).
Measurement of e by counting particles and measurement of the charge transported.
Measurement of e/m for free electrons within a metal: Tolman and Stewart, 1916.
Measurement of e/m from the cyclotron frequency ($\omega = eB/m$): the Lawrence cyclotron, 1933; Purcell and Gardner, 1950; Sommer, Thomas and Hipple, 1950.
Measurement of e/m by electron magnetic resonance: Rabi, 1938; Zavoisky, 1945.

A4.3 Quantisation of Radiated Energy

(The measurements of h are marked thus: †.)

Kirchhoff's laws of thermal radiation, 1859.
Stefan's law (1879); theoretical justification by Boltzmann, 1884.
The photoelectric effect: Hertz, 1887; Halbwachs, 1888.
The spectral distribution of thermal radiation: Wien, 1893.

Quantum explanation by Planck, 1900.
†Confirmation by Lummer and Pringsheim, 1901 (section 1.1).
Einstein's explanation of the photoelectric effect, 1905 } section 1.2.
†Confirmation and measurement of h: Millikan, 1915 } section 1.2.
(Quantum theory of specific heats: Einstein, 1907; Debye-Born and von Karman, 1912.)
†(Specific heats of solids at low temperatures: Keesom and Kamerlingh Onnes.)
†The limit of the continuous spectrum in X-ray emission: Duane and Hunt, 1915 (section 7.3).
†The Compton effect, 1922; Gingrich's measurements, 1930 (section 2.3).
†Wavelength for annihilation of an electron and a positron: Du Mond, 1952.

A4.4 Atomic Structure

(The measurements of h are marked thus: †.)
Periodic table, 1869 (volume 2, section 2.4).
Balmer's hypothesis, 1885, transformed by Rydberg, 1889 (section 1.3).
Lenard's observation of cathode rays passing through apertures, 1894.
The Zeeman effect, 1896 (sections 8.4 and 11.3).
Ionisation potentials: Lenard, 1902 (section 1.4).
The theory of paramagnetism: Langevin, 1905 (sections 9.2 and 10.4).
Wood's optical resonance experiment, 1905 (section 1.3).
The combination principle: Ritz, 1908 (section 1.3).
Measurement of the atomic number by scattering of X-rays: Barkla, 1909 (appendix 2).
Existence of the nucleus: Rutherford, Geiger and Marsden, 1911 (section 5.2).
The existence of isotopes: J. J. Thomson, 1913. The mass spectrograph: Aston, 1920.
†The Bohr atom (interpretation of Rydberg's constant), 1913 (section 6.1).
†Resonance potentials: Franck and Hertz, 1913 (section 1.4).
Moseley's law, 1913 (section 7.4).
X-ray absorption spectra: Maurice de Broglie, 1916 (section 7.1).
The theory of X-ray emission spectra: Kossel, 1917 (section 7.3).
X-ray photoelectrons: Maurice de Broglie, 1921; Robinson, 1923 (section 7.2).
Gyromagnetic experiments: Barnett, 1914; Einstein–de Haas, 1915 (section 9.3).
†The Stern and Gerlach experiment, 1921 (section 10.1).
The Paschen–Bach effect, 1921 (volume 2, section 5.4).
The Landé factor, 1923 (section 10.2 and volume 2, section 5.3).
The hypothesis of spin: Uhlenbeck and Goudsmit, 1925 (chapter 12).
Pauli's exclusion principle, 1925 (section 7.3 and volume 2, section 2.3).

The direct measurement of the spin magnetic moment of free electrons
(1) in an electron beam, polarised by scattering: Louisell, Pidd and Crane, 1954;
(2) by magnetic resonance: Dehmelt, 1958.
} chapter 12.

A4.5 Nuclear Magnetism

Existence of the hyperfine structure of spectral lines: Michelson, 1891; Fabry and Perot, 1897; Lummer and Gehrcke, 1903.
The hypothesis of a nuclear magnetic moment: Pauli, 1924; Russel, Meggers and Burns, 1927.
A detailed explanation of hyperfine structure in terms of nuclear spin: Back and Goudsmit, 1927 (volume 2, section 6.5).
The hypothesis of neutron spin: Heisenberg, 1932.
Measurement of the magnetic moment of the proton: Stern, 1933 (volume 2, section 6.1).
Nuclear magnetic resonance on atomic beams: Rabi, 1938 (section 10.5).
Measurement of the magnetic moment of the neutron: Alvarez and Bloch, 1940 (volume 2, section 6.1).
Electronic detection of nuclear magnetic resonance: Bloch, 1946; Purcell, 1946 (section 9.5).

A4.6 Wave Mechanics and Quantum Mechanics

Bohr's correspondence principle, 1923.
Matter waves $\lambda = h/mv$: Louis de Broglie, 1923 (section 4.4).
Schrödinger's equation, 1925.
Heisenberg's matrix mechanics, 1925.
Commutation laws: Born and Jordan, 1925.
Probabilistic interpretation: Born, 1926.
The uncertainty principle: Heisenberg, 1927 (section 4.2).
Diffraction of electrons by a crystal: Davisson and Germer, 1927.
Diffraction of molecules: Stern, 1932.
Fresnel diffraction of electrons: Boersch, 1940.
} (section 4.4).
The relativistic wave theory of the electron: Dirac, 1928.
The anomalous spin magnetic moment: Kusch and Foley, 1947 (sections 12.1 and 12.2).
Energy difference between the $2S_{1/2}$ and $2P_{1/2}$ levels of hydrogen: Lamb and Retherford, 1947 (volume 2, chapter 4).

Bibliography

Historical

Perrin, J. *Les Atomes*, Presses universitaires, Paris (1924).

General Books Covering Experimental History and Theory

Jacquinot, P. *Optique quantique*, Hermann.

Rouault, M. *Physique atomique*, Armand Colin.

Lopes, L. *Fondements de la physique atomique*, Hermann.

Born, M. *Atomic Physics*, Blackie, London. 8th ed. (1969).

Peaslee, D. C., and Mueller, H. *Elements of Atomic Physics*, Prentice Hall, New York (1955).

Fano, U., and Fano, L. *Physics of Atoms and Molecules: an Introduction to the Structure of Matter*, University of Chicago Press (1972).

Richtmyer, F., Kennard, E., and Cooper, J. *Introduction to Modern Physics*, 6th ed., McGraw-Hill, New York (1969).

Theoretical Books

Matthews, P. T. *Introduction to Quantum Mechanics*, McGraw-Hill, New York. 2nd ed. (1968).

Ayant, Y., and Belorizky, E. *Cours de Mécanique quantique*, Dunod.
Messiah, A. *Quantum Mechanics*, North-Holland, Amsterdam (1961–2).
Barchewitz, P. *Spectroscopie atomique et moleculaire*, Masson.
Durand, E. *Mécanique quantique*, Masson.
Merzbacher, E. *Quantum Mechanics*, Wiley. 2nd ed. (1970).
Eyring, H., Walter, J., and Kimball, G. E. *Quantum Chemistry*, Wiley, New York (1954).
Bohm, D. *Quantum Theory*, Prentice Hall, London (1960).

Discussions of the Significance of Quantum Mechanics

Bohr, N. *Atomic Physics and Human Knowledge*, Wiley, New York (1958) (Articles).
d'Espagnat, B. *Conception de la physique contemporaine*, Hermann.
Margenau, H. *The Nature of Physical Reality*, McGraw-Hill, New York (1950).

Specialised Books Giving Fuller Explanations of the Ideas Presented in This Book

The corpuscular and wave nature of light (volume 1):

Selected reprints of the American Institute of Physics:
Quantum and Statistical Aspects of Light
Mössbauer Effect
Masers and Optical Pumping

Emission and absorption of photons (volume 1, chapter 3):

Mitchell, A., and Zemansky, M. *Resonance Radiation and Excited Atoms*, Cambridge (1934).

X-rays and the spectroscopy of electrons (volume 1, chapter 7):

Compton, A., and Allison, S. *X-Rays in Theory and Experiments*, Macmillan, London (1935).
Guinier, A. *Cristallographie*, Dunod.
Siegbahn, K. *Alpha, Beta and Gamma Ray Spectroscopy*, North-Holland, Amsterdam (1965), 2 vols.
Siegbahn, K., and collaborators, ESCA. *Atomic, Molecular and Solid State Structure by means of Electron Spectroscopy*, Amqvist and Wiksell, Uppsala (1967).

Magnetism in general and magnetic resonance (volume 1):

Herpin, A. *Theorie du magnetisme*, P.U.F.
de Broglie, L. *La spectroscopie en radiofrequences*, Revue d'Optique, Paris.

Freymann, R. and Soutif, M. *Spectroscopie hertzienne*, Dunod.
Grivet, P. *Resonance magnetique nucléaire*, C.N.R.S.
Andrew, E. R. *Nuclear Magnetic Resonance*, Cambridge (1955 and 1958).
Abragam, A. *The Principles of Nuclear Magnetism*, Clarendon Press, Oxford (1961).
Pake, G. E. *Paramagnetic Resonance*, Benjamin, New York (1962).
Selected reprints of the American Institute of Physics: *Atomic and Molecular Beam Spectroscopy.*

Atomic and molecular spectroscopy (volume 2) : n.m.r. and e.p.r.

Bousquet, P. *Instrumental Spectroscopy*, Dunod, Paris (1969).
Bruhat, G., and Kastler, A. *Optique*, Masson.
Bopp, E., and Kleinpoppen, H. *Physics of One and Two Electron Atoms*, North-Holland, Amsterdam (1969).
Kuhn, H. G. *Atomic Spectra*, Longmans, London (1969).
White, H. E. *Introduction to Atomic Spectra*, McGraw-Hill, New York (1934).
Herzberg, G. *Molecular Spectra and Molecular Structure*, Volume I: *Spectra of Diatomic Molecules*, Van Nostrand, New York (1939), 2nd ed., Toronto (1950).

Two Series Presenting Modern Developments in Atomic Physics (chapter 7) :

Advances in Atomic and Molecular Physics (Edited by D. Bates and I. Estermann) (five published volumes), Academic Press, New York (1965).
Methods of Experimental Physics (Edited by L. L. Marton) (eight published volumes), Academic Press, New York (1959–).

Index